Victoria's Canadian Militia

Uniforms, Flags and Equipments of the Canadian Militia, 1837-1901

Volume 2

by

René Chartrand

Victoria's Canadian Militia: Uniforms, Flags and Equipments of the Canadian Militia, 1837 - 1901, Vol. 2
By Rene Chartrand
Cover image from Rene Chartrand

This edition published in 2023

Winged Hussar Publishing is an imprint of

Winged Hussar Publishing, LLC
1525 Hulse Rd, Unit 1
Point Pleasant, NJ 08742

ISBN 978-1-950423-57-6
LCN 2023932623

Bibliographical References and Index
1. History. 2. Canada. 3. Military

For more information
visit us at www.whpsupplyroom.com

Twitter WingHusPubLLC
Facebook: Winged Hussar Publishing LLC

Table of Contents

Dedication

To the memory of
Clive M. Law
Volunteer militiaman, law officer, senior civil servant, friend, mentor.

Preface

This is the second volume of *Victoria's Canadian Militia* whose first volume was published during October 2016 by Service Publishing in Ottawa, headed by the military historian and close friend Clive Law,. By then, most of the manuscript of the second volume was drafted as well as much of its images chosen and it was planed to be published in about mid-2017. On 10 June of that year, Clive passed away unexpectedly to the grief of the many who knew and admired him. Needless to add that publication plans for volume 2 were literally shelved.[1]

Years passed and in the early fall of 2021, military historian Gabriele Esposito in Italy contacted the author that Vincent Rospond in New Jersey (USA), a military historian and head of Winged Hussar, wished to establish contact. You now hold the result of these most cordial exchanges.

The first volume had a general introduction to Canada and its military aspects during the reign of Queen Victoria (1837-1901) with a part on the British army, the following sections presenting generals and staff departments, cavalry, artillery, infantry units in central Canada up to 1854, rifle units, the Atlantic provinces to the 1870s, colours, weapons with appendices on uniform buttons, Zouaves and independent guards in Francophone Canada.

This second volume, as seen above by its Table of Contents, rounds out the topic. All in all, it concludes what is the first extensive narrative of Canada's largely unknown military material culture from the 1830s to the early 1900s. We have paid special attention to the many illustrations, essential in such a study.

As in Vol. 1, much is drawn from archival sources and numerous printed primary and secondary sources. To orient readers wishing to know more, each chapter has a "Notes" paragraph at it end listing sources supported by many footnotes. We have also inserted in the main text many notes in brackets such as: (LAC, reel T405/868, 977, 1008) that refer to a Library and Archives Canada (LAC) microfilm reel number (T405) and page numbers. These, in effect, form the select bibliography that, presented in this way, focus better on the chapter's topic.

Of course no study is perfect. The inevitable errors, omissions or misinterpretations that do occur, even with best efforts, rest solely with the author

René Chartrand
Gatineau (Aylmer), QC
November 2021

[1] Obituary of Company Fellow Clive M. Law, SM)stJ (1954-2017), MC&H, Winter 2017.

Acknowledgements

The vast amount of data presented in this study is the result of the generous help given to the author by many individuals and institutions over the last half century. Some of the main contributors are listed below, several now regrettably departed.

Francis Back
Don Blyth
Anne S.K. Brown
Christopher Bryant
Kimberley Elizabeth Busato
William Y. Carman
Earl Chapman
Denis Darmanin
Edward Denby
John Denner
Scott Duncan
William Elms
André Gousse
Donald E. Graves
Peter Harrington
Philip Haythornthwaite
Robert Henderson
Sheldon Kasman
Anthony Kellet
Ludwig Kosche
Clive M Law
Michel Litalien
Daniel Mackay
Robert J. Marrion
Yves Martin
Michael R. McNorgan
Michael Reintjes
Barry Rich
Ron Ridley
Carol Reid
David Ross
Jack L. Summers
David Webber
Ross M. Wilson

Of the institutions, the main ones are the National Historic Sites of Parks Canada, the History and Heritage Directorate of the Department of National Defence, Library and Archives Canada (the former Public Archives of Canada), Bibiothèque et archives nationales du Quebec, the Nova Scotia, Manitoba and British Columbia archives, the Anne S.K. Brown Military Collection at the Brown University Library in Providence (USA), The National Archives (formerly the Public Records Office at Kew, UK), the Canadian War Museum, the Fort Henry Museum and a sizeable number of provincial, municipal, regimental and corps museums as well as many other types of museums and private collections that hold outstanding military objects that are credited in this work.

Since the 1960s, the author photographed objects and records whenever possible, which partly explains why so many images in these books are his own creations. These are simply credited Ph: RC for the sake of space.

Many others are credited: Ph: CL. These are images that Clive Law took or obtained and shared for this volume. The author apologises if he might have erred in some of the credits, but Clive has departed to our everlasting sorrow.

Abbreviations

AGR: Auditor General's Report (printed in yearly Sessional Papers, Parliament of Canada)

C: [and a number for] LAC, RG8, Vol. C[and a number]

CMG: *Canadian Militia Gazette*

CO: Colonial Office, series at The National Archives, Kew, United Kingdom.

DAG: Deputy Adjutant General

GO: General Order

HQ: Headquarters

JSAHR: *Journal of the Society for Army Historical Research*

LAC: Library and Archives Canada, Ottawa

MA: Manitoba Archives, Winnipeg

MC&H: *Military Collector and Historian, Journal of the Company of Military Historians*

MR: Militia Report (printed in the yearly Sessional Papers, Parliament of Canada)

PANS: Public Archives of Nova Scotia, Halifax

RG: [and a number for] Record Group

SP: Sessional Papers, Parliament of Canada

WO: War Office, series at The National Archives (TNA), Kew, United Kingdom.

Chapter 1

Infantry 1855-1870

Christopher Cheyne, officer of the Peel County Sedentary (or Non Service) Militia, mid-1850s. His dress is much as it was ordered for militia officers since 1838. A single-breasted dark blue frock coat with silver buttons and shoulder boards, a peaked cap, crimson sash and dark trousers. From an image in W.P. Bull's From Brock to Currie. Photo RC.

Up to the middle of the 19th century, Canadian militia units that were properly armed, well equipped and decently uniformed were not numerous. Those that were tended to be made up of volunteers that equipped and uniformed themselves at their own expense. But the great majority of militiamen were unarmed, unequipped, untrained and without uniforms. The 1855 Volunteer Militia Act in Canada heralded, as seen in Volume 1, a new era in the country's defence organisation thanks to arms distribution and financial encouragement that favoured drill pay and procurement of uniforms. However, these few thousand volunteers were a small minority of the country's able-bodied men liable to be mobilised for active service.

The bulk of the country's reserve consisted of many infantry battalions of "Sedentary" county militia that were simply becoming a roster of hundreds of thousand of men that could be called upon in an emergency. The old practice of forming the more enthusiastic men into "Flank" companies continued in some of these battalions. These tended to assemble for drills and might have some weapons, equipment and even a few might have somewhat similar clothing. They would gather at the annual muster, which was supposed to be attended by all sedentary able-bodied militiamen, but of course actually attended by far less men, especially in cities. The countryside county battalions would see more attendance since, as explained in Volume 1, the "June Muster" was a pleasant social occasion for men who knew each other to get together. Some flank companies went on after the passage of the 1855 Volunteer Militia Act. Although unpaid and on the whole unequipped by governments, they were obvious expressions of maintaining tradition, pride and local service in their counties. On the whole, however, the emphasis was henceforth given to the new volunteer companies subsided by the government.

Since the 1820s, officers of the sedentary units were supposed to have dark blue frock coats and peaked caps. It seems that few actually did and these tended to be in Canada West. Nevertheless, the Militia Dress Regulations issued in Toronto on 7 February 1856 updated the style of this dress as follows:

> "Frock coat blue, double breast with stand up collar, rounded off in front, cuffs and lapels all blue, two rows of buttons down the front, nine in each row at equal distances. On the left shoulder a crimson silk cord to retain the sash with a small button. Trousers, dark blue cloth with a scarlet welt down the outward seams during Autumn and Winter, and white linen during the summer. Forage cap blue cloth with black silk maple leaf lace, with the name of the Regiment and the number of the Battalion, thereof, worked in silver embroidery. The number to be one inch and a half long, sash crimson silk net, with fringe ends, united by a crimson

runner, worn diagonally over the left shoulder, and the ends of the fringe not to hang below the bottom of the coat. Waist Belt, enamelled white leather, worn over the coat, sword the same as prescribed in H. M. Army Field Officers to wear brass scabbards, Adjutants Steel scabbards, all other officers leather scabbards sword knot crimson and silver with bullion tassels."

Officer of the Sedentary (or Non-Service) Militia, c 1865. He wears an old single-breasted dark blue frock coat dating from the late 1840s or early 1850s. The crimson sash is worn over the shoulder (from 1856) and his shako (on the table) appears to be the model introduced in 1859. Courtesy Don Blyth Collection, Guelph, Ontario.

This was the same officer's blue frock coat as adopted since July 1852 in the British Army.

It further confirmed that: "The lace and buttons worn on all militia uniforms to be Silver." This provision conforming silver instead of gold was grudgingly adhered to, but never totally accepted in the infantry. Rank badges for all were updated to those adopted by the British Army in 1855: "Field Officers to have the distinction of their rank, crown and star for Colonel, crown for Lieut. Col. Star for Majors, embroidered in gold at each end of the collar. The collars of the other officers, to be plain." (Please see the chapter on rank badges for details).

This order of dress was followed by officers in some sedentary battalion companies. An example is the 5th Lincoln's No 2 Flank Company. Its officers wore the 1856 frock coat, sash trousers, sword and peaked forage cap described above with "5th Lincoln" embroidered in silver. The men do not appear to have had uniforms, but were armed with 1840s vintage British army surplus smoothbore percussion muskets, which was the case in those sedentary militia companies that had weapons from the government.

The Sedentary Militia consisting of infantry meanwhile had no uniforms whatsoever nor for that matter much weapons for its enlisted men. In 1858, a proposal was made by Adjutant General de Rottenburg to buy 1,400 coats with trousers and 5,000 forage caps that were surplus and being sold from British Army Ordnance stores at Kingston; these could be issued to the keener sedentary militiamen grouped in flank companies, but this proposal was ignored. Another try was made three years later, this time trying to obtain free red faced blue uniforms from Britain. An Order in Council of 27 February 1861 asked "the Imperial Government to transmit to Canada from the Military Stores if they can be spared" 100,000 "Red tunics with Blue facings" with an equal number of trousers "as worn by the Regular Infantry" along with fatigue caps and greatcoats. The ever-stingy Canadian government wanted all this for free and, predictably, had no success with its request.[2]

Volunteers in Red Coats

Insofar as actual infantry uniforms being worn, there were hardly any red-coated volunteer infantry units left in central Canada. In Canada East. Only the Montreal Light Infantry raised since 1837 wore scarlet coats, but it had become somewhat inactive. It was reorganised in 1856 to have six companies wearing uniforms "of a Light

[2] *Journals of the Province of Canada... 1858,* Vol. XVI, Sessional Papers No. 6; LAC, RG9, IIA6, Vol. 8, Order in Council, 27 February 1861, Clothing for Sedentary Militia.

Infantry Regiment of the Line with Blue facings" by its regulations of 18 December. Further regulations sanctioned 27 April 1857 mentioned modifications of its uniforms consisting of: "Silver lace, plate, &c., are to be worn in place of gold, and vice versa – as generally required for Militia. Shako plate and other ornaments, are to be of the patterns kept by the Quarter Master. Trousers to be of Oxford-mixture, with heavy scarlet bead down the outside seams. Officers in place of the new sling-sash, are to wear the old light infantry sash, with white patent-leather pouch-belt and black patent-leather pouch, of the pattern kept by the Quarter Master. The paymaster and Quarter Master are to be distinguished from other officers, only by wearing black belts and no sash. Cadets are to be distinguished from privates, by wearing a star of a second lieutenant on the collar of the tunic." In 1862, each volunteer of its six companies was clothed in a new "scarlet cloth tunic, blue facings – trousers grey Oxford mixture scarlet piping – shako demi regulation with green ball – winter fur cap wedge shaped – forage cap blue cloth with white band & bugle badge" that had been issued July.[3]

Canadian Militia shako, c. 1855-1861. The white metal shako plate is a general issue model with "Honi Soit qui mal y pense" on the round garter and "VR" at its centre. Fort Henry, Kingston. Ph: RC.

Otherwise, the only other unit in scarlet infantry uniform in the later 1850s appears to have been the Sorel Rifle Company that reported in July 1858 that its "present uniform is red with blue facings, grey pants" and that it would change to rifle green faced with scarlet when its members could afford it. The replacement occurred in 1860. The surge of volunteer enlistments due to the "Trent Incident" led to a desire to create, especially in Montreal, units that were clothed in scarlet that were either light or line infantry rather than rifle units. Thus, the volunteers of the 5th "Royals" Battalion were issued uniforms in May 1862 consisting of a "Scarlet cloth tunic, blue facings – regulation trousers Oxford mixture – New regulation blue cloth shako, blue cloth forage cap with red band, seal skin fur cap" for winter while the 6th "Hochelaga Light Infantry" Battalion organised in Montreal during 1862 was also dressed that year in scarlet tunics faced with dark blue facings, Oxford mixture "pants grey, scarlet stripe – blue cloth shakos."[4]

In Canada West, there were no volunteer units reported dressed in scarlet coats during the later 1850s. The first to do so appears to have been the St. Catharine's Company of Volunteer Infantry officially formed in August 1862, by actually existing as the No. 2 Flank Company of the 5th Lincoln Sedentary Militia. From about 1861 or early 1862, it adopted a uniform of "scarlet cloth, blue facings – new regulation shako", the shako obviously later since it was called of the "pattern of 1862" in the commander's report. Its officers whew the scarlet full dress like that worn in the British infantry Toronto's 10th Infantry Battalion was being organised in the fall of 1862 and was to be dressed in red coats. By November however, although armed and equipped with buff leather accoutrements and issued with grey greatcoats, it reported having "No uniforms" for its enlisted men.[5]

Meanwhile, the number of volunteer rifle companies kept rising. In February 1862, the existing companies numbered 150 with another 186 companies likely to be raised for a total of 336 companies. In cities, a few battalions of green-coated rifle companies had been timidly been formed since 1859, but these only amounted to four battalions, three of them in Montreal. More battalions had to be formed. In January 1862, the 5th Royal Light Infantry at Montreal was the first scarlet-coated infantry battalion to be organised quickly followed by the 6th Hochelaga Light Infantry also in Montreal. The Montreal Light Infantry, the 5th Royals and the 6th Hochelaga

[3] *Regulations and Standing Orders of the Montreal Light Infantry* (Montreal, 1857), pp.7, 15; LAC, RG9, IC8, Vol. 3, Commanding Officers Reports…October 1862; Earl John Chapman, *The Black Watch of Canada: the early years, 1862–1878* (Montreal: Black Watch of Canada Foundation, 2006): 11.

[4] LAC, RG9, IC8, Vol. 3, Commanding Officers Reports…October 1862.

[5] LAC, RG9, IC8, Vol. 4, Commanding Officers Reports…December 1862; R.L. Rogers, *The History of the Lincoln and Welland Regiment* (n.p., 1954): 21.

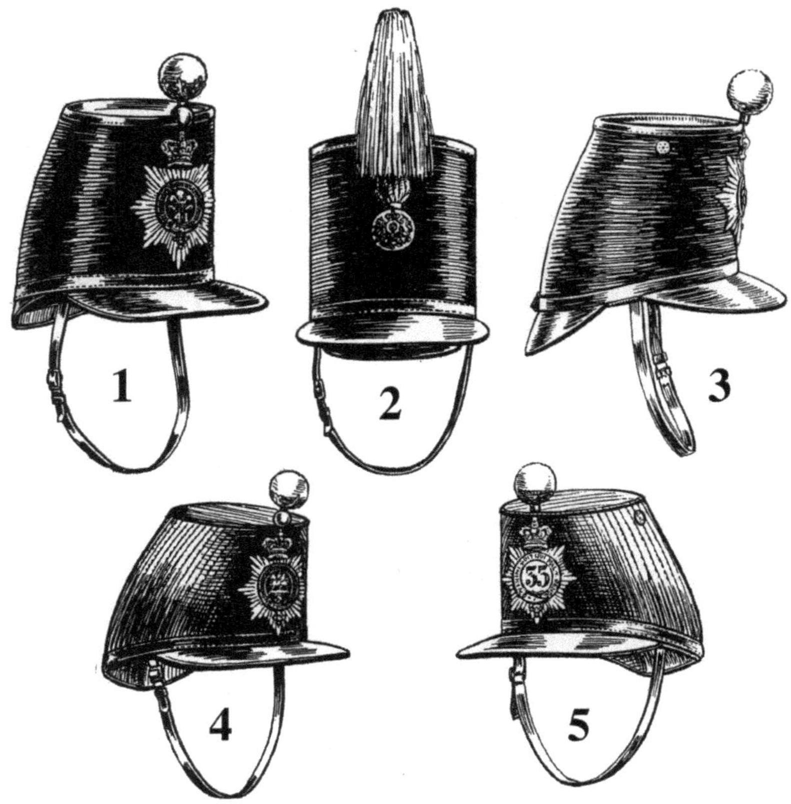

British infantry shakos, 1855-1869. 1: This was the “French Pattern Shako”, officer’s 1855-1861. 2: Fusilier units, officers 1855-1861 (. 3. men’s 1855-1861. 4: This was the “Quilted Shako”, officer’s 1861-1869. 5: men’s 1861-1861. Print after A.R. Cattley. JSAHR, Vol. XV (1936).

obtained red uniforms from England that were, except for white metal buttons instead of brass, and were exactly like those worn by the British infantry including tunics with dark blue cuff flaps and shoulder straps.

Infantry Battalions formed between 1862 and 1870:

Montreal Light Infantry raised 1837, reorganised 1856, disbanded 1868

5th Rifles (Montreal QC) 31 January 1862; 5th Royal Light Infantry 7 November 1862

6th Hochelaga Light Infantry 5 June 1863

7th Infantry (London ON) 27 April 1866; 7th London Light Infantry 15 February 1867

10th Rifles (Toronto ON) 14 March 1862; 10th Infantry 21 November 1862; 10th or Royal Regiment of Toronto Volunteers 10 April 1863

11th Infantry (St Andrews and Lachute QC) 4 June 1862, also called Argenteuil Rangers

12th York Infantry (Aurora ON) 14 September 1866

13th Infantry (Hamilton ON) 11 December 1862

15th Infantry (Belleville ON) 16 January 1863

16th Prince Edward Infantry (Picton ON) 6 February 1863

17th Levis Infantry (Lévis QC) 20 February 1863

18th Prescott Infantry (Hawkesbury and Vankleek Hill ON) 26 February 1863

19th Infantry (St. Catharines ON) 18 March 1863; 19th Lincoln Infantry 28 September 1866

20th Infantry (St. Catharines ON) 17 April 1863, disbanded 28 September 1866, companies transferred to form 19th Lincoln Infantry; 20th Halton (Milton ON) Infantry 28 September 1866, converted to 20th Halton Rifles 12 January 1872

21st Richelieu Light Infantry (Saint Jean QC) 25 May 1863

23rd Infantry (Brockville ON) 4 September 1863, disbanded 3 May 1867

23rd [originally Beauce Provisional Infantry 17 December 1869] Beauce Infantry (Sainte-Marie QC) 19 May 1871]

24th Kent Infantry (Chatham ON) 14 September 1866

Officer and private of a Montreal Volunteer Infantry unit, c. 1862. This music sheet cover of "The Canadian Volunteer Polka" probably shows members of the Montreal Light Infantry, the 5th or the 6th battalions all of which had a scarlet uniform with the cuff flap as on British infantry uniforms. Courtesy Library and Archives Canada.

25th Elgin Infantry (Waterloo ON) 14 September 1866

26th Middlesex Infantry (Strathroy ON) 14 September 1866

27th Lambton Infantry (Sarnia ON) 14 September 1866

28th Perth Infantry (Stratsford ON) 14 September 1866

29th Waterloo Infantry (Berlin [now Kitchener] ON) 14 September 1866

31st Grey Infantry (Owen Sound ON) 14 September 1866

32nd Bruce Infantry (Walkerton ON) 14 September 1866

33rd Huron Infantry (Goderich ON) 14 September 1866

34th Ontario Infantry (Whitby ON) 14 September 1866

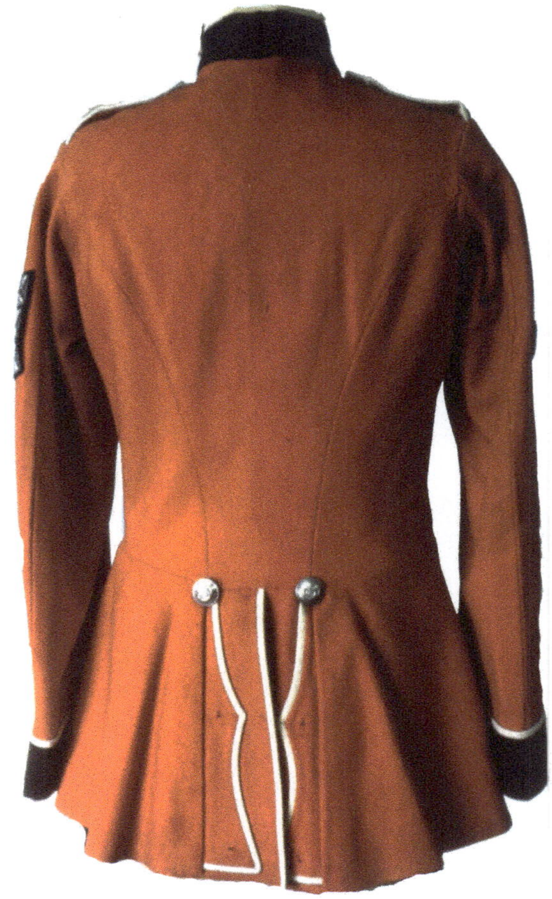

(Right) Colour sergeant's tunic, 6th Hochelaga Light Infantry Battalion, c. 1862-1867. This unit had the British pattern tunic for the first years of its existence. Scarlet with dark blue collar, cuffs, cuff flaps and shoulder straps, white piping and lace, white metal buttons, silver rank chevrons. This was one of the few units that had a distinctive unit button; it bore the numeral 6 at the centre of a bugle horn with "Canada" above and "Militia" below. Private collection. Photo RC. (Left) Colour sergeant's tunic, 6th Hochelaga Light Infantry Battalion, c. 1862-1867. Back view. Private collection. Ph: RC.

35th Simcoe Infantry (Barrie ON) 14 September 1866; The Simcoe Foresters 5 April 1867

36th Peel Infantry (Brampton ON) 14 September 1866

40th Northumberland Infantry (Cobourg ON) 5 October 1866

42nd Brockville Infantry (Brockville ON) 15 October 1866

43rd Carleton Infantry (Ottawa ON) 5 October 1866, nicknamed the Carleton Blazers

44th Welland Infantry (Clifton ON) 16 November 1866

45th West Durham Infantry (Bowmanville ON) 16 November 1866

46th East Durham Infantry (Port Hope ON) 16 November 1866

47th Frontenac Infantry (Napanee ON) 30 November 1866

48th Lennox and Addington Infantry (Napanee ON) 14 September 1866

50th Huntingdon Borderers Infantry (Huntingdon QC) 14 September 1866

51st Hemmingford Rangers Infantry (Hemmingford QC) 14 September 1866

52nd Bedford Infantry (Knowlton QC) 14 September 1866

53rd Sherbrooke Infantry (Sherbrooke QC) 21 September 1866; 53rd Melbourne Infantry 22 March 1867

54th Sherbrooke Infantry (Sherbrooke QC) 22 March 1867; 54th Richmond Infantry (Richmond QC) 10 May 1867

55th Megantic Infantry (Inverness QC) 22 March 1867

56th Prescott Infantry (Kemptville ON) 27 April 1867, converted to 56th Grenville Rifles 13 August 1871

57th Peterborough Infantry (Peterborough ON) 3 May 1857

58th Compton Infantry (Cookshire QC) 11 October 1867

59th Stormont and Glengarry Infantry (Cornwall ON) 3 July 1868

Colour sergeant's tunic, 6th Hochelaga Light Infantry Battalion, c. 1862-1867. Cuff and cuff flap detail. Private collection photo by RC

60th Mississiquoi Infantry (Dunham QC) 19 February 1869

61st [originally Montmagny and l'Islet Provisional Infantry 9 April 1869] Montmagny and l'Islet Infantry (Montmagny QC) 5 November 1869

62nd St. John Volunteer Battalion (St. John NB) 5 November 1869

66th Halifax Volunteer Infantry (Halifax NS) 5 November 1869

67th Carleton Light Infantry (Woodstock NB) 5 November 1869

68th King's County Infantry (Kentville NS) 5 November 1869

69th [originally 1st Annapolis County Infantry (Paradise NS) 8 October 1869] 1st Annapolis County Volunteers 5 November 1869

70th [originally the Champlain Provisional Infantry Battalion of 16 October 1869] Champlain Infantry (Sainte-Geneviève de Batiscan QC) 5 November 1869

71st [originally the York Provisional Infantry Battalion of 10 September 1869]; 71st York Infantry (York County NB) 12 November 1869

The 1863 Canadian Pattern Infantry Uniform

In 1862, senior military authorities rightly perceived that many Canadian volunteers had to be deployed as line infantrymen. This meant that a large number of units would need to wear scarlet tunics. This was seen

Officers of the 6th Hochelaga Light Infantry Battalion, c. 1862-1867. This unit had the British 1856 Pattern single-breasted tunic for the first years of its existence. Scarlet tunics with dark blue collar, cuffs and cuff flaps (except the ensign at right that has Austrian knots), white piping, silver buttons and lace; dark grey Oxford mixture trousers with scarlet piping; British 1855 Pattern shakos with silver crowned bugle badges with "6" at centre, white over red pompon (except for the surgeon who has a bacon and no crimson sash); colour sergeant between the colours wearing the Canadian pattern 1863 uniform with the Austrian knot above the cuff. Private collection. (Photo by RC)

as a substantial advantage: when deployed in the field with British infantry battalions, scarlet-coated Canadian volunteer infantry units could, at a distance, be taken by opponents for the redoubtable British regular "red coats". Yet nearly all units existing consisted of riflemen that were to act as skirmishers, which created a tactical imbalance in the Volunteer Militia's organisation. To remedy this situation, it was announced from early December "... that all companies to be raised in future are to be infantry." This meant, in terms of uniforms, that scarlet tunics would be worn by any new unit henceforth authorised. This, added to the fact that many of the older green-coated volunteer rifle companies raised since 1855 needed to renew their uniforms led the government to allow the purchase of some 20,000 uniforms in early 1863.[6]

Private of a Canadian Volunteer Militia Infantry unit, c. 1863-1870. He wears the 1863 Canadian Pattern uniform for infantry units. Plate by Ron Volstad. Courtesy History and Heritage Directorate, Department of National Defence, Ottawa.

By early 1863, uniforms simply had to be purchased very soon "to uniform promptly the 25,000 volunteers" that needed them. Some 20,000 pairs of trousers, grey with narrow red piping, were ordered from clothing firms in Canada. The country did not then have capacity to produce the more complicated parts of uniforms such as tunics and headdress. On 5 February 1863, both adjutants generals for Lower and Upper Canada made a common recommendation to the Minister of Militia "to procure 20,000 tunics" and headdresses "from England...requiring...13,000 scarlet tunics do. infantry – 13,000 shakos do. do...[the] tunic...should be equal in quality to those worn by sergeants of the imperial service." This meant the Canadian infantry tunics were of superior scarlet cloth. This proposal was quickly approved by an Order in Council on 7 February.[7]

Canada West Adjutant General Walker Powell looked after uniforms and he was immediately sent to England to purchase the uniforms including 13,000 infantry tunics and shakos. By 16 March, he was negotiating in London with the clothier Herbert & Company for "Scarlet cloth tunic of sergeants quality lined and made up as the regulations for Her Majesty's infantry (excepting buttons)" that, by 20 March, had "alterations in the sleeves and shoulder straps" that proved to be a "knot on sleeve" and shoulder cords instead of the cuff flaps and shoulder

[6] LAC, RG9, IC1, Vol 115, DAG to George Williamson, Quebec, 1 December 1862.

[7] LAC, RG9, IIA6, Vol. 9, de Salaberry and Powell to Minister, 5 February 1863.

Canadian Pattern 1863 infantry tunic for a private. This pattern was issued to volunteer militia infantry units until 1870. Scarlet with dark blue collar and cuffs, white piping, white cord Austrian knots at the cuffs and shoulder cords, white metal buttons. Canadian War Museum, Ottawa. Courtesy National Historic Sites, Parks Canada.

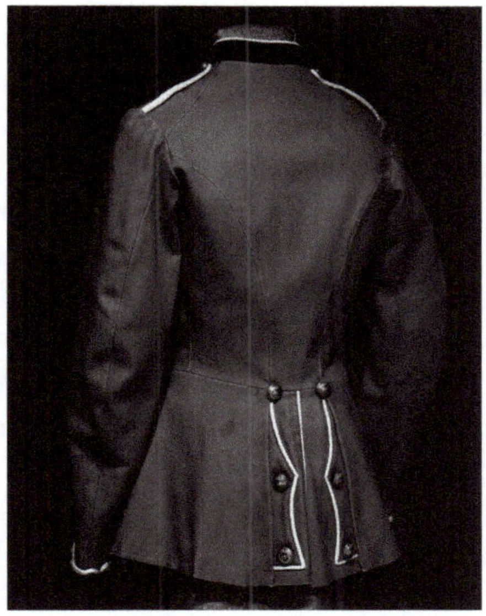

Canadian Pattern 1863 infantry tunic for a private. Back view. Canadian War Museum, Ottawa. Courtesy National Historic Sites, Parks Canada.

1863 pattern, such as this one, have brass "beaver" buttons rather than white metal. Apparently, this appeared on later supplies of this pattern's tunic. It is otherwise similar being scarlet with dark blue collar and cuffs, white piping, white cord Austrian knots at the cuffs and shoulder cords. Fort Henry, Kingston. Photo by RC.

straps seen on British infantrymen's tunics. The knot proved to be a white cord Austrian knot. Herbert lost the bid to Peter Tait & Company who were destined to supply the bulk of Canadian uniforms for the next 22 years.

The 1863 Canadian pattern uniforms were made by Peter Tait & Company within a few weeks and, by mid-May, had arrived in Canada. The issue of the 1863 pattern infantry uniforms started on 23 May with the 17th Lévis Battalion receiving 315 tunics, shakos and trousers followed on 27 May by Toronto's 10th and Hamilton's 13th receiving their uniforms. Most units received uniforms in June and July although distribution continued until the fall. The Montreal infantry: three battalions, two companies in Hamilton and one in St. Catharines that had already purchased scarlet uniforms in 1862 were compensated. Thus, by the end of 1863, the bulk of volunteers units in Canada were now clothed in scarlet coats. The uniforms, especially the tunics, were expected to last four to five years. They were paid for by deductions from the drill allowance to each volunteer.

The 1865 Canadian Pattern Serge Uniform and its Issue

The need for a more informal service dress appeared in 1864 when schools of military instructions were formed and when 30 infantry and rifle companies of the volunteer units were called on active duty at the American border from 1865. The Militia Department therefore ordered scarlet serge frocks for infantrymen from England in May 1865. These were mainly for issue to the infantry companies called up for active duty on the border within three temporary "administrative" battalions formed with detachments from various volunteer battalions (rifle companies also received green frocks; please see Volume 1). The kit for the men called up included Kilmarnock forage caps with numbers 1, 2 or 3, leather boots, haversacks and trousers. This red infantry serge frock was a loose garment with dark blue collar piped with white, plain red cuffs, white shoulder cords, four buttons in front and a patch pocket on the left breast.

In March 1866, "uniform clothing of a new pattern consisting of a tunic [actually a serge frock], trousers and shako" was offered to volunteer units for $6.50 per man instead of the more expensive cloth tunic, trousers and shakos at $10.50. This "new pattern" was a serge uniform consisting of a frock cut in the Norfolk jacket style and serge trousers. The infantry's serge frock was red with dark blue collar and pointed cuffs, white piping edging the front and the skirt. The infantry trousers offered were of dark blue serge with scarlet piping. Some 2,000 in-

Canadian Militia shako, c. 1863-1867. This general issue Pattern 1861 British shako has a white metal shako plate has "VR" with "Royal Canadian Militia Regiment" on the garter. There was no such regiment, but the British manufacturers must have had to come up with something quickly, which suggests this shako was part of the 1863 procurement. The seemingly later issues often had a beaver at the centre with "Canada Militia" around and might be of brass. Canadian War Museum, Ottawa. Photo: RC.

fantry serge uniforms were received from England in early 1866. Other items such as the shako and the forage cap remained as previously.

This new serge uniform could be more suitable as a service dress and seemed initially cheaper. The Militia Department allowed $6.00 per man every two years. On the other hand, if units waited four years or more, they made up more than the difference by ordering the better quality cloth uniforms at $10.50. As a result, these serge uniforms, which were not especially elegant, were not greeted with enthusiasm. In late April 1866, Adjutant General Walker Powell reported that "the great bulk of the Corps are asking for clothing of the old pattern" and that "not more than fifteen companies have asked for the new pattern and these are doing so with a view to have both if they can be supplied." Generally, the units that did order the new scarlet frock preferred to have Oxford grey cloth trousers. Some acquired the serge clothing for undress or fatigue purposes to save the better cloth uniforms for parades. A few battalions had some companies with cloth tunics and serge frocks for others.[8]

From the 1866-1867 issue records, the infantry companies that wore only the scarlet serge frock were at Acton, Dickinsons Landing, Napanee, Nelson, Newmarket, Norwood, Milton, Orilia, Oro, Peterborough, Portage du Fort and Williamstown. All had shakos.

Battalions that had some companies wearing cloth tunics and other companies wearing scarlet serge frocks were the 26th Middlesex, 34th Ontario, 50th Huntingdon, 51st Hemmingford, 54th Melbourne, 55th Megantic, 56th Prescott and 59th Compton.

Battalions that had both the cloth tunic and the serge frock for undress were the 20th Halton, 31st Grey, 32nd Bruce, 33rd Huron, 40th Northumberland, 44th Welland and the company at Vernon.

Later issues in 1869 showed some companies in Quebec province battalions still being issued with both cloth tunics and serge frocks, the latter obviously for undress. There were no serge frock issues in eastern Ontario, Nova Scotia and New Brunswick at that time. The serge frock was never popular and the remaining stock was declared surplus a decade later, some 2,000 being distributed to indigent western First Nations.[9]

Officers and enlisted men of the Canadian volunteer infantry, 1863-1870. From left to right: a private wearing the pattern 1863 scarlet uniform for enlisted men; a major wearing the pattern 1863 scarlet uniform for officers; a major or a lieutenant wearing the undress pattern 1856 dark blue frock coat with the officer's model peaked cap; a private wearing the 1866 scarlet serge frock and a quartermaster-sergeant wearing an undress scarlet serge frock, one of several types seen, with a peaked cap. In the background are two musicians in white tunics. A shako plate is at upper left. Plate by R. Marrion. Courtesy of the Company of Military Historians. Ph: RC.

[8]LAC, RG9, ID1, Vol. 1, Memo on service serge frocks and caps for frontier troops, 6 March 1865 by DAG Walker Powell; LAC, RG9, IC1, Vol. 110, Walker Powell to P. Tait & Co., Ottawa, 22 February 1867; LAC, RG9, IC4, Vol. 12, General Order, Ottawa, March 1865 and Regulations relating to the issue of Uniform Clothing…, Ottawa, November 1866; LAC, RG9, IC8, Vol. 5, Walker Powell to Col. McDougall, Ottawa 25 April 1866; Denis Carter-Edwards, "Border Guards: the Second Administrative Battalion and the Defence of Niagara", *Organisation of Military Museums of Canada Journal*, IX (1981-1982), pp. 15-26.

[9]LAC, RG9, IB2, Vol. 50, Return showing last issue, District No. 6, 28 November 1872 and Date of last issue, District No. 7 (for 1869-1872). The 1869 serge frock issues were: 23rd Bn 55 (& 330 cloth tunics); Rimouski Bn 55 (& 220 cloth tunics); Temiscouata Bn 55 (& 220 cloth tunics); Quebec County Bn 55 (& 165 cloth tunics); Joliette Bn 220 (& 110 cloth tunics); Trois-Rivières Bn 165 (& 165 cloth

(Left) Private of an infantry unit, 1866-1870. This figure ornamented a Province of Ontario location certificate for land grants given in 1906 to veterans of the 1866 Fenian Raids. Apart from the white shoulder cord shown as a white shoulder strap edged with scarlet piping, the uniform shown is generally correct for the c. 1866-1870 period. Private collection. Photo: RC.

(Top Right) Colour-Sergeant of the 19th Lincoln Infantry Battalion, c. 1867-1871. This superb image is correctly coloured showing the dark blue forage cap with white metal numbers (issued instead of shakos from 1867-1868), scarlet Canadian 1863 pattern tunic with dark blue cuffs and collar white piping and cords, white metal buttons, rank badges, crimson shoulder sash, dark blue trousers with scarlet piping, white infantry accoutrements with brass buckle. Courtesy of The Lincoln and Welland Regiment Museum.

Private wearing the pattern 1863 scarlet infantry tunic, but with brass buttons according to this partly coloured tintype, which is rarely seen. Such buttons appear to have been on tunics from the later 1860s. The shako badge, coloured yellow, is worn on the forage cap. The pants appear to be of a ribbed material and could be orinally civilian upon which scarlet piping has been added. Private collection. Ph: CL.

Private of a light infantry battalion, c. 1865. He wears the scarlet Canadian 1863 pattern tunic with a forage cap upon which is a stringed bugle badge indicating light infantry. Courtesy Scott Duncan.

Private of a Canadian Volunteer Militia Infantry unit, c. 1863-1870. He wears the 1863 Canadian pattern uniform for infantry units with his quilted shako on the table. Curiously, he seems to have a NCOs or officer's sword belt, but no rank badges are on his tunic. Courtesy Scott Duncan.

5th Royal Light Infantry MTL c 1866 NHS Parks Canada

Later cloth tunic and trousers issues shakos abolished

After the 1863 issue, more Canadian pattern uniforms were occasionally ordered from England. In May 1865, some 1,200 infantry tunics with shakos were ordered. According to the *Syracuse Daily Courier* citing the *St. Catharines Daily Post*, some 14,000 militia pantaloons were "in store at Quebec" and 5,000 shakos were due to "arrive by next steamer" while 5,000 more had been rejected for not being made according to contract. Officer's uniforms were expected to arrive also and all this led the Syracuse reporters to wonder "what does this all mean? Are the Canucks preparing with Father Abraham? The paragraph certainly looks warlike. One feature, however, is deserving the attention of our army officials here. The Canadian militia are not to be clothed in 'Shoddy'". There were no international incidents on that matter; in April 1866, "2,000 sergeants quality infantry tunics" with 3,500 pairs of Oxford mixture cloth trousers were expected.[10]

---

tunics); Bonaventure Coy 55; New Carlisle Coy 55; 1st & 2nd Rawdon Coys 110 (& 110 cloth tunics); St. Edouard de Gentilly Coy 55; St. Jérome 55; Ste. Gertrude Coy 55; Bécancour Coy 55; Victoriaville Coy 55; St. Florient 55; Ste. Pie 55; St. Simon 55; Wilton 55,

[10] Clothing ordered from England 12 May 1865: Artillery tunics 500; Artillery trousers pairs 500; Artillery Busbies 500; Hussar tunics 500; Hussar trousers pairs 500; Hussar Busbies 500; Infantry tunics 1200; Rifle tunics 2000; Rifle trousers pairs 2000; Rifle shakos 2500; Infantry serge frocks 1000; Rifle serge frocks 1000; Infantry shakos 1200; Infantry forage caps 1000; Rifle forage caps 1000; Great coats 1000; Ditto for officers superior make and pattern 200. LAC, RG9, IC8, Vol 5; Ordered on 12 April 1866: 2000 sergeants quality infantry tunic (scarlet); 3500 pairs Oxford mixture trousers…; 2000 pairs blue serge trousers…; 1000 Artillery forage caps, Regulation Royal Artillery; 200 Infantry chevrons sergeants; 200 Infantry chevrons corporals. W. Powell to P. Tait & Co.,(Limerick) 12 April 1866, LAC, RG9, IC1, Vol. 109.

A group of soldiers of the 3rd Administrative Battalion, Fort Wellington, May 1865. Most wear the scarlet frock issued to the companies on active service, which was similar to the frocks worn by cadets. They also had their regular pattern 1863 infantry tunic as shown by two of the soldiers. All have the white metal "3" on the dark blue forage caps. The three officers wear the dark blue double-breasted frock coat with crimson sash over the left shoulder. These officers and enlisted men are identified as being detached from the 13th Infantry Battalion. Courtesy Library and Archives Canada, PA89325.

Many new battalions had been formed, the Grand Trunk Railway Brigade was being organised and all needed new uniforms. A large issue occurred during 1866 and 1867 that, as seen above, could consist of either the "old pattern" scarlet cloth or the "New pattern" scarlet serge frock for infantry units. As it turned out, nearly all infantry units chose the scarlet cloth tunic if it was its only uniform, which was the 1863 Canadian pattern with Austrian knots at the cuffs, etc. They first issue in 1863 had white metal buttons, but it appears from surviving tunics that probably from 1866 and the later 1860s, some or all the enlisted men's scarlet cloth tunics had brass buttons. The 1866-1867 uniform issues also replaced the older British army pattern uniforms of the Montreal Light Infantry, the 5th Royals and the 6th Hochelaga which now had the 1863 Canadian Pattern scarlet cloth tunic. However, the 5th Royals had the sergeant's tunics altered since they could still be seen years later with cuff flaps and shoulder straps rather than cords.

Another aspect were the trousers that were now imported from Great Britain. In April 1866, Adjutant General Walker Powell announced that he feared a great "want of trousers" and had already "ordered 10,000" of them. Since 16 August 1865, the British regular infantry was to have "an Oxford Mixture tartan of the same shade at the cloth trousers and worn by the Infantry, having been approved for summer trousers in lieu of the blue serge." This was not the case for Canada, which ordered both the grey Oxford cloth mixture and the dark blue serge trousers. The Oxford grey cloth ones were of a very dark grey hue while the serge trousers were also a very dark hue of blue, both type having a line of scarlet piping down each outer side seam. Either type was hardly distinguishable at a distance, but some units might have scarlet cloth tunics with Oxford grey cloth trousers and others with dark blue serge trousers.[11]

[1] LAC, RG9, IC8, Vol. 5, Adjutant General to Col. McDougall, Ottawa 25 April 1866. More were ordered on 30 July 1866: 5000 sergeant's quality infantry tunics (scarlet); 2500 pairs Oxford mixture trousers assorted sizes; 6000 infantry forage caps, regulation without badges; 1000 scarlet jackets pattern 12/6 same as 2000 ordered on the 21st May; 1000 pairs blue serge trousers infantry; 1000 pairs blue serge trousers artillery. W. Powell to P. Tait & Co.,(Limerick) 30 July 1866. LAC, RG9, IC1, Vol. 109.

Infantry private of the 1st Administrative Battalion, 1865. He wears the white metal "1" on his dark blue forage cap. Courtesy National Historic Sites, Parks Canada.

Private of the 29th Waterloo Infantry Battalion, c. 1868-1871. He wears the 1863 pattern uniform. His forage cap has "29" in front. Photo by E. L. Johnson listed in the 1871 census as a photographer in Goderich, Ontario. Anne S.K. Brown Military Collection, Brown University Library, Providence. Ph: RC.

Sergeant of the 29th Waterloo Infantry Battalion, c. 1868-1871. He wears the 1863 pattern uniform with the sergeant's three white rank chevrons and the red wool sash over the right shoulder. His forage cap has "29" in front. Photo by E. L. Johnson listed in the 1871 census as a photographer in Goderich, Ontario. Anne S.K. Brown Military Collection, Brown University Library, Providence. Ph: RC

Sergeant and private of a volunteer infantry unit, c. 1863-1866. Both wear the Canadian pattern 1863 scarlet tunic. The sergeant (at left) has three rank chevrons, the crimson sash and, barely visible at the waist. Courtesy of Scott Duncan.

The 46th East Durham Infantry Battalion attending a divine service in camp, 1866. This remarkable and rare view of a military religious service was possibly at the Port Hope camp. At left, the regimental band wearing white coats and, at the centre and right, behind the Anglican churchmen, the officers and men of the battalion in scarlet coats. Courtesy Ontario Archives, Toronto, S2823.

Funeral at Toronto's St. James Cemetery of the five Canadian volunteers killed during the Fenian Raid into the Niagara Peninsula on 2 June 1866. The infantry unit in the foreground is almost certainly the 10th Royals Battalion since it wears the Canadian infantry pattern tunic and was then the only Toronto unit have the scarlet uniform. Officers wearing busbies in dark blue (artillery) and scarlet (engineers) tunics are in the middle ground. This is a coloured version of a print from a sketch after C. Fuller of Toronto published in the Illustrated London News, 30 June 1866. Courtesy Toronto Reference Library, pictures-r-1310.

Officers and men of the 15th Belleville Infantry Battalion on parade at Fort Wellington in Prescott during June 1866. Most wear the 1863 pattern scarlet infantry tunic, but a group is seen in the rifle green uniform. This was due to some rifle companies being incorporated into infantry battalions; they wore out their rifle uniforms before taking scarlet tunics. Fort Henry, Kingston. Ph: RC.

The 19th Lincoln Infantry Battalion formed in a square to repel a cavalry attack, 1866. The men wear the Canadian Pattern 1863 uniform; officers at left have the dark blue frock coat. Courtesy of The Lincoln and Welland Regiment Museum

Private and officers, 13th Hamilton Battalion, 1867. Wearing the Canadian 1863 pattern uniform, the private is armed with the newly issued Snider-Enfield infantry rifled musket. On the mock fence hangs a rolled light grey blanket and what appears to be a locally made knapsack. The officers wear the Canadian 1863 officer's pattern uniform. Courtesy Robert Marrion Collection.

Captain Edwin Isaac Parnell (seated) with his brother, 19th Lincoln Infantry Battalion, c. 1866-1867. Capt. Parnell wears the scarlet tunic with dark blue collar, cuffs and cuff flaps, white piping, silver buttons and lace, crimson shoulder sash, Oxford grey trousers with scarlet piping ; his service dress peaked cap is at his feets. Note the revolver holster slung on his sword belt. His brother, a private in the battalion, wears the standard Canadian pattern 1863 scarlet infantry tunic with an Austrian knot at each cuff. Courtesy of The Lincoln and Welland Regiment Museum.

Sergeants of Toronto's 10th Royal Infantry Battalion, 1868. All wear the 1863 Canadian pattern scarlet cloth tunic from the 1866-1867 issue when the unit received 527 tunics, 635 Oxford mixture trousers, 575 blue serge trousers, 291 shakos, 620 forage caps and 770 greatcoats. The red sergeant's sashes and chevrons were procured separately. At right is the drum major. The boy at the centre wears a white bandman's tunic and holds a flute. The numerals were probably added on the forage caps from 1867. From Chambers' History of the Royal Grenadiers. Ph: RC.

A private of the 50th Huntingdon Borderers Infantry Battalion wearing the pattern 1866 Norfolk scarlet serge tunic. In 1870, the battalion deployed against the Fenians and, on 25 May, fought at Eccles Hill (Quebec) wearing this type of tunic. Scarlet serge with dark blue collar and cuffs, white piping, white metal buttons, dark blue trousers with scarlet piping, dark blue forage cap with brass numerals. Plate by Ron Volstad. Courtesy History and Heritage Directorate, Department of National Defence, Ottawa.

A pattern 1866 Norfolk scarlet tunic for infantry battalions. This damaged tunic may be the only known example of this early and unique Canadian pattern uniform. It has white metal "beaver" buttons. Vankleek Hill Museum, Vankleek Hill, Ontario. Courtesy Michelle Campbell.

In July 1867, Adjutant General Walker Powell wrote that "we are about to abolish shakos and substitute forage caps generally as far more useful and businesslike" and that these dark blue caps would have "white metal numerals" for the battalion's number. This was followed in August 1868 by an order for "20,000 Kilmarnock forage caps" for infantry. Henceforth, if units wanted shakos, they would have to pay for them.[12]

As a result of the 1867 Confederation, the Nova Scotia and New Brunswick volunteer units were gradually integrated into the Canadian Militia. Most Nova Scotia units were issued uniforms during the summer of 1870. In New Brunswick, issues took place during a longer time between May 1869 and 1871. They consisted of the 1863 Canadian Pattern scarlet infantry cloth tunic, cloth trousers, forage caps and numerals.

Officers

Officers were also to have the Austrian knot of silver cord, but this was not the case in several battalions. A few battalions had obtained their uniforms before the middle of 1863 and the tunics had dark blue cuff flaps and shoulder straps, exactly like the British infantry tunics, rather than Austrian knots and shoulder cords. The Montreal Light Infantry, the 5th Royals and the 6th Hochelaga obtained such uniforms for their whole unit and, in the case of the 5th, still wore tunics with cuff flaps in 1870. Officers elsewhere that ordered uniforms from English clothiers rather than from the Militia Department would likely receive tunics with cuff flaps. Some officers of the 13th Hamilton, the 19th Lincoln and the 44th Welland battalions were seen, and sometimes photographed, wearing such coats well into the later 1860s.

For undress, the British army's infantry officer's all dark blue double-breasted frock coat with silver buttons and peaked forage caps was also sent out to Canada in 1863 by Peter Tait & Company for officers who wished to purchase them privately and many of them did. More were sent out three years later. When only forage caps were provided from 1867, officers could "wear the number on their forage caps in silver thread."[13]

Meanwhile, in Great Britain, a new garment called the "patrol jacket" was introduced from March 1867 replacing the 1852/1856 pattern frock coat for officers' undress; the new patrol jacket was dark blue with four black hussar-style loops in front, black Austrian knot at sleeves, the front and skirts edged with wide black lace and the back seams edged with narrow black lace with trefoil ends. Instead of buttons, it had black lace "olivets" on the chest. Officer's shell jackets continued to be used in the mess and became mess jackets while the crimson silk sash would henceforth only be worn in full dress over the scarlet coat. In 1868, the British infantry officer's dress scarlet tunic cuffs was altered to now feature elaborate lace chevrons that designated the wearer's rank. While there was no official order in Canada regarding the new 1867 patrol jacket nor for the new 1868 tunic's trim until the 1876 dress regulations for officers, Canadian officers, and the tailors that supplied them, considered it as a matter of course to conform to new British regulations. Thus, by the early 1870s, many of them would have procured the new patrol jackets and dress tunics.

Officer's gold buttons and lace almost disappeared in the infantry with the advent of the 1863 government-supplied silver laced uniform. The 4 March 1870 regulations and orders reaffirmed that "in the infantry, silver lace is substituted from gold." However, this was not the last of the matter as seen in our chapter on the infantry dealing with the 1871-1901 period.

[12]LAC, RG9, IC1, Vol. 286, Adjutant General to Col Bowen (Sherbrooke), Ottawa, July 1867 and Vol. 110, Adjutant General to Tait & Co (Limerick), Ottawa 24 August 1868.

[13]LAC, RG9, IC1, Vol 286, Adjutant General to Col Bowen (Sherbrooke), Ottawa, July 1867.

(Left) Bugler, 20th Halton Infantry Battalion, c. 1866-1869. He is shown in a rare view of the Canadian pattern 1866 infantry serge tunic with no musician's distinctions added. The battalion was issued 415 serge and 285 cloth tunics in 1866-1867. The serge tunic was scarlet with dark blue collar and pointed cuffs, white piping and white metal buttons. Dark blue forage cap with numeral "20" in white metal; dark blue serge trousers with scarlet piping. Courtesy Scott Duncan.

(Above Right) Staff sergeant and two privates of a Canadian infantry battalion, c. 1867-1870. The sergeant (left) wears the pattern 1866 Norfolk scarlet serge tunic, but the privates have an undress jacket that is not quite the same as the pattern from the Militia Department. Their jackets, which seem scarlet, don't have the white piping; the folds at each side of the breast are very faint if there at all, and no integrated cloth belt is visible. Possibly a local adaptation to emulate the frock. Regrettably, the battalion number on the sergeant's forage cap is illegible. All wear infantry accoutrements. Private collection. Ph: RC.

(Left) The battle of Eccles Hill (Quebec), 25 May 1870. This engraving in the New Dominion Monthly of July after a photo by Sawyer shows somewhat clearer the men of the 60th Mississiquoi Battalion, in pale coats, with a few men in darker coats and wide-brimmed hats that would be local "Red Sash" volunteers, firing at Fenians in the distance. Buttons shown on the skirts confirm the 60th wore its scarlet Canadian pattern 1863 tunics that day, with dark blue forage caps. The men did not wear knapsacks nor haversacks. Ph: RC.

"Warm reception at Eccles Hill" for the Fenians in May 1870. The men of No. 4 Company, 60th Missisquoi Battalion are shown firing their muskets, but this was of course a posed photo taken after the action against the Fenian. The men wear the Canadian Pattern 1863 tunic with forage caps. Courtesy Library and Archives Canada, C36668.

Canadian volunteers of the 50th Huntingdon Borderers Battalion finding a dead Fenian after the battle of Eccles Hill, 25 May 1870. The officers standing at left and right wear the dark blue pattern 1856 frock coat with peaked forage caps, the corporal at the centre wears the Canadian Pattern 1866 scarlet infantry serge frock with the NCOs and men's pattern dark blue forage cap bearing the battalion number. All three are armed with Snider-Enfield breechloading rifled muskets. The fallen Fenian appears to be wearing sky blue trousers (most likely ex-U.S. Army), possibly the dark green trimmed yellow Irish Republican Army jacket and seemingly a former U.S. Army dark blue forage cap. His weapon might be a surplus U.S. Army Springfield rifled musket. Brome County Historical Society Museum, Knowlton, Quebec. Ph: RC.

No. 4 Company, 60th Missisquoi Battalion at Eccles Hill, Québec, May 1870. This photo taken shortly after the Fenian's defeat shows the company wearing the pattern 1863 Canadian scarlet tunic with white cord Austrian knots and the dark blue forage cap with the numeral 60 in front. Captain R. McCorkill has the officer's peaked forage cap and scarlet tunic. He wears the British Pattern 1868 officer's tunic. Brome County Historical Society Museum, Knowlton, Quebec. Ph: RC.

Officer's tunic, 6th Hochelaga Light Infantry Battalion, c. 1862-1867. The star and crown indicating the rank of lieutenant-colonel is at the collar. Rear view. Private collection. Ph: CL.

Officer's tunic, 6th Hochelaga Light Infantry Battalion, c. 1862-1867. Rear view. Private collection. Ph: CL.

Field officer of the 60th Missisiquoi Infantry Battalion, c. 1862-1866. Watercolour by Derek FitzJames. Courtesy Anne S.K. Brown Military Collection, Brown University Library, Providence.

Captain John Orr, 36th Peel Infantry Battalion, c. 1865. He wears the British pattern tunic with the cuff flaps. From an image in Bull's From Brock to Currie. Ph: RC.

(Left) Tunic of Lt. Col. Sydney Barnett, 44th Welland Infantry Battalion, c. 1862. Scarlet with dark blue collar, cuffs and cuff flaps, silver crowned VR buttons and lace. The collar rank badges have been removed. Collection of the Lundy's Lane Historical Museum, Lundy's Lane, ON. Photo courtesy Kimberley Elizabeth Busato.

(Right) Infantry officer's tunic, c. 1860-1862. The buttons and collar badges have regrettably been removed. Collection of the Lundy's Lane Historical Museum, Lundy's Lane, ON. Photo courtesy Kimberley Elizabeth Busato.

(Left) Infantry officer's 1863 Canadian Pattern tunic. Scarlet with dark blue collar and cuffs, white piping, silver buttons and lace. Note the Austrian knots above the cuffs instead of the British Army's cuff flaps. The crown and lace on the collar indicates the rank of lieutenant-colonel. Fort Henry, Kingston. Ph: RC.

(Right) Captain Creasor, Owen Sound Infantry Company, c. 1865-1866. He wears the infantry officer's Canadian 1863 pattern uniform. The company formed part of the 31st Grey Infantry Battalion from 14 September 1866. Courtesy Library and Archives Canada, C49828.

(Left) Lt. Col. Sydney Barnett, 44th Welland Infantry Battalion, c. 1862. He is shown wearing his regimental coat with his crimson sash Partly coloured photo. Collection of the Lundy's Lane Historical Museum, Lundy's Lane, ON. Photo courtesy Kimberley Elizabeth Busato.

(Right) Infantry officer, 1st Administrative Battalion, 1865. He wears a locally made scarlet undress jacket with dark blue collar and cuffs, white piping, dark blue baggy trousers with scarlet piping and the white metal "1" on his dark blue forage cap. Note the black mid calf high boots that were popular in many units. This was not an official regulation uniform. Courtesy National Historic Sites, Parks Canada.

Officers of the 46th East Durham Infantry Battalion in camp at Port Hope, Ontario, during 1866. Although the image is blurry and damaged, it can be seen that about half wear the Canadian Pattern 1863 scarlet tunic faced with dark blue and trimmed with silver cuff and collar lace for officers and that a few others have an undress jacket edged with braid and that has a breast pocket, obviously in the same colours. Most wear the peaked forage cap with numeral in front. The one in the white jacket at left is probably a member of the band. Two officers near the tent have shoulder belts and forage caps and thus would belong to a light cavalry unit or a field battery rather than the 46th. Courtesy Ontario Archives, Toronto, S2822.

Officers and NCOs of Montréal's 5th Royal Light Infantry Battalion that were deployed near the Vermont border in 1870. They wore the British infantry scarlet tunic with facing colour cuff flaps edged and laced with white and shoulder straps edged with white. Corporals and privates had the 1863 Canadian pattern scarlet tunic with the Austrian knot since 1866-1867. Image in Chambers' history of the 5th Royal Scots. Ph: RC.

Militia officers, c. 1856-1860. They wear the 1856 Pattern all dark blue frock coat with silver buttons and crimson shoulder cord to hold the crimson sash. Courtesy Department of National Defence, Ottawa.

Officer's frock coat, 1856-1868. Introduced by MGO of 7 February 1856, this pattern of frock coat was widely adopted by Canadian volunteer infantry and some Sedentary Militia officers. It was all dark blue with silver buttons. There was a crimson shoulder cord on the left shoulder (invisible on this image) to hold the crimson sash worn accros the chest. Note the small percussion cap pocket at the waist. Courtesy, The Lincoln and Welland Regiment Museum. Ph: CL.

Officer, Canadian Volunteer Militia infantry, c. 1865. Possibly of the Brockville Infantry Company. Dark blue 1856 Pattern frock coat with silver buttons, dark blue or dark grey trousers with scarlet piping, crimson sash, dark blue peaked forage cap with black lace and button. Image taken by A.C. Macintyre, Brockville, Canada West. Courtesy Don Blyth Collection, Guelph, Ontario.

Officer, Canadian Volunteer Militia infantry, c. 1863-1868. Dark blue 1856 Pattern frock coat with silver buttons, dark blue or dark grey trousers with scarlet piping, crimson sash, dark blue peaked forage cap with black lace and button, badge in front. Courtesy Scott Duncan

Notes

Militia Reports for 1863 and 1866 have published issue returns; manuscript returns in LAC, RG9, IC1, Vol. 285 (1862); IC8, Vol. 5 (1866); IB2, Vol. 50 (1862-72).

Uniform issues and clothing allowance payments 1863

The pages from the Report of the Militia of Canada are reproduced here. Its Appendix No. 2 show the clothing issues from May to September 1863 to units receiving the new 1863 Canadian Pattern uniforms. Its Appendix 3 concerns the money paid for drill and for compensation clothing allowances to units that procured their own uniforms during 1862 and 1863.

Montreal Light Infantry officer, c. 1857. This damaged, but very rare image of a Canadian scarlet-coated volunteer during the 1850s, shows the officer wearing the new uniform specified in the unit's Regulations and Standing Orders of December 1856 that called for the new scarlet tunic with dark blue collar, cuffs and cuff flaps, white piping, silver buttons and lace with the new rank badges on the collar and dark blue trousers with scarlet piping. The undress dark blue shell jacket or vest with small buttons is seen underneath the unbuttoned tunic. Officers in this light infantry unit had black cartridge boxes held by a white patent leather belt worn over the shoulder as well as white sword belts with snake buckles. Private collection. Courtesy of Ross Wilson.

Officer, 32nd Bruce Infantry Battalion, c. 1870. He wears what appears to be a locally made version of the 1868 British patrol jacket. Courtesy Don Blyth Collection, Guelph, Ontario.

British army pattern patrol jacket introduced from August 1868 worn by Lt. Col. Sydney Barnett, 44th Welland Infantry Battalion, c. 1870. Dark blue with black lace. The originally black broad edging lace has turned to buff, which is not uncommon in surviving examples of this pattern of jacket. Collection of the Lundy's Lane Historical Museum, Lundy's Lane, ON. Photo courtesy Kimberley Elizabeth Busato.

APPENDIX NO. 2.

LIST of Corps, in Upper and Lower Canada, to whom Clothing has been Issued in 1863.

Date of Issue.	Post.	Corps.	Officer Commanding.	Artillery.			Infantry.			Rifles.		
				Tunics.	Busbies.	Trousers	Tunics.	Shakos.	Trousers	Tunics.	Shakos.	Trousers
May 23	Point Levi	17th Batt. Vol. Infantry, 6 Cos	Lt.-Col. Blanchet				315	315	315			
“ 23	Quebec	8th Batt. Rifles, 2 Cos	“ Reeve							133	133	133
“ 23	“	9th Batt. Rifles, 4 Cos	“ de Salaberry							189	189	189
“ 25	“	7th Batt. Rifles, 3 Cos	“ Cauchon							153	153	153
“ 27	Toronto	10th Batt. Infantry, 7 Cos	“ Cumberland				385	385	385			
“ 27	Hamilton	13th Batt. Infantry, 6 Cos	“ Buchanan				329	329	329			
“ 28	Argenteuil	11th Batt. Infantry, 6 Cos	“ Abbott				328	328	328			
“ 29	Kingston	14th Batt. Rifles, 3 Cos	“ Shaw							165	165	165
“ 30	Belleville	15th Batt. Infantry, 6 Cos	“ Ponton				330	330	830			
June 1	Picton	16th Batt. Infantry, 7 Cos	“ Ross				385	385	385			
“ 1	Storrington	1st Company Rifles	Capt. Hamilton							55	55	55
“ 1	“	2nd “	“ Spring							55	55	55
“ 1	Napanee	1st Company Infantry	“ Detlor				55	55	55			
“ 1	“	2nd “	“ Miller				55	55	55			
“ 2	Rednersville	Infantry Company	“ Anderson				55	55	55			
“ 2	Ameliasburg	“	“ Lander				53	53	53			
“ 2	St. Vincent de Paul	12th Batt. Rifles, 5 Cos	Lt.-Col. Bellerose							271	271	271
“ 3	St. Johns	21st Batt. Infantry, 6 Cos	Major Laberge				325	325	325			
“ 4	Three Rivers	1st Company of Infantry	Capt. Bernard				54	54	54			
“ 4	“	2nd “ “	“ Normand				55	55	55			
“ 4	Toronto	2nd Batt. Rifles, 3 Cos	Lt.-Col. Durie							152	152	152
“ 5	Hawkesbury Mills	18th Batt. Infantry, 6 Cos	“ Hon. Mr. Hamilton				324	324	324			
“ 6	St. Catherines	19th Batt. Infantry, 8 Cos	“ Hon. M. Currie				440	440	440			
“ 8	“	20th Batt. Infantry, 6 Cos	“ McGiverin				330	330	330			
“ 9	Montreal	1st Company Police Rifles	Capt. Flynn							47	47	47
“ 9	“	2nd “ “	“ Naigele							48	48	48
“ 9	London	1st Company Infantry	“ Taylor				55	55	55			
“ 9	“	2nd “ “	“ Chisholm				55	55	55			
“ 9	St. Hyacinthe	3 Companies, Infantry	“ Després				135	135	135			
“ 10	“	Infantry	“ Nelson				50	50	50			
“ 10	Chatham	“	“ Smith				55	55	55			
“ 10	“	“	“ Glendinning				55	55	55			
“ 10	Kamouraska	1st Company Infantry	“ I. Taché				55	55	55			
“ 10	Rivière du Loup	Infantry	“ Fraser				55	55	55			
“ 10	Acton Vale	Rifle Company	“ Cushing							53	53	53
“ 11	Lennoxville	“	“ Rawson							55	55	55
“ 11	Sherbrooke	“	Major Ibbotson							55	55	55
“ 11	“	“	Capt. Woodward							50	50	50
“ 11	Ottawa	“	Lt.-Col. Grant							55	55	55
“ 11	“	“	Capt. Beaubien							55	55	55
“ 11	Varennes	Infantry Company	Major Massue				55	55	55			
“ 11	Toronto	10th Battalion Infantry	Capt. Sherwood				55	55	55			
“ 11	Prescott	Rifle Company	Lt.-Col. Jessup							55	55	55
“ 11	Morrisburg	“	Capt. Holden							55	55	55
“ 12	Cornwall	“	“ Bergin							55	55	55
“ 12	“	“	“ Oliver							55	55	55
“ 12	Brockville	Infantry Company	“ Buell				57	57	57			
“ 13	Cornwall	Infantry	“ Pringle				57	57	57			
“ 13	Gananoque	Rifle Company	“ McCrum							55	55	55
“ 15	Barrie	“	“ M'Kenzie							55	55	55
“ 15	Collingwood	“	Major Stevens							55	55	55
“ 15	Whitby	“	“ Wallace							55	55	55
“ 15	Cookstown	“	“ Ferguson							55	55	55
“ 15	Scarborough	“	Capt. Norris							55	55	55
“ 15	Oshawa	“	“ Warren							55	55	55
“ 15	“	Infantry	Major Fairbanks				55	55	55			
“ 17	Guelph	Rifles	Capt. Higinbotham							55	55	55
“ 17	Owen Sound	“	“ McNab							53	53	53
“ 17	Galt	“	“ Dato							50	50	50
“ 17	Elora	“	“ Donaldsan							55	55	55
“ 17	Mount Forest	“	Lieut. Mitchell							47	47	47
“ 17	Stratford	Infantry	Capt. Imlach				55	55	55			
“ 17	Owen Sound	“	“ Brodie				55	55	55			
“ 18	Peterborough	Rifles	“ Ed. Poole							55	55	55
“ 18	Bowmanville	“	Lt.-Col. Cubitt							55	55	55
“ 18	Lindsay	“	Capt. Smith							40	40	40
“ 18	Port Hope	“	“ Kirckhoffer							49	49	49
“ 18	“	Infantry	“ Williams				55	55	55			
“ 18	Peterborough	“	“ Kennedy				55	55	55			
“ 18	Lakefield	“	“ Vizard				55	55	55			
“ 22	Villa Nova	Rifles	“ McLaren							55	55	55
“ 22	Simcoe	“	“ Tisdale							55	55	55
“ 22	Port Rowan	“	“ Pitt Mabee							55	55	55
“ 22	Walsingham	“	“ Morgan							40	40	40
“ 22	Waterford	“	“ Green							55	55	55
“ 22	Komoka	“	“ Atwood							40	40	40
“ 22	Southampton	“	“ Sproat							25	25	25
“ 22	Richelieu	Infantry	“ Dorion				55	55	55			
“ 22	North Ridge	Rifles	“ Wagstaff							55	55	55
“ 22	Embro	“	“ Wallace							55	55	55
“ 23	L'Islot	Infantry	“ S. Gamache				55	55	55			
		Carried forward.										

List of Corps, in Upper and Lower Canada, to whom Clothing has been Issued in 1863.—*Continued.*

Date of Issue.	Post.	Corps.	Officer Commanding.	Artillery.			Infantry.			Rifles.		
				Tunics.	Busbies.	Trousers	Tunics.	Shakos.	Trousers	Tunics.	Shakos.	Trousers
		Brought forward										
June 23	Rimouski	Infantry	Capt. Michaud				55	55	55			
" 23	Trois Pistoles	"	" Bertrand				55	55	55			
" 24	Huntingdon	"	" Hall				55	55	55			
" 24	"	"	" Whyte				55	55	55			
" 24	Durham	"	" McEachern				55	55	55			
" 24	Athelstan	"	" Leonard				55	55	55			
" 24	Napierville	"	" Morrison				55	55	55			
" 24	"	"	" Trudeau				55	55	55			
" 25	Berthier (en haut)	"	" Chalut				55	55	55			
" 25	R. du Loup (en haut)	"	" Ed. Caron				55	55	55			
" 25	Gore, 2nd Co. 11th Bt.	"	" Wm. Smith				55	55	55			
" 25	Buckingham	"	" McNaughton				55	55	55			
" 25	Aylmer	"	" W. A. Campbell				55	55	55			
" 25	Philipsburg	"	" Moore				45	45	45			
" 25	Granby	"	" Miller				50	50	50			
" 25	"	"	" J. Chalmers				50	50	50			
" 25	Waterloo	"	" Charles Maynes				55	55	55			
" 25	"	"	" Reynolds				55	55	55			
" 26	Perth	"	" Morris				55	55	55			
" 26	"	"	" Haggart				55	55	55			
" 26	Almonte	"	" McIntosh				55	55	55			
" 26	Bell's Corners	"	" W. F. Powell				55	55	55			
" 26	Fitzroy	"	" Fraser				55	55	55			
" 27	Kenyon	"	" A. K. McDonald				55	55	55			
" 27	Glengarry	"	" L. McDougald				55	55	55			
" 27	Morrisburg	"	" Rose				55	55	55			
" 27	Lyn	"	" Wilson				55	55	55			
" 27	Dixon's Corners	"	" Lowery				55	55	55			
" 30	Ashburnham	"	Major Rogers				55	55	55			
" 30	Mill Brook	"	Capt. R. Howden				55	55	55			
" 30	Campbellford	"	" Tice				55	55	55			
" 30	Cobourg	"	" Elliott				55	55	55			
" 30	Whitby	"	" Woodward				55	55	55			
" 30	Cobourg	"	" Railton				55	55	55			
July 1	Lloydtown	"	" Ed. Bull				50	50	50			
" 1	Prince Albert	Infantry	" Forman				55	55	55			
" 1	Brampton	"	" Stork				55	55	55			
" 1	Greenwood	"	" Warren				55	55	55			
" 1	Bradford	"	" McMaster				55	55	55			
" 1	Barrie	"	" Russell				55	55	55			
" 2	Seaforth	"	" Coleman				55	55	52			
" 2	New Hamburg	"	" Goodman				52	52	52			
" 2	Orangeville	"	" Buckham				55	55	55			
" 2	Kincardine	"	" Alex. Shaw				55	55	55			
" 2	"	"	" Daniel				55	55	55			
" 2	Paisley	"	" Bruce				55	55	55			
" 3	Dundas	"	" Crossland				55	55	55			
" 3	Georgetown	"	" Young				55	55	55			
" 3	Thamesford	"	" Dawes				55	55	55			
" 3	Stewart Town	"	" Murray				55	55	55			
" 3	Norval	"	" Kyle				55	55	55			
" 3	Ingersoll	"	" Oliver				55	55	55			
" 3	St. Ann la Pocatière	"	" Deguise				55	55	55			
" 4	Mount Pleasant	"	" Ellis				40	40	40			
" 4	Drumbo	"	" Howell				40	40	40			
" 4	Wardsville	"	" Henderson				55	55	55			
" 4	Vienna	"	" Gustin				55	55	55			
" 4	St. Johns	"	Major Jackson				55	55	55			
" 4	Harrietsville	"	Capt. McMillan				55	55	55			
" 6	Windsor	"	" J. G. Macdonald				55	55	55			
" 6	Sandwich	"	" Chas. Casgrain				55	55	55			
" 6	"	"	" Moynahan				55	55	55			
" 6	Amherstburg	"	" Wilkinson				55	55	55			
" 6	Blenheim	"	" McMichael				55	55	55			
" 6	Leamington	"	" Fox				55	55	55			
" 7	Rockburn	"	" Blackwood				55	55	55			
" 7	Franklin	"	" Rogers				55	55	55			
" 7	Hemmingford	"	" McNaughton				55	55	55			
" 7	St. Remi	"	" Hugh McGill				55	55	55			
" 7	Roxham	"	" Lucas				55	55	55			
" 7	Beauharnois	"	" Martin				55	55	55			
" 7	Levis	Engineers	" Freer				55		55			
" 8	Kamouraska	Infantry	" V. Taché				55	55	55			
" 8	Burritt's Rapids	"	" G. Shephard				55	55	55			
" 8	Farmersville	"	" Booth				55	55	55			
" 8	Mallorytown	"	" Fields				40	40	40			
" 8	Albion	"	" Evans				55	55	55			
" 8	Derry (West)	"	" Grafton				55	55	55			
" 9	St. Catherines	Artillery	" Stoker		50							
" 9	Napanee	Infantry	" Embury				55	55	55			
" 9	Uxbridge	"	" Spears				55	55	55			
" 9	Grahamsville	"	" Graham				55	55	55			
		Carried forward										

LIST of Corps, in Upper and Lower Canada, to whom Clothing has been Issued.—*Continued.*

Date of Issue.	Post.	Corps.	Officer Commanding.	Artillery. Tunics.	Artillery. Busbies.	Artillery. Trousers	Infantry. Tunics.	Infantry. Shakos.	Infantry. Trousers	Rifles. Tunics.	Rifles. Shakos.	Rifles. Trousers
		Brought forward										
July 9	Caledonia	Rifles	Capt. Wm. Jackson							55	55	55
... 9	York	"	" Davis							55	55	55
... 9	Cayuga	"	" Stewart							55	55	55
... 10	Ste. Scholastique	"	" Valois							55	55	55
... 10	St. Eustache	"	" Marsil							55	55	55
... 10	Baie St. Paul	Infantry	" Duperé				55	55	55			
... 11	Kamouraska	"	" Martineau				55	55	55			
... 23	Easton's Corners	"	" Ferguson				55	55	55			
... 23	Consecon	"	" Clark				55	55	55			
... 23	Aurora	"	" Ashton				55	55	55			
... 23	King	"	" Garden				55	55	55			
... 23	Alton	"	" Riddel				55	55	55			
... 23	Lucan	"	" Hedgins				55	55	55			
... 23	Morriston	"	" McFarlane				55	55	55			
... 23	Durham	"	" McNab				55	55	55			
... 24	Cobourg	Rifles	" Wm. Smith							48	48	48
... 24	Belleville	"	" C. G. Levesconte							55	55	55
... 24	London	"	" R. Lewis							55	55	55
... 24	St. Thomas	"	Major Stanton							49	49	49
... 24	London	"	Capt. Barber							55	55	55
... 24	Dunnville	"	Major Amsden							55	55	55
... 25	Stratford	"	Capt. R. S. Service							53	53	53
... 25	Oakville	"	Lt.-Col. Chisholm							55	55	55
... 25	Pakenham	"	Capt. Dunnett							55	55	55
... 25	Carleton Place	"	" James Poole							55	55	55
... 25	Merrickville	"	" Gwyne							55	55	55
... 25	"	"	" Wm. Wright							55	55	55
... 27	Columbus	"	" George Prentice							55	55	55
... 27	Brooklin	"	" Th. Hodgson							55	55	55
... 27	Goderich	"	" Seymour							55	55	55
... 27	Meaford	"	" Pollard							51	51	51
... 27	Leith	"	" J. Cannon, Sen.							47	47	47
... 27	Lucknow	"	" McDonald							52	52	52
... 27	Dundas	"	" Gibson							55	55	55
... 28	Beachville	"	" C. Greig							55	55	55
.. 28	Princeton	Rifles	Capt. Th. Cowan							40	40	40
... 30	Port Hope	Engineer	" Ridout				55	55	55			
... 31	Brampton	Rifles	Major Wright							55	55	55
Aug. 14	North Oxford	"	Capt. Henderson							40	40	40
... 14	Delaware	"	" Graham							55	55	55
... 14	Wolverton	"	" Cole							50	50	50
... 21	Goderich	Artillery	" Alex. Ross	50	50	50						
... 24	Quebec	"	" N. H. Bowen	55	55	55						
... 24	"	"	" Grant	50	50	50						
Sept. 2	Hinchinbrooke	Rifles	" McWilliams							45	45	45
... 3	Quebec	Artillery	" Lindsay	50	50	50						
.. 4	"	Rifles	" Burns							55	55	55
... 19	Fergus	"	Lieut. Strathmore								40	
... 20	Gannanoque	Artillery	Capt. Ferd. Jones	50	50	50						
... 21	Prescott	"	" Roebuck	50	50	50						
... 21	London	"	" Buckley	50	50	50						
... 21	Montreal	"	" Alex. Wand	50	50	50						
... 22	Sherbrooke	"	" Bowen	50	50	50						
... 22	Port Colborne	Field Battery	" Verner	69	69	69						
... 23	*Hemmingford*	*Rifles*	" *Shields*							15	15	15
... 23	Morrisburg	Artillery	" Rubidge	50	50	50						
... 23	Iroquois	"	" Alex. McDonell	50	50	50						
Oct. 12	Perth	Rifles	" Fraser							55	55	55
... 20	Hamilton	Field Battery	Major Villiers	50	50	50						
... 23	Paris	Rifles	" Patton							55	55	55
Aug. 29	*Quebec*	*Rifles*	*Capt. Alleyn*								55	
... 30	"	*Rifles*	" *Pellitier*								55	
Sept. 8	St. Johns	Infantry	Sgt.-Major Labranche				1	1	1			
July 22	Ottawa	Rifles	No. 1 Co.							1	1	1
... 24	*Toronto*	*Rifles*	*Lt.-Col. Durie*								355	
Sept. 5	*London*	*Infantry*	*Capt. Taylor*				12	12	12			
		Total		674	724	674	9897	9842	9897	4927	5422	4917

Those in *italics* have paid for the clothing issued to them.

THOS. WILY, Lieut.-Col.
Superintendent of Stores.

APPENDIX NO. 3.

STATEMENT shewing the sums paid to the different Corps of Volunteers for Drill and Clothing Allowance, in the year 1863.

To whom Paid.	Designation of Corps.	For what Service.	$ cts.	$ cts.
Lt. Duff	Cavalry	For drill of Troop for (1862) Frontenac	480 00	
Major McLeod	do	do do York	447 00	
Capt. Burgess	do	do do London	511 20	
				1438 20
Major Stevenson	Field Battery	For drill of Battery for (1862) Montreal	930 00	
Lt.-Col. Notman	Foot Company of Artillery	For drill of Company for (1862) Dundas	186 00	
				1116 00
Capt. Smith	Rifle Company.	For drill of Company for (1862) Toronto	330 00	
Lt.-Col. Jessup	do	do do Prescott	300 00	
Capt. Moir	do	do do Montreal	330 00	
Major Barwis	do	do do Megantic	318 00	
Capt. Levesconte	do	do do Belleville	168 00	
" Hunt	do	do do Sorel	286 50	
" Lewis	do	do do London	318 00	
Major Amsden	do	do do Dunnville	291 00	
Capt. Rawson	do	do do Lennoxville	145 82	
" Higinbotham	do	do do Guelph	285 50	
" Woodward	do	do do Sherbrooke	306 00	
Major Stanton	do	do do St. Thomas	228 00	
Capt. Hanson	do	do do Montreal	492 00	
				3798 82
		Total for Drill		$6353 02
		Clothing Allowance.		
Capt. Audet	Chasseurs Canadiens, 4th Battalion Rifles.	Allowance for his Company in lieu of clothing for (1862) Montreal	330 00	
" Emond		do do do	330 00	
" Beaudry		do do do	330 00	
" Cinq-Mars		do do do	330 00	
" Spénard		do do do	330 00	
" D'Orsonnes		do do do	324 00	
" L. Spénard		do do do	330 00	
" Normandeau		do do do	330 00	
Lt. Bissonnette		do do do	330 00	
Capt. Hon. J. O. Chauveau		do do do	246 00	
" Scott	5th Battalion, or Royal Light Infantry.	do do do	330 00	
" Allan		do do do	330 00	
" Mackenzie		do do do	300 00	
" Kirby		do do do	330 00	
" Hopkins		do do do	330 00	
" Campbell		do do do	276 00	
" Grant		do do do	330 00	
" Lyman		do do do	294 00	
" Johnson	3rd Battalion Victoria Volunteer Rifles.	do do do	336 00	
" Nivin		do do do	300 00	
" Ogilvie		do do do	300 00	
" Henderson		do do do	222 00	
" Crawford		do do do	300 00	
" McGrath		do do do	312 00	
" Bossé	Rifle, 9th Bat.	do do do Quebec	354 00	
" Thomson	do do	do do do do	324 00	
Major Bernard	do	do do do do	366 00	
" Barretto	do	do do do Mount Forest	222 00	
Capt. Forsyth	Engineer	do do do Montreal	330 00	
" Galway	Rifle	do do do Ottawa	180 00	
" Skinner	Inf. 13th Bat.	do do do Hamilton	204 00	
" Wallace	Rifle (Highl'd).	do do do Whitby	222 00	
" Ince	Rifle	do do do Toronto	204 00	
" Gibsone	do 8th Bat.	do do do Quebec	432 00	
" Shepherd	do	do do do Vaudreuil	240 00	
" Yule	do	do do do Lennoxville	182 75	

STATEMENT shewing the sums paid to the different Corps, &c.—*Continued.*

To whom Paid.	Designation of Corps.	For what Service	$ cts.	$ cts.
		Clothing Allowance.—Continued.		
Capt. Wall	6th Battalion or Hochelaga Light Infantry.	Allowance to his Company in lieu of clothing for (1862) Montreal	378 00	
" Binmore		do do do Montreal	342 00	
" Reinhardt		do do do do	348 00	
" Jackson		do do do do	342 00	
" Porter		do do do do	342 00	
" Ibbotson		do do do do	342 00	
" Crispo		do do do do	342 00	
" Martin		do do do do	342 00	
" St. Germain	Rifle	do do do St. Hyacinthe	282 00	
Major Teeter	Cavalry	do Troop do Grimsby	210 00	
Capt. Herring	Rifle, 9th Bat	do Co'py do Quebec	330 00	
" Forest	Foot Artillery	do do do Ottawa	240 00	
" Macdougall	Troop	do Troop do Montreal	138 00	
" Drummond	do	do do do Sherbrooke	300 00	
Lieut. Heming	do	do do do Guelph	150 00	
Major Button	do	do do do Markham	168 00	
" Bate	do	do do (1863) St. Catharines	186 00	
Capt. Cook	do	do do do Cookshire	288 00	
" DesRivières	do	do do do St. John's	240 00	
Major Teeter	do	do do do Grimsby	216 00	
" Smart	do	do do do Port Hope	228 00	
Capt. Smith	do	do do do Montreal	186 00	
" Perry	do	do do do do	180 00	
Lt. Duff	do	do do do Kingston	186 00	
Capt. Sinclair	Engineer	do do do Ottawa	300 00	
" A. G. Forrest	Foot Artillery	do do do do	312 00	
Major Evans	Battalion of Artillery.	do do do Montreal	300 00	
Capt. Ferrier		do do do do	300 00	
" Hobbs		do do do do	300 00	
" Shaw		do do do do	300 00	
" Scott		do do do do	300 00	
" Drum		do do do do	300 00	
" Stoker	Artillery	do do do St Catharines	235 50	
" Pelletier	Rifle, 9th Bat	do do do Quebec	254 65	
" Galway	do	do do do Ottawa	300 00	
Major Alger	do	do do do Brantford	360 00	
Capt. Richardson	Oxford Rifles	do do do Woodstock	390 00	
" "	Rifle	do(special) do do do	260 00	
" Grant	Highland Rifle.	do do do Brantford	330 00	
Major Bernard	Rifle	do do do Quebec	360 00	
Capt. Hanning	do	do do do Danville	300 00	
" Thomson	do 9th Bat.	do do do Quebec	330 00	
" Shields	do	do do do Hemmingford	216 00	
" Alleyn	do 8th Bat.	do do do Quebec	240 65	
" Herring	do 9th Bat.	do do do do	330 00	
" Litchfield	do	do do do Portsmouth	360 00	
" Murray	Queen's Own R.	do do do Toronto	370 40	
" Paterson	do	do do do do	248 65	
" Harris	do	do do do do	254 65	
" Shepherd	Rifle	do do do Vaudreuil	276 00	
" Minnes	do 14th Bat.	do do do Kingston	444 00	
" White	do	do do do Prescott	240 00	
" Gardiner	Queen's Own R.	do do do Toronto	324 00	
" Hodgins	do	do do do do	224 65	
" Croft	do	do do do do	254 65	
" Draper	do	do do do do	224 65	
" Johnson	Vic. Vol. Rifles	do do do Montreal	318 00	
" Nivin	do	do do do do	234 00	
" Ogilvie	do	do do do do	270 00	
" McGrath	do	do do do do	276 00	
" Gibsone	Rifle, 8th Bat	do do do Quebec	432 00	

Chapter 2

Infantry 1871-1901

Infantry School Corps, C Company on parade at Stanley Barracks in Toronto, 1884. The company is in marching order wearing undress scarlet serge jackets, serge trousers, black short leather leggings and boots, armed with Martini-Henry rifles and equipped with British pattern Valise equipment. Apart from forage caps, Canadian volunteer militiamen were not issued any of these item, only regulars. Courtesy Library and Archives Canada, C31372.

The last three decades saw many changes in the organisation, dress and arms of the Canadian infantry. From the 1870s, the nation's infantry become progressively the worst armed such force in the Americas, tens of thousands of volunteer infantrymen still carrying antiquated Snider-Enfield rifled breech-loading muskets into the late 1890s, with the exception of a few regular companies. Equipment was equally ancient and training not given to all. It could be said that only the scarlet cloth tunics kept appearances up and even those were replaced for serge frocks for rural battalions from 1896. Nevertheless, the volunteer infantrymen looked amazingly good and soldierlike, which was largely due to officers and men often giving their meagre "drill pay" to purchase uniforms, extra kit and often dress helmets. As will be seen in this chapter, the infantry remained nevertheless a viable-looking force as a result.

It was also a time when small regular units of regular infantry were raised in the nation. The first such regular unit was organised in 1872 in Winnipeg from the two rifle battalions sent there in 1870 as a "Provisional Battalion of Canadian Light Infantry and Artillery" and disbanded in 1878.

The battalions of volunteer militia organised across the expanding nations needed military instruction from professional soldiers. The British army having withdrawn its garrisons in Canada in 1871 (except at naval bases Halifax, NS, and Esquimalt, BC), two batteries of artillery had been formed late that year whose duties included training volunteer gunners. Volunteer cavalry and infantry were somewhat left to their own resources until late 1883 when "School Corps" organised as regular units were established to extend the scope of the training, which was provided especially at the annual summer training camps. They also served as garrison troops. For the infantry, the Infantry School Corps was created in December 1883 and evolved into becoming the Royal Canadian Regiment from 1892 and the nation's senior regular infantry unit ever since.

The volunteer militia battalions meanwhile were formed from the early 1860s and considerably expanded especially to counter the Fenian's menaces of 1866 and 1870, the preference being given to forming scarlet-coated line infantry units. Another important factor was the 1867 Confederation of Canada as an independent self-governing nation amalgamating the several British "colonies" on the North America continent as provinces and territories. The federal government was responsible for defence and, from 1868, amalgamated the existing units into what became the Canadian army. As a result, there were over 80 infantry battalions by the mid-1870s, the majority being of scarlet-coated infantry (the others being rifle units) and more were organised later on.

Uniforms of the Royal Canadian Regiment, 1890s-1901. From left: officer, dress uniform; private, South Africa 1900-1901; private and officer, service dress, 1896-1901; officer, winter dress; private, service dress, c. 1890-1895. Detail of a print in Fetherstonhaugh's 1936 The Royal Canadian Regiment 1883-1933. Ph: RC.

Infantry School Corps, C Company, 1885. Pattern 1878 scarlet tunic, dark blue trousers with red piping, white helmet with brass plate, spike and chinscales. In front, a sergeant (three white chevrons on sleeve) and at the centre an officer holding a sword (gold buttons and lace). This image is especially interesting because it is an English uniform print in Richard Simkin's 1891 Our Armies that was adapted by Canadian Adjutant General Walker Powell for his "Scapbook" to depict the company in 1885. He painted a crow's foot at the cuffs of the privates and NCOs and also painted every (originally blue) helmet white. The British valise equipment and Martini Henry rifles would be correct for the permanent infantry. Courtesy Library and Archives Canada, C12034.

The units making up the Canadian scarlet-coated infantry from 1871 to 1901 were:

Regular infantry:

Provisional Battalion of Canadian Light Infantry and Artillery (converted from Rifles at Winnipeg on 23 October 1872, disbanded July 1877).

Infantry School Corps 21 December 1883; Canadian Regiment of Infantry May 1892; The Royal Regiment of Canadian Infantry May 1893; The Royal Canadian Regiment of Infantry April 1899; The Royal Canadian Regiment November 1901.

Permanent (regular) Infantry Units

Provisional Battalion, Manitoba, 1871-1877

Following the 1870 Red River Expedition's successful and peaceful occupation of Winnipeg's Fort Garry and Lower Fort Garry, two companies detached from the two Canadian rifle battalions were left on permanent duty to garrison Manitoba. The following year, the Fenians again threatened to invade Manitoba and were rumoured to be assembling a force south of its border. On 12 October 1871, an Order in Council authorised the immediate dispatch of 215 officers and enlisted men formed into two rifle companies to reinforce the detachment "on duty at Fort Garry" in Manitoba. Each rifleman was armed with the Short Snider Enfield rifle with its sword-bayonet, black accoutrements and clothing that was issued on 21-22 October at Collingwood before leaving. The reinforcement arrived at Fort Garry on 18 November. The 88 officers and men of the two companies in garrison were merged with the reinforcements to form the "Provisional Battalion of Rifles" of 16 officers and 300 NCOs and men.[14]

Infantry School Corps soldiers in a guardhouse, c. 1890. All are wearing the corps' undress scarlet serge jacket with five buttons in front with dark blue collar, cuffs and shoulder straps edged white, trousers with red piping and dark blue Glengarry forage caps. Courtesy Library and Archives Canada, C31393.

Part of the battalion came in aid of civil power on 19 September 1872, the day of the Manitoba provincial elections, when a mob in Winnipeg was "becoming unruly." Some reputedly drunk rioters who were trying to break into the Dominion Savings Bank ran off at the cry of "here come the soldiers!" after which guards were posted at the bank and other buildings. On 25 September, a draft of 184 replacement men enlisted for three year's service arrived. This included an officer and 25 gunners detached from A an B batteries with two 6-pdr field guns to provide a cadre of artillery in Manitoba.

Up to the fall of 1872, the battalion's foot troops were classed as riflemen. This meant that the uniform of the battalion was of the rifle green pattern. The troops that had arrived in 1870 and stayed in Manitoba had, with some additions and modifications, had the pattern 1863 rifle uniform with the Austrian knot at the cuffs. The 1871 issue would have been of the 1870 pattern rifle uniform with the chevron cuffs already described in Volume 1. By Order in Council of 10 September 1872, the name of the unit was changed to "Provisional Battalion of Canadian Light Infantry and Artillery" to reflect its detachment of gunner and to henceforth transform its riflemen into scarlet-coated infantrymen. According to Ernest Chamber's history of the militia, it was "a concession to the respect and veneration held by indigenous warriors for the British troops their forefathers had fought along with in olden days."

For the battalion, this meant that the officer's and men's uniforms would be the Canadian Militia's pattern 1870 scarlet cloth tunic for infantry with the white chevron cuff (see below). The troops that arrived on 25 September 1872 appear to have come with scarlet infantry uniforms for all. As for the gunners, they wore the blue artillery uniform already described in volume 1. However, there remained 12 officers who had previously purchased rifle pattern uniforms. By Order in Council of 25 January 1873, a compensation of $135 was approved for each officer to obtain: 1 scarlet cloth tunic, 1 scarlet cloth patrol jacket, 1 serge patrol jacket, 2 pairs cloth trousers, 1 pair serge trousers, 1 shako, 1 forage cap, 1 mess jacket, 1 mess vest, 1 sword belt and 1 crimson sash. These items were already "in possession of the officers" when the order was announced.

---

14 LAC, RG9, IIB1, Vol. 518, Adjutant General Robertson-Ross to Governor General, 6 November 1871. The battalion's commander was Maj. A.G. Irwin.

Major Henri Beaufort Vidal, Infantry School Corps, c. 1890. His cap bears the corps 1884-1894 bullion badge and his rank badge is worn on the shoulder cords of his dark blue and black patrol jacket. Lithograph, private collection. Ph: RC.

Rifle battalions did not have colours, but some were issued to the battalion after its conversion to infantry. An order of 17 January 1876 mentioned a "Guard of Honour furnished from the Provisional Battalion of Infantry…and the Regimental Colour…"[15]

On 6 July 1873, another 100 NCOs and men from New Brunswick and Nova Scotia enlisted for two years joined the battalion. As previously, each volunteer for service in the battalion received: 1 cloth tunic, 1 pair cloth trousers, 1 serge jacket, 1 pair serge trousers, 2 pairs of boots and 1 forage cap. The free kit consisted of: 2 flannel shirts, 1 pair braces, 2 pairs socks, 2 linen towels, 1 knife, fork and spoon with hold-all, 1 cloth brush, 2 blacking brushes, 1 box blacking, 1 tin waterproof blacking and 1 soap. For winter: 2 knitted undershirts, 2 pairs knitted drawers, 1 muffler, 1 fur cap, 1 pair of mitts and a greatcoat. Other clothing was also worn. 304 pairs of moccasins were requested in November 1871, leggings for sentries were approved in December, white or black felt hats were mentioned in June 1873, serge blouses and trousers requested in April 1874 and canvas trousers in May.[16]

A march to Qu'Appelle and back to Winnipeg by two companies from 17 August to 5 October 1874 revealed a number of features to Lt. Col. Osborne-Smith. "Besides the ordinary field kit, each man was provided with moccasins… On the line of march, the men were compelled to carry their rifles, waist belts, ball-bags, canteen, bayonet and haversack." Marching on the Prairies was a difficult experience; the boot's soles "became so polished in marching over the grass that the wearer slips as though on ice, all times the hard wiry grass cuts through the toe of the boot, as though with a knife." The "best ammunition boots" were to be issued with "the toes capped with fine sheet copper" and an "ample supply of moose moccasins should be taken with the quartermaster's stores." Furthermore, "drums and fifes and bugles should always be taken for marching on prairie land: nothing is so monotonous or fatiguing as a wide expanse of plain unbroken either by tree or shrub." Finally, Osborne-Smith considered the "Bell tent" with its "high wall" was "the best for troops on a prairie march."

By that time, it was fairly clear to the federal government that there was little threat in Manitoba. Apart from the few outings, the troops had been mostly called upon to occasionally intervene in aid of civil power and help put out a large fire in Winnipeg. The Fenians had largely vanished, the Métis and First Nations were peaceful and settlement of the province was creating a strong volunteer militia force; there was already a field battery, a company of mounted rifles and three rifle companies organised. Thus, on 14 November 1874, the battalion was advised to reduce its strength to 12 officers and 188 NCOs and men including two artillery officers and 43 gunners. This was scaled down further on 1 June 1875 to 11 officers and 100 NCOs and privates including two artillery officers and 20 gunners. From 30 June 1876, the artillery was raised from 20 to 25 NCOs and gunners, but the infantry was reduced from 78 to 25 NCOs and men. From 21 August 1876, the number of officers consisted of one captain, two lieutenants and two ensigns. The "battalion" was thus reduced to two small companies. This force, after a detachment enforced quarantine regulations for six months due to an epidemic of smallpox between the shores of Lake Winnipeg and the District of Keewatin, was disbanded on 3 August 1877. Thus came to an end Canada's first regular infantry unit.[17]

[15] MA, MG6, B5, District Orders, 1874-1876, 17 January 1876

[16] *The Volunteer Review*, 20 May 1873; MA, MG6, B1, Military District No. 10, letter book 1, pp. 6, 21; ibid, letter book 2, 9 May 1873; ibid, letter book 3, 11 April 1874; ibid, letters received, book No. 3, 12 May 1874. Officer's servants wore civilian clothes.

[17] MA, MG6, B5, District Orders, 1874-1876, 19 June 1876.

The Infantry School Corps/Royal Canadian Regiment

Corporals and officer, Royal Canadian Regiment, c. 1897-1900 The corporal wears the 1896 infantry frock with the regiment's badge added on the collar and shoulder straps. His frock's shoulder have edging at the end, which would indicate that it was made before 1898. He also has the new brown leather Oliver equipment. The officer has the dark blue peaked cap with a red band and the 1892 officer's scarlet frock. Watercolour by "H.J.W." Ph: CL.

Between 1877 and 1883, the only regular Canadian army units were the two batteries of artillery posted at Quebec City and Kingston. There was also a small staff cadre for the instruction of the cadets at Kingston's military college. All other officers and men were volunteers in the many militia units existing all over the country. The financially conservative federal government was not especially keen to create more regular units that would lead to increased military spending. On the other hand, something had to be done about training infantry and cavalry units and for that to happen, a core of professional regular soldiers that knew "soldiering" was needed. Thus, on 23 December 1883, the "Infantry School Corps" as well as a similar cavalry unit. The infantry school actually functioned as a regular infantry battalion with companies A at Fredericton, NB, B at Saint-Jean, QC and C at Toronto, ON. D Company raised on 18 August 1887 was at London, ON.

The dress uniform of all ranks of the regiment was generally the same as that of the volunteer infantry (described below). It was, however, fully equipped like British army soldiers with various orders of dress and kit including undress uniforms, greatcoats, boots, short black leather leggings, grey and cotton shirts, and dress white helmets that were not issued to volunteers, only to regulars. As in the British army, being regular officers, they wore, from the outset, gold buttons, lace and gilded badges rather than silver (reserved to volunteer militia officers).

The annual issue to NCOs and enlisted men from 1884 included the dress cloth tunic and the serge jacket (called "Tunic, Serge" in 1884, and later "Frock") until May 1892 when two serge jackets were served on the second year, the cloth tunic now being issued every second year. The undress scarlet serge jacket was based on the one used in the British infantry from 1873 and featured dark blue cuffs edged with narrow white lace ending in a trefoil (like on the dress tunic), a dark blue collar with the bottom edged with white, dark blue shoulder straps edged with narrow white lace, five brass Canadian militia buttons in

Royal Canadian Regiment officer's service dress, c. 1900. Scarlet frock with dark blue collar, cuffs and shoulder straps, gilt buttons and badges Brown leather "Sam Browne" equipment with revolver holster and sword frog. White helmet with gilt plate and fittings. Courtesy Michael Reintjes.

Privates and buglers, Royal Canadian Regiment, c. 1898-1901. All have the 1896 Canadian Pattern scarlet frock (some with the 1898 pocket pleats) and the dark blue field service caps worn on service dress. Detail from a photo taken at Frederiction, New Brunswick. Courtesy New Brunswick Provincial Archives.

front, no white piping edging from collar to skirt in front and no pockets on the breast. With this serge jacket were worn undress dark blue serge trousers with red piping and a dark blue Glengarry forage cap. This serge jacket remained unchanged until the advent throughout the Canadian militia of the pattern 1896 infantry serge frock, which was generally similar to the one previously worn by the Royal Canadian Regiment, but with the addition of breast pockets and additional buttons down the front (see description below).

The white helmets of the Infantry School Corps had yellow metal star-shaped plates with a beaver at the centre surrounded by a garter bearing "INFANTRY SCHOOL CORPS" surmounted by the imperial crown for NCOs and privates. Officer's helmets plates had the same general design, but with laurels surrounding the garter whose designation read "ROYAL SCHOOL OF INFANTRY" and the beaver in silver. They also had small gold bullion badges of the same design with a light blue velvet scroll bearing the "PRO PATRIA" unit motto worn on their peaked forage caps (with a red band) and tunic collars.

"Calling the Roll" of the Royal Canadian Regiment's detachment of the Yukon Field Force leaving Vancouver in May 1898. Courtesy City of Vancouver Archives.

By Militia General Order of May 1894, the insignia for the regiment was henceforth the "Royal and Imperial - cypher (V.R.I.) surmounted by the [Queen's] Imperial Crown" without unit designation and the same, but smaller, for forage cap badges and buttons, in gilt for officers; collar badges of the "Canadian Beaver on a scroll inscribed the motto Pro Patria" in frosted silver for officers. Their gilt waist plates had the VRI cipher and crown "in silver relief on pebbled gilt centre, with outer circle the designation 'Royal Canadian Regiment' in relief." The officer's pouch ornament also had the cipher and crown in gilt, the pouch itself having a brown leather belt like the waist belt. "Staff Sergeants, the Royal and Imperial cypher surmounted by the Imperial Crown to be embroidered on the strap in gold of special dimensions, on blue cloth shoulder-strap." NCOs and men had the same pattern helmet plates and buttons "as described for officers; *sergeants,* water gilt ; other ranks, gilding metal." The Queen's crown appear to have been worn by the regiment until 1907 (LAC, reel T405/868, 977, 1008).

Yukon Field Force 1898-1900

In 1898, the gold rush that drew tens of thousands of would-be prospectors looking to make their fortune in the wilderness of the Yukon Territory worried the Canadian government. In order to insure that law and order would be respected in a potentially turbulent populace of over 30,000 adventurers as well as to protect the storage of gold found, the government ordered the formation and immediate dispatch of the 198 strong Yukon Field Force to Dawson, the territorial capital. This included 133 members of the Royal Canadian Regiment. By 1899, the rush was over and tens of thousands of prospectors were going back to their homes as did about a hundred soldiers. The rest of the Yukon Field Force leaving in June 1900. Their weapons and ammunition were left at Dawson City in the care of the North West Mounted Police. This rather epic Yukon episode marked the first time a Canadian regular force had gone north of the 60th parallel.[18]

[18] R.J. Marrion and J.L. Summers, "The Yukon Field Force 1898-1900", MC&H, Vol. XXII, No. 4 (Winter, 1970) pp. 123-126; David Ross and Robin May, *The Royal Canadian Mounted Police 1873-1987* (London: Osprey, MAA 197, 1988), pp. 14-16, 20-21; "The Yukon Field Force 1898-1900", *Canadian Army Journal,* November 1950, pp. 30-34; according to an order in council of 21 March 1898, the Yukon Field Force was to have five staff personnel, 16 of the Royal Canadian Dragoons, 49 of the Royal Canadian Artillery and 133 of the Royal Regiment of Canadian Infantry (now the Royal Canadian Regiment).

The Yukon Field Force, summer dress, 1898. At left, a bombardier of the Royal Canadian Artillery. The private wearing the white helmet, the sergeant and the two privates, one in service dress and the other in fatigue dress were from a detachment of the Royal Canadian Regiment. A brimmed hat and high, laced boots were added for service dress in the Yukon's summer environment. All had the new brown leather Oliver equipment, which was first issued to the Yukon Field force. As can be seen, the valise could be replaced by a kit bag with the lower braces used to carry the rolled blanket. Plate by R. Marrion. With the kind permission of the artist and of Company of Military Historians. MUIA 345. Ph: RC.

For summer duty in the Yukon, the men were issued high brown leather boots that laced in front. The "Specifications for Beef Mocassin Boots laced in front" given out on 7 April 1898 with an order for 250 pairs stated:

Length of leg from heel: 16 1/2 in to 17 1/2 inch (depending on length of leg).
The leather is to be similar…as sample pair… both upper and sole and insoles as well.
The thread…eight strands or cord…of strong linen…to be heavily waxed…
The inside is to be…absolutely smooth for the foot.
The back of the leg is to be shaped for the calf of the man's leg.
The heels…one inch high…two inches long…

This was followed on 8 June by an order for: "Hats, wide brim…, Sweaters, jersey…, Gloves…, Mocassins, buck…, Boots ankle…, Boots, beef or moccasins…[see above], Boots, felt…" and ordinary boots. The brown felt hats had pugarees and their wide brims turned up on the left side. They also had their white helmets for guards and parades and used the serge frocks for service dress. The above was also intended for issue to the gunners and cavalrymen in the Field Force. A special and substantial winter dress was also issued.[19]

Militia (no-permanent) Infantry Units

Volunteer infantry:

1st Battalion, Governor General's Foot Guards (Ottawa ON) 7 June 1873; Governor General's Foot Guards (Ottawa ON) 16 September 1887.

1st Rifles (Montreal QC) 17 November 1859; 1st Prince of Wales Rifles 7 September 1860; 1st Prince of Wales Fusiliers 2 May 1898.

5th Rifles (Montreal QC) 31 January 1862; 5th Royal Light Infantry 7 November 1862; 5th Royal Fusiliers 19 November 1875; 5th Royal Scots Fusiliers 27 February 1880; 5th Royal Scots of Canada 29 February 1884.

6th Hochelaga Light Infantry (Montreal QC) 5 June 1863; 6th Hochelaga Fusiliers 3 December 1875 and 6th Fusiliers 28 January 1876. Amalgamated into the 1st on 2 May 1898.

[19] LAC, RG9, IID1, Vol. 12, p. 3805, Specifications for Beef Mocassin Boots laced in front (7 April 1898); Return showing supplies purchased for the Yukon Field Force…(8 June 1898), Vol. 13, p. 4126.

Volunteer militia infantry cloth tunic for enlisted men, 1870-1877. This pattern was ordered in Great Britain in September 1870 was issued in Canada from 1871 until replaced by the 1878 pattern. Scarlet, dark blue collar, cuffs and shoulder straps, white piping and lace, white metal crown buttons. Courtesy Canadian War Museum, Ottawa. Photo National Historic Sites, Parks Canada.

Volunteer militia infantry cloth tunic for enlisted men, 1870-1877. Curiously, the sergeant's chevrons are fixed on the left upper sleeve rather than on the right. Scarlet, dark blue collar, cuffs and shoulder straps, white piping and lace, white lace chevrons on dark blue background, brass crown buttons. Fort Henry, Kingston, Ontario. Ph: RC

Canadian Pattern 1870 tunic for enlisted men. This example, like most tunics seen of this pattern, has brass crown buttons. York-Sunbury Museum, Fredericton, New Brunswick. Ph: Courtesy Department of National Defence Library, Ottawa.

Volunteer militia infantry cloth tunic for enlisted men, 1870-1877. Back view. Fort Henry, Kingston, Ontario. Ph: RC.

7th Infantry (London ON) 27 April 1866; 7th London Light Infantry 15 February 1867; 7th Fusiliers 16 January 1880.

10th Rifles (Toronto ON) 14 March 1862; 10th Infantry 21 November 1862; 10th or Royal Regiment of Toronto Volunteers 10 April 1863; 10th Royal Grenadiers 5 August 1881.

11th Infantry (St Andrews and Lachute QC) 4 June 1862, also called Argenteuil Rangers.

12th York Infantry (Aurora ON) 14 September 1866; 12th York Rangers [Infantry] 10 May 1872.

13th Infantry (Hamilton ON) 11 December 1862.

15th Infantry (Belleville ON) 16 January 1863; Argyll Light Infantry 2 June 1871.

16th Prince Edward Infantry (Picton ON) 6 February 1863.

17th Lévis Infantry (Lévis QC) 20 February 1863, disbanded 1 July 1901 and reactivated 1902.

18th Prescott Infantry (Hawkesbury and Vankleek Hill ON) 26 February 1863, disbanded c. 1892.

18th Saguenay Infantry (Chicoutimi QC) 1 February 1900.

19th Infantry (St. Catharines ON) 18 March 1863; 19th Lincoln Infantry 28 September 1866; 19th St. Catharines Infantry 1 October 1897.

20th Infantry (St. Catharines ON) 17 April 1863, disbanded 28 September 1866, companies transferred to form 19th Lincoln Infantry; 20th Halton (Milton ON) Infantry 28 September 1866; 20th Halton Rifles 12 January 1872; 20th Halton Battalion Lorne Rifles 11 November 1881; 20th Halton Regiment Lorne Rifles 8 May 1900.

21st Richelieu Light Infantry (Saint Jean QC) 25 May 1863, disbanded 24 November 1882.

21st Essex Infantry (Windsor ON) 12 June 1885; 21st Essex Fusiliers 4 February 1887.

23rd Essex Light Infantry (Windsor ON) 14 September 1866, disbanded 11 March 1870.

23rd [originally Beauce Provisional Infantry 17 December 1869] Beauce Infantry (Sainte-Marie QC) 19 May 187[illegible]]; amalgamated into the 92nd on 1 July 1899.

24th Kent Infantry (Chatham ON) 14 September 1866, disbanded 25 November 1892; reactivated as 24th Kent Infantry (Chatham ON) 1 January 1901.

25th Elgin Infantry (Waterloo ON) 14 September 1866.

26th Middlesex Infantry (Strathroy ON) 14 September 1866; 26th Middlesex Light Infantry 24 March 1880.

27th Lambton Infantry (Sarnia ON) 14 September 1866; 27th Lambton Infantry [and] St. Clair Borderers 1 March 1872; 27th Lambton Regiment (St. Clair Borderers) 8 May 1900.

28th Perth Infantry (Stratsford ON) 14 September 1866.

29th Waterloo Infantry (Berlin [now Kitchener] ON) 14 September 1866.

31st Grey Infantry (Owen Sound ON) 14 September 1866.

32nd Bruce Infantry (Walkerton ON) 14 September 1866.

33rd Huron Infantry (Goderich ON) 14 September 1866.

34th Ontario Infantry (Whitby ON) 14 September 1866.

35th Simcoe Infantry (Barrie ON) 14 September 1866; The Simcoe Foresters 5 April 1867.

36th Peel Infantry (Brampton ON) 14 September 1866.

40th Northumberland Infantry (Cobourg ON) 5 October 1866.

42nd Brockville Infantry (Brockville ON) 15 October 1866; 42nd Lanark and Renfrew Infantry 1 December 1897.

43rd Carleton Infantry (Ottawa ON) 5 October 1866, nicknamed the Carleton Blazers, disbanded 3 December 1875.

44th Welland Infantry (Clifton ON) 16 November 1866; 44th Lincoln and Welland Infantry 1 October 1897.

45th West Durham Infantry (Bowmanville ON) 16 November 1866; 45th Victoria Infantry (Lindsay ON) 1897.

46th East Durham Infantry (Port Hope ON) 16 November 1866; 46th Durham Infantry 8 May 1900.

47th Frontenac Infantry (Napanee ON) 30 November 1866.

48th Lennox and Addington Infantry (Napanee ON) 14 September 1866, disbanded 3 September 1880.

48th Highlanders (Toronto ON) 16 October 1891.

50th Huntingdon Borderers Infantry (Huntingdon QC) 14 September 1866; 50th Huntingdon Borderers and Hemmingford Rangers 28 July 1899.

51st Hemmingford Rangers Infantry (Hemmingford QC) 14 September 1866, amalgamated into 50th on 28 July 1899.

52nd Bedford Infantry (Knowlton QC) 14 September 1866; 52nd Brome Light Infantry 31 November 1872.

53rd Sherbrooke Infantry (Sherbrooke QC) 21 September 1866; 53rd Melbourne Infantry 22 March 1867.

54th Sherbrooke Infantry (Sherbrooke QC) 22 March 1867; 54th Richmond Infantry (Richmond QC) 10 May 1867.

55th Megantic Infantry (Inverness QC) 22 March 1867; 55th Megantic Light Infantry 8 May 1900.

56th Prescott Infantry (Kemptville ON) 27 April 1867; 56th Grenville Rifles 13 August 1871; 56th Lisgar Rifles

Men of the 29th Waterloo Battalion milking a cow at the summer training camp near Gooderich, Ontario on 28 June 1871. This photo is is possibly the earliest showing militiamen wearing the Pattern 1870 Canadian infantry tunic. Courtesy Library and Archives Canada, PA92322.

Members of the 77th Wentworth Battalion at the summer training camp in Dundas, Ontario, in September 1873. Courtesy Ontario Archives, S11869.

Private of the 69th Battalion, c. 1875. He has the battalion's number pined on the front of his Glengarry that also seems to be lacking the cockade on the left side. Courtesy Andrew Moss collection. Ph: CL.

29 August 1871; 56th Grenville 8 May 1900.

57th Peterborough Infantry (Peterborough ON) 3 May 1857; 57th Peterborough Rangers 16 January 1880.

58th Compton Infantry (Cookshire QC) 11 October 1867.

59th Stormont and Glengarry Infantry (Cornwall ON) 3 July 1868; 59th Stormont Infantry 22 June 1883; 59th Stormont and Glengarry Infantry 23 March 1888.

60th Missisquoi Infantry (Dunham QC) 19 February 1869, disbanded 6 June 1898.

61st [originally Montmagny and l'Islet Provisional Infantry 9 April 1869] Montmagny and l'Islet Infantry (Montmagny QC) 5 November 1869.

62nd St. John Volunteer Battalion (St. John NB) 5 November 1869, disbanded 4 August 1871.

62nd St. John Infantry (St. John NB) 22 March 1872; 62nd St. John Fusiliers 14 April 1882.

66th Halifax Volunteer Infantry (Halifax NS) 5 November 1869; 66th Princess Louise Fusiliers (Halifax NS) 14 November 1879.

67th Carleton Light Infantry (Woodstock NB) 5 November 1869.

68th King's County Infantry (Kentville NS) 5 November 1869.

69th [originally 1st Annapolis County Infantry (Paradise NS) 8 October 1869] 1st Annapolis County Volunteers 5 November 1869. Amalgamated in the 72nd on 1 September 1898.

70th [originally the Champlain Provisional Infantry Battalion of 16 October 1869] Champlain Infantry (Sainte-Geneviève de Batiscan QC) 5 November 1869, disbanded 1 February 1900.

71st [originally the York Provisional Infantry Battalion of 10 September 1869]; 71st York Infantry (York County NB) 12 November 1869.

72nd The 2nd Annapolis County Infantry (Wilmot NS) 14 January 1870. Amalgamated in the 69th on 1 September 1898.

73rd Northumberland Infantry (Northumberland County NB) 25 February 1870.

74th Infantry (Sussex NB) 12 August 1870.

75th Lunenburg Infantry (Lunenburg NS) 12 August 1870.

77th Wentworth Infantry (Dundas ON) 23 May 1872.

78th [originally the Colchester and Hants Provisional Infantry 6 April 1871; 78th Colchester, Hants and Pictou or Highlanders Infantry (Truro NS) 5 September 1879.

79th Shefford Infantry Highlanders (Waterloo QC) 31 May 1872.

80th Infantry (Gentilly QC) 25 June 1875.

81st [originally Porneuf Provisional Infantry 9 April 1869]; 81st Porneuf Infantry (Pont Rouge QC) 6 July 1878.

82nd [originally the Queen's County 28 June 1875, and Charlottetown 2 July 1875 provisional infantry battalions]; 82nd Queen's County Infantry (Charlottetown PEI) 30 June 1876 and 4 October 1878.

83rd [originally the Joliette Provisional Infantry Battalion of 13 January 1871]; 83rd Joliette Infantry (Joliette QC) 27 December 1878.

84th [originally the St. Hyacinthe Provisional Infantry Battalion of 24 March 1871]; 84th St. Hyacinthe Infantry (St. Hyacinthe QC) 12 December 1879.

85th Infantry (Montreal QC) 4 June 1880.

86th [originally Three Rivers Provisional Infantry, 24 March 1871]; 86th Three Rivers Infantry (Trois-Rivières and Louiseville QC) 4 June 1880.

87th [originally Quebec Provisional Infantry 9 April 1869]; 87th Quebec Infantry (Quebec QC) 27 May 1881.

88th [originally the provisional infantry battalions of Kamouraska 9 April 1869 and Charlevoix 19 October 1870]; 88th Kamouraska and Charlevoix Infantry (Kamouraska and Rivière Ouelle QC) 29 September 1882, disbanded 1 May 1901.

Two men of the 77th Wentworth Battalion, c. 1871-1878. Photo of the period. Courtesy National Historic Sites, Parks Canada.

Sergeant, infantry, Volunteer militia infantry, c. 1871-1877. Photo of the period. Courtesy National Historic Sites, Parks Canada.

89th [originally the provisional infantry battalions of Temiscouata and Rimouski 9 April 1869]; 89th Temiscouata and Rimouski Infantry (Fraserville and Rivière du Loup QC) 12 January 1883.

91st Winnipeg Light Infantry (Winnipeg MN) 10 April 1885, disbanded 23 November 1888.

91st Manitoba Light Infantry (Winnipeg MN) 4 January 1889, disbanded 11 November 1892.

92nd [originally Dorchester Provisional Infantry 9 April 1869] Dorchester Infantry (Saint-Anselme QC) 12 June 1885.

93rd [originally Cumberland Provisional Infantry 6 April 1871]; 93rd Cumberland Infantry (Amherst NS) 12 June 1885.

94th [originally Victoria Provisional Infantry 13 October 1871; Victoria Highland Provisional Infantry 12 December 1879; Victoria Provisional Infantry, Argyll Highlanders, 9 April 1880; Victoria Infantry, Argyll Highlanders (Amherst NS) 12 June 1885.

95th Winnipeg Infantry (Winnipeg MN) 10 April 1885; 95th Manitoba Grenadiers (Portage la Prairie MN) 21 May 1886, disbanded 19 August 1892.

Canadian infantry battalion private, c. 1878-1885. He wears the Canadian Pattern 1878 tunic with the white lace trefoil crow's foot edging the cuffs, the most visible feature of the new uniform. The collar was now edged only at the bottom. The shoulder strap continued to have its end edged with white although this was eventually omitted. Plate by Ron Volstad. Courtesy History and Heritage Directorate, Department of National Defence, Ottawa.

Unnumbered:

King's Provisional Infantry (Georgetown PEI) 25 June 1878, disbanded 17 October 1884.

From 8 May 1900, all battalions were henceforth designated regiments.

Volunteer Infantry 1871-1885

Just as during the 1860s, the uniform and related equipment issue to a volunteer infantryman remained limited to a few basic items, namely a scarlet tunic, a pair of trousers and a forage cap. Everything else was to be provided by the volunteer at his own expense. As will be seen below notably concerning headdress and in the chapter concerning equipment, certain other items might be procured in a concerted effort by the officers and men of a battalion. Footwear was not necessarily important for occasional evening drills, but good shoes or low boots were essential during a summer training camp. Otherwise, like grey greatcoats (see the winter uniforms chapter), footwear and short gaiters would be supplied by government only when a unit was mobilized on full duty in the field for a fairly prolonged period of time such as the 1885 North-West Campaign, which was the only instance when thousands of volunteer militiamen went on active service for several months in a remote part of the nation. Otherwise, militia units were usually mobilized in aid to civil power and such duty usually took place in towns as a result of labour disputes.

Regarding footwear, some city units made efforts to privately procure items such as the British army's short black leather leggings, at least according to some period images. In some photos, it is difficult to clearly see if the volunteers might not be wearing low boots or half boots with the bottom of the trouser legs tucked in. Such boots, including "jack boots", were popular with civilian workmen. Furthermore, following the German army's

Infantry private's pattern 1878 tunic. Scarlet with dark blue collar, cuffs and shoulder straps, white piping edging the bottom of the collar, front, shoulder strap and cuffs forming a decorative "crow's foot" above each cuff, brass Canada Militia "crown" buttons. Fort Henry, Kingston, Ontario. Ph: RC.

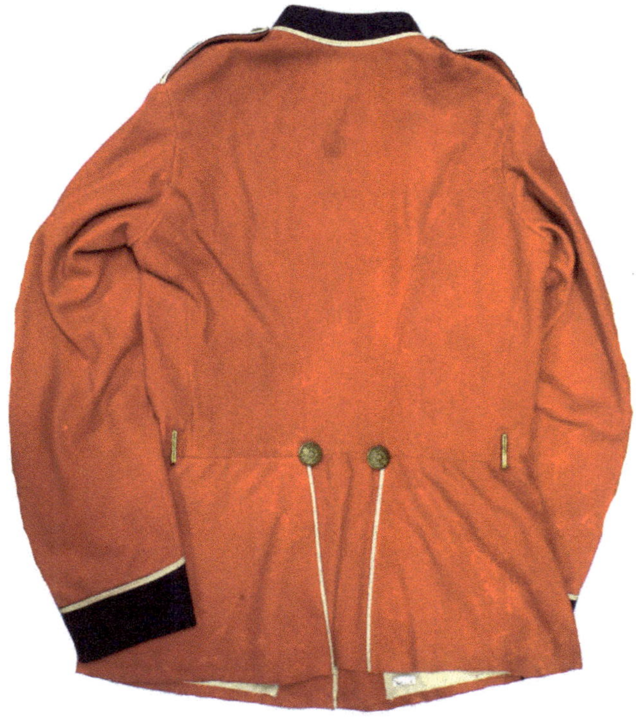

Infantry private's pattern 1878 tunic, back view. The back skirt is simply edged with white piping from the waist to the edge. Fort Henry, Kingston, Ontario. Ph: RC.

1870-1871 triumph, many military men including volunteers sought to emulate the look of the German troops; trousers tucked into jack boots had a definite martial look. However, most Canadian volunteers appear to have kept to wearing trousers long.

Canadian Pattern 1870 Infantry uniform

Replacement of newer patterns of government issue clothing did not occur overnight. The average clothing issue cycle was five years. For example, a battalion that had been clothed in, say, 1869, was not likely to receive new clothing until 1874. Thus, it would be wearing the Canadian Pattern 1863 infantry tunic with the Austrian knot on the cuffs until at least 1874 when the replacement 1870 tunic (first issued in 1871) with the chevron cuff would be served to NCOs and men. Photographs can be misleading as they might show older tunics still in use as secondary clothing. As late as the North-West Campaign of 1885, some members of the 66th Halifax battalion could be seen with the old Pattern 1870 tunic. Another militiaman posed wearing the Pattern 1866 infantry frock while his two companions have the pattern 1878 tunic with the crow's foot cuffs. Perhaps the most curious instance regarding uniform issues during the 1870s concerned the 59th Stormont and Glengarry Battalion, designated since 1868 as an infantry rather than a rifle unit. It was only from 14 June 1878 that the 59th was officially henceforth to be issued scarlet tunics rather than the dark green rifle uniforms that this battalion had been issued until then. A remarkable example of bureaucratic bungling and probably not the only one.[20]

In the mid-1870s, questions were increasingly asked regarding the procurement of uniforms from Canadian manufacturers rather than import them from Britain. From 1875 to 1877, the government ordered uniforms from Canadian firms, but they turned out to be below the quality expected. One of the main problems concerned infantry tunics because there were difficulties obtaining good scarlet dye that would be steadfast on the cloth and not be significantly altered by the elements. Since 1863, the Canadian uniforms imported from the United King-

[20] H.M. Jackson, *The Roll of the Regiments (The Active Militia)* (Ottawa, 1959), p. 99b.

A volunteer militiaman most likely of the 82nd Queen's County Battalion at Prince Edward Island, c. 1883. Scarlet tunic with dark blue collar, cuffs and shoulder straps, white piping and braid, brass buttons, dark blue trousers with scarlet piping and, remarkably, the dark blue British Home Pattern helmet. This carte de visite photo was taken at Charlottetown. Anne S. K. Brown Military Collection, Brown University Library, Providence, USA. Ph: RC.

Corporal and private of the 42nd Brockville Battalion, c. 1880. Both have the pattern 1878 tunic with the trefoil at the cuffs, but also wear the older pattern forage cap. Courtesy Carleton Place Museum, Carleton Place, Ontario.

The Midland Infantry Battalion marching to Humbolt, Saskatchewan, in the spring of 1885. The Midland Battalion was a temporary unit formed for active service in the North-West with officers and men of the 15th, 40th, 46th infantry and the 49th Rifles) battalions. This detail from an exceptional image by O.B. Buell shows the battalion wearing the "Home Pattern" British dark blue spiked helmet. This, and several other 1885 campaign photos and prints, appears to be the only evidence of the use of the Home pattern helmet being worn on campaign anywhere in the British Empire. Courtesy Library and Archives Canada, PA118748.

dom were made with superior quality "sergeant cloth" so Canadian manufacturers certainly had a hefty challenge. In the meantime, uniforms were needed and, in 1878, large uniform orders were again made in Great Britain. The 1878 public accounts reports reveal some $35,207 spent on "Militia clothing" in Britain, nearly all at Messrs Dolan & Co in London. This was a handsome sum and what it bought was certainly a substantial part of the Canadian militia's supply, most of which would be for infantry.

Canadian Pattern 1878 Infantry uniform

British patterns for other arms of the service had not changed since the last major uniform order from Britain in 1870, but scarlet tunics for the British line infantry had changed. Since 1871-1872, the chevron cuff had been replaced by the so-called "crow's foot" (a trefoil) in white lace edging the pointed cuff on British army tunics. The Canadian policy was to obtain tunics that were nearly identical to the British patterns. No contracts nor detailed descriptions have so far been found, but what evidence seen points to 1878 as the era when "crow's foot" (or trefoil) lace massively appeared on the cuffs of Canadian infantry tunics. No major purchases of infantry uniforms were made from Britain after 1871; minor orders may well have been made following the older 1870 pattern. When, from 1875, stores would need to be replenished after the five year cycle, the main purchases were made in Canada for the next two years, until, as seen above, large orders were made in Britain in 1878.

The 1878 Canadian pattern scarlet infantry tunic had some differences with the one worn by regulars of the imperial army. The facing colour was dark blue for all units and applied at the pointed cuffs, the collar and the shoulder straps. White piping edged the bottom of the collar, the coat down the front, all around the shoulder straps and down the two seams of the back skirts below the waist. Narrow white tape edged the pointed cuff ending in the "crow's foot" (or trefoil) above. Button were of brass. It had no badges or numerals on the collar and shoulder straps.

This pattern has remained the Canadian infantry's enlisted men's dress tunic to this day. There were no major purchase of uniforms in Britain after 1881 when the British infantry tunic was again changed to the straight "jam pot" cuff. Thus, this British pattern was never worn by Canadian infantry. By the mid-1880s, Canadian manufacturers were able to produce textiles for uniforms of equivalent quality to those previously imported from Britain and, from 1885 when the next large purchases had to be made, orders were placed with Canadian firms. The only infantry pattern tunic readily available was that of 1878 and it continued to be used by the Militia Department for decades.

Canadian Volunteer Militia during the 1885 North-West campaign. From left: drummer, 90th Winnipeg Rifles; drummer, 7th Fusiliers; sergeant, Governor General's Foot Guards; private, 46th Infantry; private, 66th Fusiliers wearing an old pre-1878 tunic. In the background, a trooper in field dress of the Governor General's Body Guard. Plate after Robert Marrion. Courtesy Company of Military Historians, MUIA 366.

1885 North-West Campaign

Order of Battle (scarlet coated Infantry units)

Private of the 91st Winnipeg Light Infantry at Calgary, 1885. This figure is based on group photo of this unit in 1885 showing the men still wearing the older 1870 pattern tunic and the round "pork pie" older type forage cap. The 1860s equipment is typical of that issued to the troops deployed during the North-West Campaign. Watercolour by R. Marrion. Private collection. Photo by Jack L. Summers.

Batoche Column, Maj. Gen. Middleton: Infantry School Corps, detachment of C Company (2 officers, 40 other ranks); 10th Royal Grenadiers (17 officers, 250 other ranks); Midland provisional battalion, one infantry company each from 15th, 40th, 45th, 46th, 47th and 57th battalions (34 officers, 342 other ranks)

Battleford Column, Lt. Col. W.D. Otter: Infantry School Corps, detachments from C Company (1 officer, 50 other ranks); Governor General's Foot Guards, detachment of shapshooters (3 officers, 48 other ranks).

Alberta Field Force, Maj. Gen. T.B. Strange: 91st Winnipeg Light Infantry (29 officers, 298 other ranks).

Along lines of communications: York and Simcoe provisional battalion formed from 12th and 35th infantry, at Humboldt (27 officers, 342 other ranks); 7th Fusiliers, at Clarke's Crossing (20 officers, 237 other ranks); 95th Winnipeg Infantry, at Qu'Appelle (24 officers, 408 other ranks); Halifax Provisional Battalion, included detachments of the 63rd along the Canadian Pacific Railway line; Yorkton Militia, at Yorkton (1 officer, 63 other ranks).

Worn and torn

The infantrymen went out on the Prairies dressed in their scarlet tunic, dark blue trousers, forage caps and in some cases Home Pattern helmets (see below). On 9 May, a resident of Batoche noted that "we see the red coats forming the line of battle all around the mission" which confirms that the infantrymen wore their scarlet tunics in the last battle of the North-West Campaign. By then, they were quite worn and the troops had assumed a much more campaigning appearance. The sharpshooters detachment of the Governor General's Foot Guards "went off [in early April 1885] to the front in their ceremonial tunics, 'with the buttons two by two,' their ordinary civilian shoes, and very little else but what they stood up in, and 'hoped for the best'...and a heavy greatcoat completed their outfit. A few days later at Winnipeg the stores of the Hudson's Bay Company were able to furnish suitable boots, socks, underclothing, and other essential articles of kit, as well as blankets" recalled Colour Sergeant Winters.

Campaigning on the Prairies soon transformed the appearance of the men and their uniforms and Winters added: "Our uniform, originally scarlet and blue, was worn and faded beyond recognition, often patched with oat bags and sometimes abandoned for any clothes we could procure. Many pairs of trousers had been worn out and replaced by others, bought in a Hudson Bay store or some other trading post, of a very loud grey plaid pattern which did not harmonise very well with the Queen's uniform. On our heads we wore service caps which we had manufactured ourselves out of oat bags and dyed with coffee." Alexander Campbell of the 7th Fusiliers later related that: "A Guard of Honour composed of the 7th and the Artillery with Busbies, bright scarlet and blue tunics, pipeclayed belts, etc., were at the [London, Ontario, train] station to meet us, a great contrast to those of the 7th, who had just returned with their soiled and faded tunics, tea stained belts and helmets with many of their trousers patched with flour sacks..."[21]

[21] Fourmond, quoted in Walter Hilderbrand, *The Battle of Batoche* (Ottawa: Parks Canada, 1985), p. 43; Lt. J. Preston, Midland Battalion, "Diary 1885", *Saskatchewan History*, VIII: 3 (Autumn 1955), p. 107; Alexander Campbell, "An Account of the Advances of the 7th Fusiliers of London to aid in the suppression of the North West Rebellion of 1885" University of Saskatchewan Library, MSS 49 #17 at: http://library.usask.ca/northwest/campbell/campbell.html

A sergeant and a private of the Midland Infantry Battalion in a skirmish line at Batoche, May 1885. This print shows both wearing the dark blue "Home Pattern" British spiked helmet in battle, but this is so far unsupported by witnesses' accounts. Illustrated War News, 30 May 1885. Courtesy Bruce Peel Special Collections, University of Alberta, Edmonton.

Bayonet charge of the Midland Infantry Battalion at Batoche, 12 May 1885. In this print, the battalion charges wearing its dark blue "Home Pattern" British spiked helmets, but the is no certainty that it was actually the case as it is not mentioned in battle reports. Illustrated War News, 30 May 1885. Courtesy Bruce Peel Special Collections, University of Alberta, Edmonton.

Infantry skirmishing at Batoche, May 1885. The head dress is the more commonly worn Glengarry or the folding winter cap. Note the musician with a water pail for thirsty skirmishers. The little church at Batoche is seen in the distance. Illustrated War News, 30 May 1885. Private collection. Courtesy Bruce Peel Special Collections, University of Alberta, Edmonton.

A "look out on the Qu'Appelle Trail", April or early May 1885. Infantryman sketched on the spot by E.W. Morrison for a plate in the Illustrated War News of 23 June 1885. Courtesy Bruce Peel Special Collections, University of Alberta, Edmonton.

Three members of the Canadian Volunteer Militia in the North-West Campaign, 1885. This photo shows two men wearing home-made forage caps white the third, sitting, has the issue Glengarry. The private at left wears the 1878 Pattern tunic and fairly long boots. The sergeant standing in the back has his sash over the shoulder. The most surprising feature is the 1866 Pattern infantry serge frock worn by the private sitting. This was several years after the remaining supply had been disposed of by the Militia Department, some units may have still have some for service dress and fatigues. Courtesy Library and Archives Canada, C24335.

Officers sleeping in the 'zareba' at Batoche, about 10 May 1885. This photo by Capt. James Peters shows the real appearance of men on campaign trying to secure some comfort to rest. Most have wool caps, one has a hat and one wears his regulation Glengarry. The rest of the clothing also seems to be a mixture of civilian and military items, for instance check shirts. The weapons, three swords and a rifle, are nearby. Courtesy Library and Archives Canada, C4522.

The 1885 temporary York and Simcoe Battalion and the Midland Battalion deployed during the North-West Campaign had contributing units wearing Home pattern helmet. Although some photos show the Midland Battalion marching through Saskatchewan wearing blue Home Service helmets, they do not generally appear to have actually been worn in battle, at least from written accounts of the engagements. The undress cap was obviously the popular choice on those occasions. However, two illustrations in *The Canadian Pictorial and Illustrated War News* published in Toronto on 30 May 1885 shows men identified as being of the Midland Battalion in a skirmish line at Batoche wearing their dark blue Home pattern helmets on and a bayonet charge, also at Batoche, of infantrymen — the unit is not identified — also wearing dark blue Home Pattern helmets (the officers are shown

Company A, 15th Argyll Light Infantry Battalion: "The Heroes of Batoche" 1885. This remarkable image shows the unit's members who took part in the North West Campaign wearing their homemade forage caps while in the field and probably new tunics and trousers issued when they came back to Belleville. However, the colour sergeant at left has a scarlet Norfolk serge jacket and the wounded officer at right wears his dark blue patrol jacket. Little numbers are superimposed on the men's breast referring to their names in a caption. Hastings & Prince Edward Regiment Museum, Belleville. Ph: CL.

wearing undress peaked caps and dark blue patrol jackets). Whether this detail was from information received in the field or from illustrators in Toronto who assumed the Midland Battalion would be wearing its helmets in battle remains unknown. Other prints of engagements (including Batoche) in that campaign show forage or fur caps being worn.

Boots were necessary on campaign and, to keep them comfortable, Sergeant Evans of the 10th Royal Grenadiers noted on 23 April that the men would "make a pillow of our boots at night to keep them from freezing stiff." Temperatures fortunately rose thereafter during the day and many volunteers made forage caps out of burlap and linen sacks as a protection against the sun although nights were still cool on the prairies.[22]

Not everyone had uniforms. The hundred men of the Fort Battleford Home Guard had none during the fort's siege. They were, according to the *Saskatchewan Herald* of 28 April 1885, "exempt from drill, but [obliged to] do a full share of guard and other duty, and are always on hand when any real work is to be done" and "Every man's waist encircled with a belt brisling with cartridges, a rifle in his hands and a revolver by his side."[23]

[22] Thomas Edward Champion, *History of the 10th Royals and of the Royal Grenadiers* (Toronto, 1896), p. 174.

[23] Quoted in Charles P. Mulvaney, *The History of the North-West Rebellion of 1885* (Toronto, 1885), p. 370.

Private of a Canadian infantry battalion, 1890s. This rendering was probably made from notes taken of the detachment sent to London for the 1898 Jubilee. Cutting from an unknown period publication. Vinkhuizen collection, Draper fund. Courtesy New York Public Library.

Private, 69th Annapolis Battalion, c. 1896-1898. He wears the new 1896 Canadian Pattern serge frock for infantry. The breast pockets on this image have no pleats indications it was made between 1896 and 1898. Private collection. Ph: CL.

Canadian volunteer militia Pattern 1896 infantry serge frock (also called serge jacket), c. 1898-1901. Scarlet with dark blue collar, cuffs and shoulder straps, white edging at the bottom of the collar, cuffs (with trefoil), shoulder straps (and invisible back skirt seams), brass buttons. Argenteuil Museum at Carillon, QC. Ph: RC

Canadian volunteer militia Pattern 1896 infantry serge frock (also called serge jacket), c. 1898-1901. Back view. Argenteuil Museum at Carillon, QC. Ph: RC.

Militiaman wearing the Pattern 1896 infantry serge frock with seemingly pleated pockets introduced from late 1898, early 1899. Courtesy Scott Duncan.

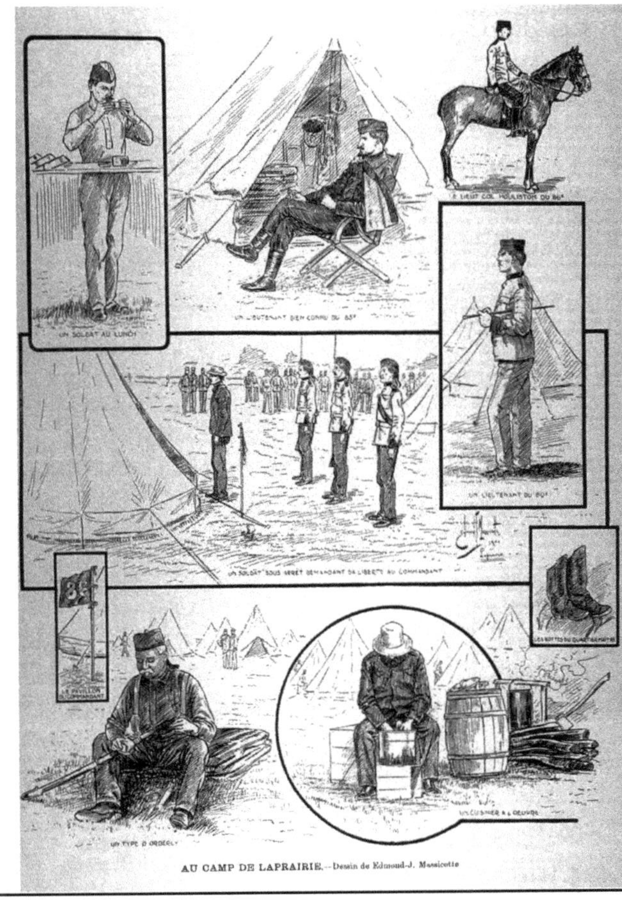

Scenes at the volunteer militia summer camp at Lapraire, QC, during July 1900. These sketches by Edmond-J. Massicotte show various and not commonly seen activities. Top row, from the left: a soldier having lunch, probably on a break from some fatigue as he is wearing a shirt; a lieutenant of the 85th; Lt. Col. Houliston of the 86th and a lieutenant of the 80th below both wearing the scarlet serge frock. Centre: three soldiers escort a soldier in civilian dress under arrest. Bottom: the 86th's command pennant; a soldier orderly; a cook at work and the quartermaster's boots. Some of the men still wear the Glengarry cap. Plate in Montréal's Le Monde Illustré, 22 July 1900.

Officer's patrol jacket, c. 1880s. Dark blue patrol jacket with (now somewhat faded) black lace and acorns. This garment was likely worn by an officer of the 67th Carleton Light Infantry Battalion in western New Brunswick. Carleton County Historical Society Museum, Woodstock, NB. Ph: RC.

Infantry 1886-1901

Forage cap of the 46th East Durham Battalion, c. 1880. Dark blue with brass numbers and a yellow cloth button on top. This example is somewhat more rigid than some shown in photos and might be meant for NCOs or officers. However, examples of the c. 1860s-1880s round forage caps are practically unseen so comparisons cannot presently be made. Courtesy National Historic Sites, Parks Canada.

Following the 1885 North-West Campaign, the infantry's standard issue uniform for NCOs and enlisted men continued to be the pattern 1878 scarlet tunic with the crow's foot cuff. For most units, there were no substantial changes until the late 1890s. As seen above, many battalions then adopted the pattern 1896 scarlet frock from that year onwards. Trousers and Glengarry forage caps remained the same. Except for certain infantry units such as the Royal Canadian Regiment, Governor General's Foot Guards, Highland, Fusilier and Grenadier units, the shoulder straps and collars of the NCOs and enlisted men did not generally have badges for nearly the whole pre-1901 period. Only in 1897 did the department order brass numerals for shoulder straps and this appears to have been a very gradual process. On 29 January 1898, instruction from the contracts branch in Ottawa specified that the "shoulder strap on the Infantry tunics may be set into the shoulder seams as in the Infantry jackets [frocks]" which meant that the white lace or piping edge would no longer be needed at the end. Regimental collar badges were only authorised to be issued from the summer of 1909 and most unit received them in the next two years. (LAC, reel T384/253, 399)

The 1896 serge frock for NCOs and private men

As seen in volume 1 for other arms of the service, the serge frock was introduced in 1896 for NCOs and enlisted men for all units of the Canadian armed forces, be they regulars or volunteers. For the infantry, it consisted of scarlet serge with dark blue collar, cuffs and shoulder straps, white piping (or narrow lace) edging the bottom of the collar, the shoulder straps and the pointed cuffs ending in a trefoil. There were two patch pockets on the breast and seven buttons in front. Initially, the pockets were initially unplaited, but all were made with plaits from late 1898 or early 1899.

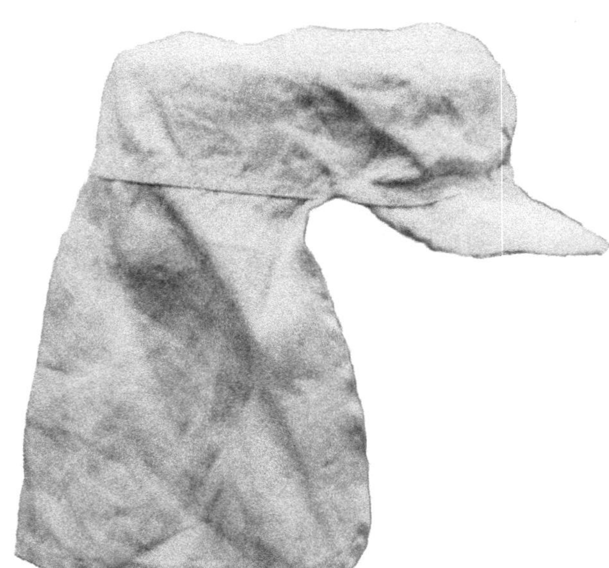

Canadian Militia Havelock, c 1870s-1880s. This white linen cover was procured by some units. It was put over the round forage cap. The Army Museum, Halifax Citadel. Ph: RC.

Although the scarlet cloth tunic, which now was to last nine years, continued to be worn by the regular Royal Canadian Regiment and city volunteer infantry units, the scarlet serge frock henceforth clothed all battalions and, for rural units, this was their only uniform coat. The scarlet serge frock was used for service dress by regular and city volunteer battalions; their scarlet cloth tunics were reserved for more formal occasions.

Glengarry pattern infantry forage cap, 35th Simcoe Battalion, 1880s. Dark blue with red tourri (or pompon) and black shiny and somewhat wide leather band. Brass unit badge on black silken material. The black ribbons that hung at the back are missing. Courtesy National Historic Sites, Parks Canada.

Headdress

From 1867 until 1902, the only official government issue headdress was the forage cap. For infantry battalions, this initially was the round all dark blue "pork pie" cap. As during the 1860s, the enlisted men's cap had no peak and indicated the battalion by metal numbers.

Since the 1860s, some units added a linen "Havelock" cover — named after a British hero of the 1857 Sepoy mutiny in India — over their round forage caps to provide better protection from the summer sun. These usually had a padded linen peak and a "curtain" at the back to protect the neck. Most covers appear to have been white, but the 37th Haldimand Rifles had them in grey during 1885-1888. In June 1885, the militia gazette commented that Havelock cap covers could be "of a different pattern for every company, crushed and dirty before they had been worn three days, or cheap grey cotton caps with peaks before and behind, looking like neither helmets or Havelocks. Such headdress make us respect those corps that let their faces peel under Glengarrys or forage caps in preference to looking untidy or un-uniform."

Lt. Col. Brown Chamberlin, 60th Missisquoi Battalion, 1878. He wears his full-dress uniform for state occasions that includes the British 1869 shako and the gold and crimson shoulder sash (adopted in the British army from 1868). His medal is the C.M.G. (Companion of the Order of St. Michael and St. George) awarded on 23 September 1870 for his role in the repulse of the Fenians at Eccles Hill. It was suspended on a blue ribbon. He retired from the force in 1871 and was allowed to retain his rank as a special case. Courtesy Library and Archives Canada, PA 145315.

In 1871, the first 3,000 Glengarry style infantry forage caps arrived in Canada. This type of forage cap of Scot's origin had been adopted by the British infantry three years earlier. It could be folded flat, was all dark blue except for a black leather band edging the bottom (and a red and white diced band above for Scottish units) with a small round red tuft (called a tourie) at the centre of the crease on top. Two narrow black ribbons hung from its back, at least in theory since this is not always apparent in illustrations and surviving caps. The replacement of the round forage caps by the Glengarry was gradual and took many years. The majority of battalion probably had it only by the later 1870s, but some still wore the round forage cap as late as 1885, sometime covered by a Havelock as seen above. Furthermore, it was only on 16 July 1880 that a Militia General Order allowed officers to have "the blue Glengarry with scarlet tuft, and black silk cockade on the left side. Numbers or badges to be worn on cockade, with a scarlet edging" at drills and summer camps.

From 1881, the "Torin" cap described as a

"Service Cap for Active Service and Peace Manoeuvres" was introduced in the British army and some were soon worn by Canadian volunteers. It was a flat folding dark blue cloth cap with a crown that could be of a contrasting colour and edged with narrow lace or piping. Period images show that officers and some units wore it instead of the issue Glengarry. During the 1885 North-West Campaign, some of these caps were homemade in the field by the soldiers out of coarse linen bags.

The Glengarry was officially replaced by the dark blue field service cap with two small buttons in front from August 1898. The Canadian contingents sent to South Africa had a khaki field service cap imported from England. Eventually, the field service cap gradually gave way to the peaked cap from 1903.

Although there was no full dress headgear provided by the government, units were allowed to procure them at their own cost. In many cases, the officers bought their own shakos, fur caps or helmets. In other instance, especially in cities, whole units would organise various fundraising events to buy dress headgear for the whole regiment. The patterns followed were those of the British army except for plates or badges. Thus, in the 1870s, photos show that some officers purchased the 1869 British infantry shako. It does not appear to have been especially very popular with Canadian volunteer enlisted infantrymen. However, the 1st Prince of Wales Rifles was seen wearing its "neat cloth shakos" in 1876, but it seems that NCOs and privates of the whole scarlet-coated infantry battalions did not adopt it.[24]

The officer's forage cap described in the 1876 Canadian dress regulations was of dark blue cloth, with "black leather peak and chin strap; black netted button and braided figure on the crown" and a 1 1/4 inch wide band of red cloth with the battalion's number embroidered in silver. The red band for all was unusual; only regiments with a "royal" title in the British army had it. This lapse or error in the Canadian regulations was corrected ten years later. While it is likely that forage caps ordered by officers from England would have the "correct" black lace band, others would have had locally made forage caps with red bands until, in theory, 1886.

The "Home Pattern" helmet introduced in the British army from 1878 was soon noted by Canada's volunteers. As early as May 1878, there was an enquiry to the Deputy Minister of Militia about such helmets for the Canadian Militia and, the following month, an entry in the ledger mentions "Helmets for 66th Battn" that suggests the Halifax unit was the first to order them. The ledger also mentions helmets for the 53rd in May 1879. This followed a Militia General Order of 3 January 1879 that there would be no objection to helmets being procured by units as long as they were "similar to those worn by the corresponding arm in the Imperial Regular Army. with the exception" of the Canadian ornament and devices. Then again, not everyone in Canada was charmed by this Germanic apparition; after all, it was really the old Prussian army's spiked helmet. On the other hand, in military circles worldwide following Germany's 1870 triumph over France, shakos were suddenly unfashionable and spiked helmets popular in many countries. Thus, in Canada, a number of officers, and, sometimes, whole units raised enough money to have such helmets as dress items. The presently known units to have worn Home Pattern helmet, either just officers or by all ranks, were the 12th York, the 13th Hamilton, the 17th Lévis (probably officers only), the 19th Lincoln, the 26th Middlesex Light Infantry, the 31st Grey, the 44th Welland, the 45th West Durham, 47th Fronteenac, the 51st Hemmingford Rangers, the 53rd Sherbrooke, the 54th Richmond, the 57th Peterborough, the 59th Stormont and Glengarry, the 62nd St. John Fusiliers, the 66th Princess Louise's Fusiliers (probably officers only), the 82nd Queen County (PEI) and the 85th Montreal Battalion (probably officers only). There were certainly more.[25]

From the 1880s, the British army's white "pith" tropical helmet turned out to be the favourite choice of Canadian soldiers. For dress, they also had yellow metal spikes and plates The reason for this preference is undocumented, but it must have had something to do with having a distinctive appearance that looked handsome and different. It also seems to have been less expensive and some were imported from the United States where both the British "six panels" or the American "four panels" models could be bought. With units holding charity picnics, diners and dances to buy their helmets, cost was a factor. As a result, a few battalions — notably the 16th, 32nd

[24]William J. Patterson, *Soldiers of the Queen: The Canadian Grenadier Guards of Montreal 1859-2009* (Montreal: Canadian Grenadier Guards Corporation, 2009), p. 41

[25]LAC, reel T4175/449, Deputy Minister's ledgers, June 1878; David Ross, "The Home Pattern service helmet in Canada", *Organisation of Military Museums of Canada Bulletin*, I, No. 2 (1972), pp. 14-17.

1

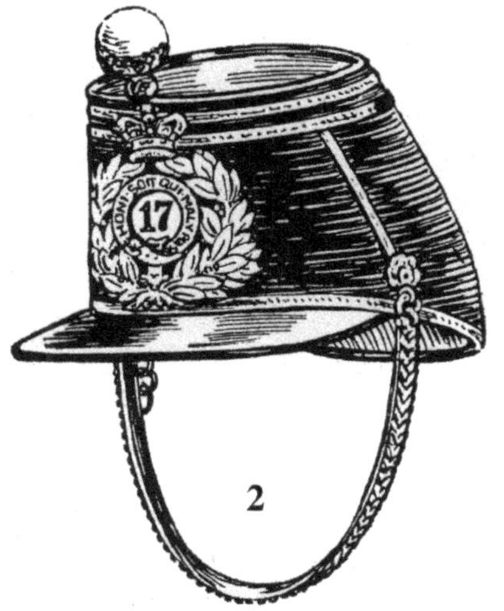

2

(Above) The Pattern 1869-1878 British infantry shako. 1: officer's. 2: men's. From 1867, shakoes were not issued by the government. Some Canadian officers obtained it for full dress, but it seems very few if any battalions wished to purchase it for NCOs and enlisted men. Print after A.R. Cattley. JSAHR, Vol. XV (1936).

(Left) Officer's blue Home Pattern spiked helmet, 17th Lévis Battalion, 1880s-1890s. The silver-plated plate is a standard one of the Canadian Militia whose centre has been replaced by a "17" in silver laid on dark blue velvet. Courtesy National Historic Sites, Parks Canada.

(Below) Men of the 44th Welland Battalion in line at Camp Niagara in October 1891. They still wear their dark blue "Home Pattern" helmet with, a most unusual detail, the artillery's round ball finial on top rather than the infantry's spike; possibly the result of a mistaken order from a British supplier. Courtesy Lincoln and Welland Regiment Museum, Niagara-On-The-Lake.

Dawson Rifle Company, 1902. Close up of a corporal, Color Sergeant F. E. Davis and a sergeant. Color Sergeant Davis was a veteran of the Royal Canadian Regiment of Infantry detachment with the Yukon Field Force whose small brass collar badge is seen on his scarlet frock's collar. Courtesy City of Vancouver Archives.

(Left) Enlisted men's white helmet, 67th Carleton Light Infantry, 1890s. Carleton County Historical Society Museum, Woodstock, NB. Ph: RC.

(Bottom left) Enlisted men's white helmet, 16th Prince Edward Battalion, 1890s. This is the American four panel model. Courtesy National Historic Sites, Parks Canada.

(Bottom right) Enlisted men's white helmet, 46th Durham Battalion, 1890s. This is the American four panel model. Courtesy National Historic Sites, Parks Canada.

and 46th — obtained white American helmets.[26]

Volunteer Militia Infantry Officers

Gold versus Silver

The first detailed Canadian dress regulations for officers were issued as the Militia General Order of 4 March 1876. It generally repeated almost word for word the imperial army's dress regulations, but continuing to replace the British regular infantry officer's gold with silver buttons and lace, to the ongoing displeasure of many Canadian infantry officers who, as would be seen later, absolutely wanted gold and rejected silver.

The matter of gold versus silver buttons and lace for the Canadian Militia's officers was raised at various times from 1838. Since the early 1860s, the cavalry and artillery had been confirmed in their right to wear gold, but the matter was not quite settled in the infantry. Indeed, some of the enlisted men's scarlet tunics had brass "beaver" buttons in the later 1860s, which must have made their officers wonder why they could not have gold, especially since such a practice could be traced back, not only to the June 1814 dress regulations for Upper Canada, but even to Lt. Governor Simcoe's 1794 orders for the Upper Canada Militia. Lower Canadian units had used gold or silver. Many units in the pre-Confederation Maritime provinces had also used gold buttons and lace on their uniforms.

The 4 March 1870 Canadian regulations and orders reaffirmed that "in the infantry, silver lace is substituted from gold." Silver for infantry was confirmed in the 1876 dress regulations. Meanwhile, an Order-in-Council of 22 December 1874 had given a "special sanction" for "Gold instead of Silver Lace" for the newly raised Governor General's Foot Guards who wore the uniform of the British Coldstream Guards. In 1875-1876, the officers of Montreal's 5th and 6th battalions's secured a similar derogation to wear gold lace when their units received permission to wear fusilier bearskin caps. The 7th and 10th fusilier battalions followed suit within a few years. The implicit message was that gold was reserved for the rich and privileged grenadier-like urban "elite" units. As a result, resentment persisted amongst many battalions whose officers felt that gold was the proper and traditional metallic hue that should be worn by Canadian volunteer infantry officers. Newspapers got involved and questions were asked in the House of Commons in Ottawa. In 1880, the matter was referred to Lord Kimberley, Secretary of State for the Colonies in Britain, who answered on 31 March "suggesting that silver lace should be adopted instead of gold lace."

Infantry colonel's tunic, sash and belt, c 1868-1883. Scarlet tunic, dark blue collar and cuffs, white piping, silver lace, cords and Canada Militia "beaver" buttons, gold collar rank badges. Silver state occasions dress sash with two crimson stripes. Crimson leather belt with silver lace, silver buckle with gilt centre. Courtesy Michael Reintjes.

[26] The ultimate and lavishly illustrated study on headdress is: Clive Law, *Fuss & Feathers 200 Years of Canadian Military Headdress* (Ottawa: Service Publications, 2013).

Lt. Col. J. Brosseau, 85th Montreal Battalion, c 1885-1890. He led the battalion from 1880 to 1892. Private collection. Ph: RC.

This was not the answer that the Canadian government wanted. In an Order-in-Council of 19 July 1880, the minister of Militia and Defence, Sir Alexander Campbell, noted that the "uniform clothing worn by the Militia of Canada is similar to that supplied in England to the Regular army and was adopted at the suggestion of the Imperial Authorities, but that in all cases the distinguishing devises on the badges and buttons are especially Canadian." After observing that the cavalry, artillery, engineers and the Governor General's Foot Guard "wear gold lace, the Infantry Silver lace", he stated that the use of gold lace had been "so long in existence, [that] such a change could not be made without causing considerable irritation" and it would be "undesirable to attempt to do so at present." The governor general, the Marquess of Lorne, could only concur and nothing further was done.[27]

This, in effect, opened the whole issue. Sir John A. Macdonald's conservative government wanted to stay clear of it and, on the contrary, coveted the support of the militia officers since many of the senior ones were also influential men in their communities. Indeed, in April 1881, Trois-Rivières's 86th Battalion obtained "the right to wear gold lace" for its officers (LAC, reel T4176/100). So, wishing to avoid controversy, the Department of Militia and Defence did not change the regulations with a formal announcement. Instead, in the militia regulations and orders published in 1883, the high command simply left out the requirements regarding lace…! And that year, the School of Infantry (later Royal Canadian Regiment) was created and, being a regular unit, its officers wore gold buttons and lace. A Militia General Order of 9 January 1885 mentioned that they would wear full dress "at balls, leves and State occasion…trousers with gold lace stripes (Infantry Pattern)". In effect, gold lace was "in" and it was confirmed for all infantry officers by the 1 May 1886 dress regulations.

Another matter was where the officer's rank badges should be worn on the uniform. The British moved them from the collar to the cuffs in 1880, but no formal orders to that effect were given in Canada until November 1885 (please see Chapter 10 on rank badges for details).

Officer's Service Dress

Officer's service dress uniforms that were not the regulation dark blue patrol jacket existed more or less officially in some units. The 22 January 1885 standing orders of the London (ON) 7th Fusiliers mentioned that the battalion had "previously been authorised to wear a scarlet cloth patrol jacket" that was now to be discontinued in favour of the regulation blue patrol jacket. The pattern may have been that of the 1872-1874 British scarlet patrol jacket. In the wider area of Kingston, ON, the officers of the 47th Frontenac Battalion adopted a scarlet patrol (or possibly fatigue) jacket with four pockets that had dark blue collar and cuffs edged white (see illustrations) during the 1870s.

As observers noted, uniformity amongst officers was often wanting regarding service dress. The militia gazette published the remarks of a correspondent on 29 December 1885 that nicely summed up the officer's service dress in many battalions: "There are…many varieties of patrol jackets, and they include all shades of blue and even green and black, with poor quality braid that fades during a single camp, and as many rows, and drop loops, and eyes as the wearer fancies; large badges of rank and small ones, and so on. The writer has seen officers

[27] LAC, RG9, IIA6, Vol. 15, Order in Council, 19 July 1880.

Officer, 44th Welland Battalion, c. 1890. Courtesy Lincoln and Welland Regiment Museum, Niagara-On-The-Lake.

Lt. Col. James Brock, 11th Argenteuil Rangers, c. 1890. His white spiked helmet bears the numeral "11" and he wears the standard infantry officer's scarlet tunic. Musée d'Argenteuil, Carillon. Ph: RC.

(Left) Lt. Col. R.R. McLennan, 59th Stormont and Glengarry Battalion of Cornwall, ON, November 1898. He wears the dress uniform of infantry officers with the full dress ceremonial sash and the white helmet. Courtesy Library and Archives Canada, C46602.

(Right) A captain of the 44th Welland Battalion. c. 1880. His dress uniform includes dark blue "Home Pattern" helmet with the spike on top. and white gloves. Courtesy Lincoln and Welland Regiment Museum, Niagara-On-The-Lake.

on parade with sky blue patrols, and [wear] the men's tunic with a little common tinsel for lace, and it is no uncommon thing to see officers wearing [enlisted] men's serge trousers..." The "sky blue" patrol jackets were originally a blue-black hue that had faded to a sort of blue grey. Surviving Canadian examples of this garment are usually somewhat blue-black and nearly all have their originally black braid and lace faded to a brown or an olive hue.

The 1 May 1886 officer's dress regulations were more complete than those of 1876 and, with a great many additions and amendments, remained the standard reference until the advent of the 1907 dress regulations. It introduced the "Scarlet Patrol Jacket" for officers, to "be used in place of the tunic on all occasions except those of state or ceremony, and in place of the blue patrol jacket, provided uniformity in Corps is observed." It had a dark blue collar, rounded in front, and dark blue cuffs, white piping "same as tunic", five gilt buttons and "two outside breast pockets." For rank badges, the collar and cuffs were trimmed with "one row of gold tracing braid at top of each." Field officers were to wear an additional row of quarter inch braid: for major, one; for lieutenant-colonel, two. The badges of rank for all officers was "embroidered in gold on the shoul-

Officer, 47th Frontenac Battalion, c. 1875. He wears a scarlet service undress jacket with dark blue collar and cuffs, white piping, silver buttons and shoulder cords seemingly adopted by the battalion's officers. Dark blue cap with "47" in silver and a black peak, dark blue or grey trousers with scarlet piping. Courtesy Anthony Gero.

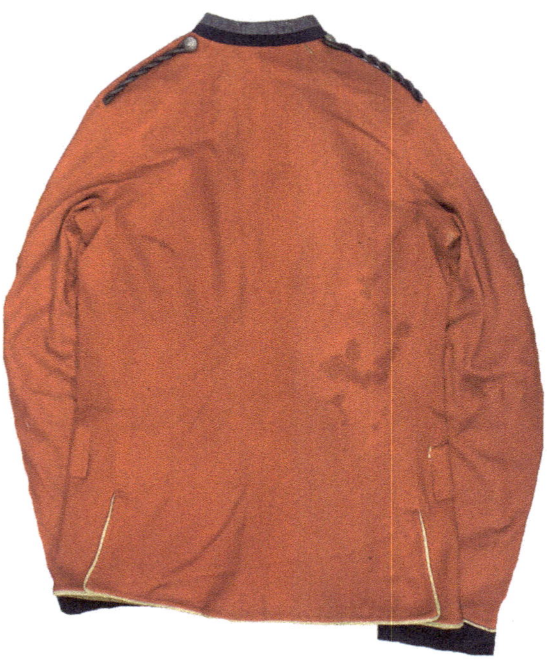

Officer's service scarlet undress jacket, c. 1875. Rear view. Fort Henry, Kingston. Ph: RC.

Officer of the 46th Durham Battalion, 1885. The mannequin shows the typical dress of infantry officers during the 1885 North-West Campaign. A dark blue forage cap with a black peak edged with gold, a red band and the numeral 46 in gold on dark blue. Dark blue patrol jacket with (now faded) black lace and acorns, dark blue trousers with scarlet piping, black leather belts, binocular case and revolver holster. All items original artefacts. Batoche National Historic Site, Parks Canada. Photo: David Ross.

der straps which are of same material as garment." Judging from images of the time, this pattern does not seem to have been widely adopted by infantry officers. It must also be borne in mind that volunteer officers hardly wished to have yet another expense because of their military interest.

Instead, most officers obviously continued wearing their dark blue patrol jacket with black cords and flat lace. For instance, members of the 86th Trois-Rivières Infantry Battalion were seen in July 1891 at Shawinigan by the parish priest who noted an officer wearing a "black" uniform with black lace (the patrol jacket), two gold stars (for a lieutenant) on the shoulder straps and white belt while his men had scarlet tunics and "black" trousers with red piping.[28]

Much more popular, judging from extant images and garments, was the British army's officer's serge patrol jacket that was adopted by GO of 24 March 1892 for field service dress for Canadian officers. For infantry (and engineers), it was scarlet with five buttons, two pockets with pointed flaps on the breast, dark blue collar, pointed cuffs and shoulder straps with gold rank badges. No collar badge was to be worn. The forage cap band was henceforth of "black lace, maple leaf pattern" with the scarlet band now reserved only for corps with the "Royal" designation. Furthermore, in corps "in which the Glengarry cap is worn by the men, officers will wear the same."

The scarlet-coated Dawson (Yukon) Rifle Company 1900-1905

With the Yukon Field Force departing from Dawson City in 1900, guarding the gold securely as well as maintaining the law and order in the territory became a concern as the small detachment of the North-West Mounted Police would not suffice. Local volunteers were called for to assist and, on 1 July 1900, the Dawson Rifle Company was created by a General Order. It was to have three officers and 39 other ranks. Its organisation was finally completed by early 1902. For its uniform, the 15 April *Victoria Daily Colonist* reported that the "coat is scarlet with blue facings, handsomely trimmed, and ornamented with bright military buttons, and the head dress a jaunty little field service cap of blue" that, presumably, had been issued in late 1900 or 1901. A rather startling dress since rifle corps were to be uniform in their regulation dark green uniforms. This was confirmed by a photo of the assembled company on the "Queen's Birthday" in 1902 uniformed with "Serge tunic, scarlet, trousers black with red piping, black boots, F.S. [Field Service] caps." Obviously, scarlet had been approved of, which perhaps made it unique in the British Empire. The reason was surely because it would have taken many months and been quite expensive to ship from Ottawa arms, ammunition and rifle pattern clothing for the company. In 1900, the departing Yukon Field Force left many weapons and equipments in Dawson. The rifle company obtained uniforms, rifles and Oliver equipment left by the Yukon Field Force and used them until disbanded in 1905.[29]

[28] *La Normandie et le Québec vus du presbytère*, N.-J. Chaline, R. Hardy and J. Roy, ed. (Rouen, 1987), p. 149.

[29] We thank Michael Gate and Derek Cooke, formerly of Parks Canada in Whitehorse and Dawson City, for certain additional details. Captain John Herbert Donald Hulme was a barrister in Dawson. In 1986, construction workers building a dike on the shore of the Yukon River at Dawson discovered many Lee-Metford rifles in the river's silt bed. It appears to have been some of those left behind with a lot of equipment by the Yukon Field Force that, after the company's disbandment in 1905, were consigned to the North-West Mounted Police who lacked facilities to keep them and eventually disposed of them in the river according to an old gentleman in Dawson who had been a young policeman there in the 1920s and 1930s. A fuller article on this unit by the author can be found in MC&H, Vol. 73, No. 3 (Fall 2021), pp. 260-264.

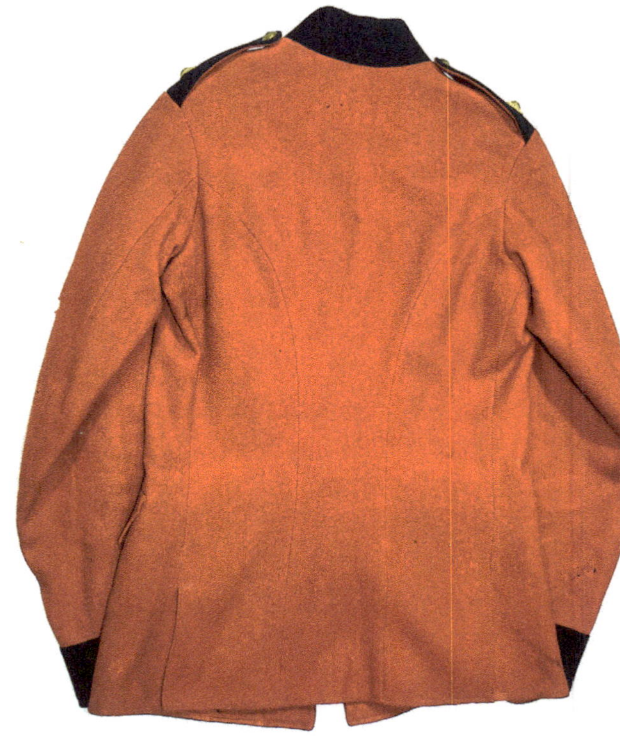

Infantry officer's scarlet serge frock, 1892-1901. Dark blue (actually black in this and some other instances) collar, cuffs and shoulder straps, gilt crown buttons and "pip". Courtesy Michael Reintjes.

Infantry officer's scarlet serge frock, 1892-1901. Back view. Courtesy Michael Reintjes.

Dawson Rifle Company, 1902. Close up of a corporal, Color Sergeant F. E. Davis and a sergeant. Color Sergeant Davis was a veteran of the Royal Canadian Regiment of Infantry detachment with the Yukon Field Force whose small brass collar badge is seen on his scarlet frock's collar. Courtesy City of Vancouver Archives.

Chapter 3

Distinct Dress of the Foot Guards, Fusiliers and Grenadiers

Governor General's Foot Guard

The Governor General's Foot Guards, 1874. This detail from a large cover print by William Brymner for the unit's regimental march published in Ottawa during 1874. It has the figures officers shown whose details are not always accurate (the collars and cuffs should be dark blue, no epaulette to the mounted officer, etc.). Those in scarlet have the Fusilier's pattern busby with a red feather while those in blue would be wearing the patrol jacket. It appears to be the earliest print showing the unit and its colours (not shown here but see Vol. 1, p. 254). Courtesy Anne S. K. Brown Military Collection, Brown University Library, Providence, USA.

In 1872, it was felt that the Governor General of Canada should have a foot guard unit to attend him in Ottawa. Thus was the Governor General's Foot Guard raised in Ottawa from 7 June of that year. Although a volunteer militia unit and not a regular corps, its uniform was totally distinct from the rest of the volunteer militia's infantry. Except for badges, it adopted the same uniform as the 2nd Coldstream Regiment of Foot Guards of the Queen's royal guard in London. "The Uniforms for the Non Commissioned Officers and Men...were ordered in the usual manner...on approval of the Honourable Minister of Militia..." in April 1873.[30]

For the enlisted men, this consisted of a scarlet tunic with dark blue collar, cuffs, cuff flaps and shoulder straps, white piping, brass buttons set in pairs, white lace at the cuff flaps and rear pocket flaps buttonholes, dark blue trousers with scarlet piping. The collar had a regimental badge on each side of the collar, which was exceptional on Canadian infantry enlisted men's uniforms in the early 1870s. It consisted of a six-pointed silver star with an oval bearing a blue cross at its centre. The uniform was issued every five years. However, for the first issue, there were no guards pattern uniforms available and the guardsmen had to settle with whatever could be obtained; these would have been the standard 1870 Pattern tunic although it seems there were tailoring attempts to have them look somewhat like the Coldstream Guards. The first issue fur caps were not the bearskin grenadier's

[30]LAC, RG9, II B1, Vol. 519, Memorandum, Lt. Col. J. Robertson Ross to Minister, Ottawa, 26 March 1872 and note by same, 9 April 1873.

pattern, but rather of the fusilier pattern and of racoon fur; its red plume was on its left side (the Coldstream Guards had theirs on the right side). The second issue of caps in 1878 was said to be of the proper Coldstream Guards pattern bearskin with the red plume of the Governor General's Foot Guard continuing to be on the left side ever since for all ranks. By the early 1880s, the tunic's shoulder straps also had a maple leaf badge. Outside of dress parades, the men had the Glengarry. This cap was worn on campaign during the 1885 North-West Rebellion.

In May 1889, it was further permitted to award its NCOs and guardsmen with a six-pointed gold star "having in centre a St. George's Cross in blue" for every five year's service badge to be worn on the centre of the scarlet tunic's left sleeve, which was unique to this unit. Corporals were allowed to carry side arms, a privilege otherwise granted only to sergeants; lance corporals had two chevrons instead of one.

The enlisted men of the Governor General's Foot Guard were not immediately issued the infantry's 1896 Canadian pattern serge frock. A "special pattern" had to be prepared that respected the regiment's peculiar button set in pairs placement and this was done in the spring of 1898 and made during May (LAC, reel T384).

Some officers of the Governor General's Foot Guards, 1876. Seated at left, Captain and Quartermaster the Honourable George Grant, at right Major and Paymaster the Honourable Horatio Asprey Wicksteed, standing is Captain Henry Stuart Weatherley. Being field officers on the regimental staff, Grant and Wicksteed had bicorn hats and sabretaches. Weatherley led a company and thus had the regimental fur cap. As can be seen, the first uniforms had the buttons were spaced evenly rather than in pairs. Private collection. Ph: CL.

Officers had the Coldstream Guards dress uniform with gold lace made official by Order-in-Council of 22 December 1874 although already in wear. The use of gold lace at a time when silver lace was the official Canadian infantry lace is said to have enjoyed strong support from Governor General Lord Dufferin. The officer's trousers had a wide scarlet stripe. The ceremonial and Levée dress sash for officers of the Governor General's Foot Guards was ordered to be "blue and gold" by Order-in-Council of 18 March 1875. On 9 April 1873, it was reported that the officers ordered their uniform "by themselves according to the usual custom observed by the Militia of Canada, from their own tailors in England and entirely at their own expense." In spite of this, they had difficulties in obtaining uniforms that were exactly identical to Coldstream Guards patterns. This was especially seen on the tunic's cuff flaps that, as early photos show, the buttons and lace are not being set in pairs, but seen as equidistant. The problem lingered on until 1889 when a dress committee was established and uniformity for officers achieved (see also chapter 12 on Mess Dress).

A note of 16 October 1900 kept by retired Adjutant General Walker Powell summed up nicely the dress of the regiment that "had been patterned after this distinguished regiment (Coldstream Guards)...upon the shoulder strap our men wear the maple leaf instead of the Coldstream Rose, and the cross in our badge is blue instead of red... Our bearskin [cap] plume is red while the Coldstream is white and the officers full dress sash, blue and gold instead of their red and gold" (LAC, Powell's Scrapbooks, Vol. 5, p. 116).

Undress, drill, mess and patrol uniforms also followed the Coldstream Guards patterns rather than those of the British line infantry that were also used by the Canadian infantry. The drill jacket was white, like that worn by enlisted men of Britain's Brigade of Guards. The undress cap was a dark blue pillbox with a white band with the regimental badge in front. Officers had the guards pattern dark blue patrol jacket with flat and wide black lace. They had a dark blue peaked cap. The militia gazette reported on 7 February 1889 that the unit would adopt the "frock coat, of the same pattern as the [British] brigade of guards, also the new regulation brown gloves for drill

Lance corporal, Governor General's Foot Guards, 1878. Scarlet tunic with dark blue collar, cuffs and shoulder straps, white lace and piping, brass buttons set in pairs, dark blue trousers with scarlet piping, black bearskin cap with red plume and brass plume holder and chinstrap. Courtesy Library and Archives Canada, C43484.

Private, Governor General's Foot Guards, 1878. Watercolour by Derek FitzJames. Courtesy Anne S. K. Brown Military Collection, Brown University Library, Providence, USA.

Four senior NCOs of the Governor General's Foot Guard that were part of the Ottawa Sharpshooters that took part in the 1885 North-West Campaign. The officer belongs to the 43rd Rifles. The guardsmen wear Glengarry caps or a homemade burlap cap rather than their distinctive pillbox caps. Courtesy Governor General's Foot Guard, Ottawa.

Three privates of the Governor General's Foot Guard, c. 1875-1885. They wear the regiment's distinctive scarlet tunic with cuff flaps and buttons set in pairs like Britain's Coldstream Guards with pillbox caps with a white band. Courtesy Library and Archives Canada.

Private's tunic and fur cap of the Governor General's Foot Guard. Although these items date from the later 20th century, their basic patterns have remained essentially unchanged since the 1870's. Private collection. Ph: RC.

Colour-Sergeant, Governor General's Foot Guard. c. 1900. Vinkhuizen collection, Draper fund. Courtesy New York Public Library.

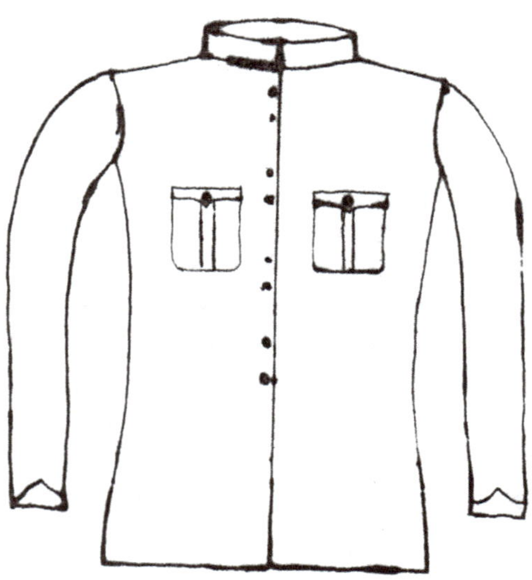

Basic design for the pattern 1898 frock for enlisted men of the Governor General's Foot Guard, 1898. Note the buttons in pairs. Courtesy Library and Archives Canada.

Enlisted man's 1898 pattern frock of the Governor General's Foot Guard, c. 1898-early 1900s. Governor General's Foot Guard Museum, Ottawa. Ph: CL.

(Right) Officer, Governor General's Foot Guard, 1898. An officer of the engineer unit at Brighton, Prince Edward Island, is in the background. Detail from a large plate by A.H. Hilder that was a supplement to 1898 "Christmas Globe" published in Toronto. Private collection. Ph: RC.

(Leftt) Officer, Governor General's Foot Guard, 1893. Scarlet tunic with dark blue collar and cuffs, white piping, gold buttons (set in pairs) and lace, dark blue trousers with red stripe, crimson sash, black bearskin cap with red plume. Dress identical to Britain's Coldstream Guards except for badges. Print in Adjutant General Walker Powell's Scrapbooks. Courtesy Library and Archives Canada, C12069.

(Bottom Left) Sergeant, Governor General's Foot Guard, c. 1880. He wears the distinctive white drill jacket worn by the Britain's foot guards. Courtesy Governor General's Foot Guard, Ottawa.

Lt. Hamilton Gray, Governor General's Foot Guard, 1885. He commanded the Sharpshooter detachment at the battle of Cut Knife Hill during the North-West Campaign. He wears the guard's pattern patrol jacket. Courtesy Governor General's Foot Guard, Ottawa.

purposes." The Governor General's Foot Guards continues today to wear the same basic dress uniform.[31]

Fusiliers

In Montreal, the 5th Royal Scots and the 6th Hochelaga Light Infantry, which became the 6th Fusiliers from 28 January 1876 and adopted the British fusilier's bearskin cap. The 6th also procured white grenade badges that were added on the men's shoulder straps. The 5th then evolved towards a Highland uniform (please see the chapter on Highland Dress). The 6th remained a Fusilier unit and, during 1890, decided to change its facings from the standard dark blue to white after having obtained permission to do so. When inspected on 21 June, the battalion "looked particularly smart and soldier-like on parade, what with their bearskins and new white facings and clothing, and their well cleaned arms and accoutrements... The Corps has a good Brass Band as well as Drums and Fifes, and also an Ambulance and a Signalling detachment." According to the *Canadian Military Gazette* of 12 February 1891, "its dress now corresponds in every particular with that of the fusiliers of Her Majesty's army." The model followed by the Canadian 6th Fusiliers was the British 5th Fusiliers, which had white facings and the same red over white plume on the left side of its bearskin cap. From 1896, its serge frocks had white collars and cuffs instead of the standard dark blue. The 6th was the only battalion to be granted a change in the facings of its uniform. This white facing, however, did not last for long. As seen below, the 6th was merged with the 1st Prince of Wales Rifle Battalion in May 1898, which meant another change in uniform for both units.[32]

Fusilier's bearskin Busby, 5th Royal Fusiliers of Montreal, 1875-1881. White plume at the left side, gilt flaming grenade badge with VR at centre and THE ROYALS around, gilt chin scales. This Busby belonged to Ensign James Alexander McLennan commissioned on 19 November 1875, lieutenant 30 November 1877, retired 1881. Black Watch of Canada Museum, Montreal. Ph: Earl Chapman.

Fusilier battalion's collar detail, c. 1875. Fusilier regiment added a flaming grenade badge to the collar. This detail is from Ensign James Alexander McLennan's doublet. Black Watch of Canada Museum, Montreal. Ph: Earl Chapman.

One of the most notable battalion change of uniform occurred with the dress of the 1st Prince of Wales Rifle Battalion in Montreal. The *Canadian Military Gazette* of 1 November 1892 even published an editorial article promoting that the "Fighting First" should change its rifle green uniform to scarlet, arguing that the "first regiment of the Canadian infantry service should, we think, be of imposing aspect, and a model to its junior battalions in appearance as well as in other matters." The uniform of the Grenadier Guards was suggested. Nothing occurred immediately, but the idea had been raised to change the unit into a more prestigious looking guard style unit. It also

---

[31] *History of the Governor General's Foot Guards* (Ottawa, 1948), pp. 237-241; LAC, RG9, IIA6, Vol. 12, Orders in Council, 20 June, 22 December 1874 and 18 March 1875; Robert M. Foster, *Steady the Buttons Two by Two* (Ottawa, 1999), pp. 12-15, 17-21, 32, 63. The Guard's Pattern red and gold sashes were finally adopted in 1939.

[32] Patterson, *Soldiers of the* Queen, pp. 42, 479; F.S. Meighen, "The Canadian Grenadier Guards", *Canadian Defence Quarterly*, VII (October 1929-July 1930), p. 550; MR 1890.

Private of a Fusilier regiment, most likely Montreal's 6th Fusiliers, c. 1880. Note the Busby with the brass grenade in front on the table at left. Contemporary photo. Courtesy National Historic Sites, Parks Canada.

had to be a substantial unit and, on 2 May 1898, the 6th Fusiliers was amalgamated with the 1st Rifles to form the 1st Prince of Wales Fusiliers. This meant that the 6th lost its white facings, but the new consolidated unit would now wear a dress uniform of scarlet faced with dark blue and have bearskin caps. Eventually, the 1st Prince of Wales' Fusiliers were re-designated Canadian Grenadier Guards in 1912 and then assumed the uniforms of the British Grenadier Guards.

The 7th London Light Infantry of London, Ontario, became the 7th Fusiliers on 16 January 1880. Its aim was to emulate, as far as possible for volunteer militiamen, the City of London, England, 7th Royal Fusiliers Regiment in the regular British army. That was a tall order, but, by 1884, the Ontario 7th Fusiliers had obtained fusilier pattern bearskin caps with a brass grenade badge in front and brass chinstrap, but without a side plume. Its January 1885 Standing Orders mentioned wearing "tunic, busby, complete kit and equipment" for marching order and the forage cap for Drill Order. It officially prescribed the issue Glengarry with a brass grenade badge on its left-front side for NCOs and privates. The brass grenade badge was also worn on the collar. The men's scarlet tunic was the standard Canadian infantry 1878 pattern.

Watercolours by Alexander Campbell of 1885, a soldier of the 7th and veteran of the North-West Campaign, show the cuff's edging and crows foot as yellow rather than white. Since other details are accurate, it would appear the 7th might have indulged in creating this non-official distinction at the time. The regiment was reorganised in 1899 and continued to be a fusilier unit that paraded with busbies in January 1900. In October of 1901, it was part of a gathering in Toronto of some 12,000 volunteers for a royal review by His Royal Highness the Duke of York then on a visit in Canada. During the march past occurred a peculiar incident. The duke, who was wearing his uniform of honorary colonel of the Royal Fusiliers of London, England, saw something that intrigued him, leaned over in his saddle and asked Gen. O'Grady-Hay, the commander-in-chief at the review:

"But what Regiment is this?"
"The Seventh Fusiliers of London, Sir."
"But do you mean—?"
"Of London, Canada"

At this reply, the duke "seemed immensely pleased to see approaching a regiment named after his own, wearing the same uniform and from a city bearing the same name." He was indeed so pleased that, a few days later, both the Duke and the Duchess of York decided to travel to London, Ontario, and there present the new colours of the 7th Fusiliers.[33]

The 21st Essex Infantry, designated Fusiliers in 1887, do not appear to have procured bearskin fusilier caps nor, after 1896, scarlet cloth dress tunics for its enlisted men. As late as 1904, NCOs and men paraded at the St. Louis (Missouri) exhibition dressed in the scarlet serge frock and white helmets as the unit's full dress. The

[33] *Standing Orders of the 7th Battalion Fusiliers* (London, ON, 1885), pp. 26-34; Alexander Campbell, "An Account of the Advances of the 7th Fusiliers of London to aid in the suppression of the North West Rebellion of 1885", http://library.usask.ca/northwest/campbell/campbell.html; University of Saskatchewan, Special Collections, MSS 49#17; Francis B. Ware, *The Story of the Seventh Regiment, Fusiliers of London, Canada 1899 to 1914* (London, Canada, 1945), pp. 21, 23-24.

The 6th Fusiliers' Signalling Section, 1890. Bearskin fusilier cap with brass grenade and chin scales, scarlet tunic with white collar and cuffs (worn until 1898 when changed to dark blue), white cuff trefoils and piping, brass buttons, dark blue trousers with scarlet piping, white accoutrement belts. They hold blue and white signal flags. Contemporary photo. Courtesy Canadian Grenadier Guards Museum, Montreal.

Private of the 7th Fusiliers of London, Ontario, 1885. At the centre, a private in marching order. He wears the Glengarry with a brass grenade badge, the standard scarlet tunic but with yellow grenade at the collar and yellow trefoil cuff lace. The insets are, clockwise, the "grey summer" helmet, "full dress" fur cap with grenade badge, "off duty" wearing the Glengarry and the "winter" fur cap. Pencil drawing by Alexander Campbell in his diary of the North-West Campaign. Courtesy of the University of Saskatchewan Library, Saskatoon.

Sergeant of the 7th Fusiliers of London, Ontario, 1885. This pencil drawing may well represent the artist, Sergeant Alexander Campbell, in his diary of the North-West Campaign. He wears the fur cap with its brass grenade and is also shown with a yellow trefoil on the cuffs. The three gold chevrons and crimson-red sash of his rank, which was also Campbell's rank. Courtesy of the University of Saskatchewan Library, Saskatoon.

Two officers and a private of the 7th Fusiliers of London, Ontario, 1885. They are shown at Clark's Crossing on the west bank of the Saskatchewan River. The tents of a military camp are seen in the distance on the east bank. The officers wear a service dress jacket that would have been scarlet with dark blue collar and cuffs, five or six gilt buttons in front. The service dress caps were likely dark blue with a yellow band with black peaks edged with gold lace. Pencil drawing by Alexander Campbell in his diary of the North-West Campaign. Courtesy of the University of Saskatchewan Library, Saskatoon.

Officer of the 7th Fusiliers of London, Ontario, 1885. He wears the dark blue patrol jacket with black cords and lace. The dark blue cap has a black band, gilt badge with silver "7", black patent leather peak with gold lace. Pencil drawing by Alexander Campbell in his diary of the North-West Campaign. Courtesy of the University of Saskatchewan Library, Saskatoon.

Field officer of a Fusilier battalion, c. 1885. Dark blue forage cap with a red band and gold grenade, dark blue patrol jacket with black cords, dark trousers and black boots. Watercolour by R.W. Rutherford. National Historic Sites, Parks Canada. Ph: RC.

Sergeant, 62nd St. John Fusiliers, 1898. This sergeant was certainly part of the Canadian detachment sent to London for Queen Victoria's Jubilee. He wears the unit's fusilier's full dress with the bearskin cap adopted since 1882. Cutting from an unknown British publication. Vinkhuizen collection, Draper fund. Courtesy New York Public Library.

The 1898 Christmas and centennial card of the 62nd St. John Fusiliers. An 1898 private is shown in a rare back view facing a private of 1798 wearing the uniform of the time. Ph: CL.

Officer, 62nd St. John Fusiliers, c. 1880. He wears a Glengarry with his dark blue patrol jacket with black cords.

(Right) Second lieutenant of the 66th Halifax Volunteer Infantry, c. 1878-1879. This remarkable photograph shows a dark blue Home Pattern helmet that may first have been worn by this unit; its plate bearing the battalion's number 66 is of the infantry pattern. This indicates that this image was taken between the late summer of 1878 and November 1879 when the battalion was designated as a Fusilier unit. The Home Pattern helmets were kept, but a new flaming grenade badge, a distinction of fusilier units, replaced the original badge and collar grenade badges were added, probably in 1880. Image by Millman & Ray in Halifax. Courtesy Andrew Moss. Ph: CL.

(Left) Lt. Col. J. J. Bremner, 66th Princess Louise Fusiliers, c. 1880. He wears the fusilier's grenade plate on his Home Pattern helmet, which would have been procured following the November 1879 granting of the Fusilier title. Image in Collections of the Nova Scotia Historical Society, XVII (1913). Ph: RC.

Private H.R. Roberts, 10th Royal Grenadiers, 1883. This image appears to be the earliest dated photo that shows the bearskin fusilier cap worn by an enlisted man of the battalion. Courtesy The Royal Regiment of Canada Museum, Toronto.

Private, 10th Royal Grenadiers, in marching order, c. 1883-1885. Fusilier cap with red over white plume and brass grenade badge stamped with numeral "10"; Canadian pattern scarlet infantry tunic and dark blue trousers. Armed with a Snider-Enfield rifled musket. White accoutrements with a haversack whose linen belt slung over the right shoulder gives the impression the soldier wears 1840s style cross belts. The haversack is of the 1860s pattern. Courtesy Scott Duncan.

Firing party of the 10th Royal Grenadiers at the funeral of Lt. W. Fitch (upper inset) and Pte T. Moore, 1885. The men wear their full dress including bearskin caps. From the inset portraits, it can b seen that only officers have collar grenades at this time. Illustrated War News, 6 June 1885. Private collection. Ph: RC.

brass fusilier's badge was, however, worn on the collar and the helmet. The officers had the scarlet cloth tunic with gold buttons and lace.

The 62nd St. John Battalion of Infantry became the "62nd St. John Fusiliers" on 14 April 1882 and received its fusilier bearskin caps ordered from England during the summer. The officers "received permission to wear cross-belt and pouch instead of the sash, being, probably, the first regiment in Canada to make this change." Furthermore, they were allowed "to wear gold lace instead of silver" at that time. From about the mid 1880s, the men's tunic had the brass grenade at the collar and the brass numeral "62" on the shoulder straps.[34]

The 66th at Halifax became the "Princess Louise Fusiliers" on 14 November 1879, but it did not acquire bearskin caps. From about 1879, the unit's officers took into wear the British dark blue Home Pattern that was eventually replaced by white helmets.

Grenadiers

The 10th "Royals" Battalion of Toronto was officially assigned a distinctive unit helmet badge as per General Order of 1 August 1879: "This Battalion is hereby permitted to adopt and use on the helmet plate, the following description of badge: In centre on shield, figure 10 with crown on top; behind both, and showing above crown and 10, a sheaf of spears; on dexter side, Thistle and Shamrock; on sinister side, Roses, and in base. Maple Leaves. The shield surrounded with garter, bearing the motto, 'Ready, aye Ready,'which is surrounded by a wreath of laurel leaves, behind which and extending outside wreath, a military star. The whole surmounted by a Royal Crown."

Sergeant, 10th Royal Grenadiers, 1885. He is photographed in marching order with the dress fusilier bearskin, short Snider-Enfield rifled musket and 1860s accoutrements. His scarlet sash is visible under the belts worn over the pattern 1878 infantry tunic. Courtesy Library and Archives Canada, PA1992-2.

The 10th Royals' dark blue helmet was replaced by fusilier's pattern bearskin caps after 5 August 1881 when the unit became the "Royal Grenadiers". Its badge, henceforth worn on officer's forage caps with a red band, now had a cipher RG at centre with 10 and the motto below, a grenade above and side floral wreaths. Up until 1893, the 10th looked like a fusilier unit in full dress, but on 12 November 1893, the 10th Royal Grenadiers "wore their new 'bearskins' for the first time." These new bearskin caps were larger and more rounded like those of the British Foot Guards in Britain. They had a white plume with red tip. The 10th also kept wearing the round dark blue pillbox forage cap, ornamented with a wide red band and a brass grenade in front. During the 1885 North-West Campaign, the men of No. 2 Company had a homemade peaked linen cap. The grenades worn on the tunic collar appears to have been added during the later 1880s. The tunic and trousers otherwise remained according to the standard Canadian patterns.

The service dress for officers consisted of the round peaked cap with a scarlet band and, from 1879, an embroidered version of the authorised helmet plate that continued to be worn on the forage cap into the 20th century. The dark blue patrol jacket was initially the same as other infantry battalions with four cords on the chest. From the 1880s, this was changed to the Guards' style dark blue patrol jacket with five white black laces ending in a point whose ends were always left hanging down; the collar had a gold embroidered grenade at each side.[35]

The short-lived 95th Manitoba Grenadiers (1886-1892) did not have fur caps, but its officers did have

[34] E.T. Sturdee, *Historical record of the 62nd St. John Fusiliers (Canadian Militia)* (St. John, NB, 1888), p. 89; tunic in the Canadian War Museum No. 19870190-004.

[35] Thomas Edward Champion, *History of the 10th Royals and of the Royal Grenadiers* (Toronto, 1896), p. 237.

(Left) A group of NCOs and privates of the 10th Royal Grenadiers of Toronto, c. 1898. From left to right: sergeant Noble in parade dress; bandsman Herbert; pioneer McCleary; colour sergeant Philips of the bicycle squad in service dress frock and forage cap; private Cordingly in dress marching order with tunic, white helmet and Oliver accoutrements; drummer Witheridge in parade dress; private ambulance man White of the medical detachment. This image appeared in the British The Army and Nay Illustrated, 2 April 1898.

Private, 10th Royal Grenadiers, c. 1900. Note the distinctive red on white plume. Vinkhuizen collection, Draper fund. Courtesy New York Public Library

Officer, 10th Royal Grenadiers of Toronto, 1898. Detail from a large plate by A.H. Hilder that was a supplement to 1898 "Christmas Globe" published in Toronto. Private collection. Ph: RC.

Mounted field officer of the 10th Royal Grenadiers of Toronto, c. 1900. He is shown in parade dress. The sabretache hanging below the waist belt was abolished from 1901. Coloured photo print. Private collection. Ph: RC.

(Right) Officer's patrol jacket for Toronto's 10th Royal Grenadiers, c. 1886-1895. Courtesy National Historic Sites, Parks Canada.

(Bottom) Members of the 95th Manitoba Grenadiers Battalion, c. 1886. They are lined up on the centre of Portage la Prairie for this image. The men wear the standard scarlet tunic with the Glengarry, the Snider-Enfield rifled muskets having fixed bayonets. Capt. H.J. Woodside at left wears the dark blue peaked cap and patrol jacked. This photo has often been dated at 1885, but Capt. Woodside's patrol jacket has grenades at the collar that could only have been added after 21 May 1886 when the 95th became the "Manitoba Grenadiers" with its HQ at Portage la Prairie. It was disbanded in 1892. Courtesy Library and Archives Canada, C79107.

grenades on their collars.[36]

Notes

MA, MG6, B5, Journal of the Provisional Battalion of Canadian Light Infantry and Artillery, 1871-1874 by Lt. Col. Osborne-Smith; LAC, RG9, IIA6, Vols. 11 and 12, Orders in Council, 1871-1876; Clive M. Law, *Fuss & Fashion 200 Years of Canadian Military Headdress* (Ottawa: Service Publications, 2013); Charles H. Stewart, *The Concise Lineages of the Canadian Army 1855-date* (Toronto: author, 1982); Michael Barthorp, *The British Army on Campaign 1816-1902* (London: Osprey,1988, MAA 198 and 201)

[36]A final and surprising note regards a proposal found in the January 1896 issue of *The Indian Magazine* published in Brantford, Ontario (III, No. 4, pp. 1-3). This called for the creation of a six company battalion of Infantry to be called "The Six Nation's Regiment" that, notably would draw its traditions from the services of the Canadian Iroquois against the Americans during the War of 1812. A distinct uniform was proposed that called for a red tunic, a blue or black kilt, a cap of deerskin with a "fan-shaped panache" of feathers, brown breeches, leggings, shoes and accoutrements with a scarlet blanket. This proposal probably came at the worst time for the small staff of the Defence Department officials in Ottawa then labouring to introduce the new serge frock and seeking to obtain modern rifles as a priorities. Furthermore, permission to have the battle honours "of *Queenston Heights*, and *Beaver Dams* — or at least *Niagara*" on its colours were sought. Nothing came of this amazing proposal and the battalion never formed, but, well over a century later, the idea of having War of 1812 battle honours for Canadian units emerged again and was approved by the government of Canada in 2012.

Chapter 4

Highlanders

Early Highland Military Dress

During the early 1820s, the cultural "Highland Revival" that was taking place in Britain had repercussions in eastern Upper Canada. The counties of Glengarry and Stormont had been settled by a substantial number of Highlanders from Scotland that had retained many traditional aspects. It was not uncommon for some of them to wear their traditional costume and, as early as 1792, "Highlanders in their national dress" at Glengarry came to greet Governor Simcoe and his wife. Nearly four decades later, Eleanora Hallen noted in her diary that in "the evening, Mr Joseph put on his kilt, it is a Highland dress. It is of plaid with a scarf and bow of the same. It comes above his knees, and his legs are not covered… He had short stockings with very low shoes."

The question of what was the tartan used by Scots settlers and their immediate descendants has often been raised and remains much debated. We have no pretension or wish to settle any part of these somewhat sensitive debates, but merely evoke what some tartans might have been for early Canadian volunteers with a Scot's heritage. As with nearly every clan, there were several setts (patterns) possible for the tartan of a clan and although, since the later 19th century there is a certain uniformity — if not quite unanimity — is settling on one predominant tartan, this was not necessarily the case earlier. The Macdonell of Glengarry clan, whose members made up the majority of early Scots settlers in eastern Upper Canada, can be a good example. The mid-18th century portrait of Alasdair Ruadh Macdonnell of Glengarry shows him clothed in a predominantly red tartan. Sir Henry Raeburn's 1812 portrait of the 15th Chief, Alexander Ranaldson Macdonell shows him dressed in a predominantly dark green and dark blue tartan with many red and white lines, which has been basically retained as the clan's primary tartan in its darker version for there is also a paler sett of it.[37]

From the 1820s, officers of some Upper Canada Glengarry county militia units looked somewhat like this officer. From 1821, the 1st and 3rd Glengarry Regiments were allowed tartan trousers and sashes, Highland bonnets with plumes with the regulation scarlet coatee faced with dark blue and trimmed with gold buttons and lace. Prior to 1856, Highland infantry units wore kilts or trews, plaid sashes, bonnets or shakos, but their coats were generally the same as other infantry units except that they had short tails and officers usually had wings instead of epaulettes. This detail of a print of an officer of the 72nd Highlanders in 1825 after "I.N. McKs" was transformed digitally by the author to have the proper dark blue facings and gold lace; the red tartan of the 72nd is not quite the "red" Macdonnell tartan, one of the clan's severals, but it can be taken as an evocation, especially as setts of tartans were then not fixed permanently as they later came to be. Ph: RC.

During the War of 1812, militiamen from the eastern counties of Upper Canada were once seen wearing kilts, which would have been their civilian dress. Thus, it is not too surprising

[37] The modern 59th Battalion now The Stormont, Dundas and Glengarry Highlanders did not officially adopt Highland features until 1904 with doublets, Glengarry caps and trews of the Sutherland tartan as the Militia Stores had those available for issue. The Glengarry tartan was adopted during the First World War and kilts of that tartan in 1922 when the unit became Highlanders as per Boss, pp. 108, 125, 134.

Colonel Alastair Ranaldson Macdonell of Glengarry (1771-1828). This portrait painted in 1812 by Sir Henry Raeburn shows the 15th clan chief, who wore only Highland clothing and was an outstanding promoter of Scots heritage. It shows him wearing one of the Macdonell's darker setts of tartan. This was most probably also worn by many Glengarry volunteers militiamen and was officially adopted in 1922 by the Stormont, Dundas and Glengarry Highlanders of the present Canadian Militia. Print after Raeburn. Ph: RC

to read that, on 5 June 1821, the officers of the 1st, 2nd and 3rd Glengarry regiments and the Stormont regiment were permitted to wear "Highland bonnet with black plumes, the established [1814 scarlet faced with dark blue] uniform coat with [gold] epaulettes, a crimson silk sash, waist belt, sabre, and tartan trousers. On 15 July 1822, the 2nd Glengarry's regimental uniform was changed to: "Regimental coat of green cloth, with black velvet collar, cuffs and lapels, with gold epaulettes and gold lace. The Highland plaid to be worn across the left shoulder. The remainder of the uniform to continue as heretofore ordered." However, on 29 March 1823, officers of the new 2nd Stormont Regiment were not allowed to assume Highland dress.[38]

Clan Macdonnell was not the only one to emigrate to Canada. In November 1823, Archibald MacNab (1781-1860), 13th chief of the MacNab clan obtained a land grant of 81,000 acres in Upper Canada on the shores of the Madawaska River in Renfrew County. He convinced a number of Scots to settle at the township of McNab. Years later, in 1836, he was authorised to form a "Regiment of Scotch Militia", the 1st Renfrew Regiment of the Upper Canada Militia, which he commanded as lieutenant-colonel and whose officers may have worn elements of Highland dress. What this was in detail is so far unknown, but an early photo and print of his 1840s portrait shows him wearing a jacket, kilt and hose of the red and black McNab tartan over a high collared scarlet jacket with silver buttons, buckles, a large brooch and a seemingly black Glengarry with eagle feathers. One would suppose that this could have been the unit's dress and that some officers of the unit also had it. In early December 1837, Col. McNab's militia was embodied, boarded three steamers that conveyed "Colonel MacNab's and other bodies of militia to the Toronto" where they arrived on the 9th presenting "a motley appearance. Only the chief officers were mounted and in uniform; the rank and file were un-uniformed; they had a sort of extemporised military band, and were preceded by the two field-pieces from the City Hall." McNab and his militiamen eventually went back home. However, his excessive greed and dictatorial ways drove his settlers to have his grant revoked by the government in the years after the rebellion. He resigned his commission in 1846 and now ruined, McNab eventually left Canada. As for the militiamen of McNab Township, they do not appear to have carried on with Highland traditions bearing in mind that many other settlers from various origins now joined them.[39]

The Glengarry and 1st Stormont Militia regiments likely continued wearing plaid and, when the Rebellions broke out in November 1837, the officers may well have had the Highland uniform. In April 1838, the Williamstown detachment of Glengarry Highlanders received grey greatcoats, trousers, mitts, moccasins and a

[38] *Mrs. Simcoes's Diary*, Mary Quayle Innis, ed. (Toronto, 1971), p. 67; *Eleanora's Diary*, Caroline Parry, ed. (Richmond Hill, 1994), p. 142 (11 January 1836); LAC, RG9, IB1, Vol 67, Adjutant General of Militia Upper Canada to colonels of the 1st, 2nd and 3rd Glengarry regiments and the Stormont Militia, York, 5 June 1821; LAC, RG9, IB1, vol. 73, Adjutant General to Col. Donald Macdonell, 2nd Glengarry, York UC, 15 July 1822; LAC, RG9, IB1, vol. 74, Adjutant General to Col. Anderson, 2nd Stormont, York UC, 29 March 1823. LAC, RG9, IC1, vol. 21, J. Fraser to Adjutant General , Embro, 3 October 1836. A proposed unit at Embro that was not raised was to wear the "Sutherland tartan. Coat, Vest and Trowsers with Blue Bonnet."

[39] McNab was later described in *Chamber's Edinburgh Journal* of 17 November 1849 as "usually dressed in a blue coat and trousers, with a whole acre of McNab tartan for a waistcoat — at great dinners he wore a full suit of tartan. On the jacket were large silver buttons, which his ancestors wore in the rising in 1745." *History of the County of Peterborough, Ontario* (Toronto, 1884), p. 173.

true Highland Scot item: some bonnets. Col. Fraser's 500 Glengarry Highlanders that went to Montreal in May 1838 "created a sensation as they marched through our streets to the martial music of the spirit-stirring bagpipes" reported the *Montreal Herald*. That same month, a volunteer company at Esqueesing was also allowed to "assume the Highland garb" which indicates that Scots traditions in military fashion were favourably looked upon. In late November, the Adjutant General of Upper Canada approved the formation of a "Scotch Company" in the 4th Battalion of the North York Militia from "the Highlanders throughout the Township of Vaughan...provided that they were permitted to form a Company of themselves officered by their own Clan..." One may suppose they had so far unknown features of Highland dress.[40]

Private, The Highland Company, 1837-1838. "Paid by Government as Regulars. 80 Men." Dark blue Balmoral cap with tartan band, white thistle badge and black ribbons, dark blue greatcoat with tartan collar and cuffs, dark blue buttons, tartan trousers, light brown mitts. The tartan is shown as red and black and blue, somewhat reminiscent of the Rob Roy sett. Plate by Peregrine Pouchbelt and Rodrick Ramrod. Anne S.K. Brown Military Collection, Brown University Library, Providence, USA. Ph: RC.

The 4th Provisional Battalion of Upper Canada of Incorporated Militia was largely recruited in the Scots Glengarry counties during the fall of 1838 and its men had "black plumes and tartan caps" that were most likely dark blue bonnets with tartan or diced headbands and black plumes. This would have been for undress as the battalion since it was also issued with shakos. (Please see the chapter in volume 1 on the 1816-1854 infantry.)[41]

Following the 1837-1838 Rebellions, the features of Highland dress were retained in the Glengarry and Stormont militia units. The Militia General Order of 16 April 1852, specified: "The Glengarry Battalions:— To wear the Highland bonnet, tartan scarf and trousers, as heretofore." This was followed by the Militia General Order of 16 June 1852 that further ordered: "The 3rd & 4th Battalions of Stormont Militia, shall wear the Highland Uniform, as prescribed for the Glengarry Battalions in the General Orders of 16th April 1852. Their Commanding Officers to select the pattern of the Tartan for their Battalions." What tartans were chosen remain unknown, but some would have been Macdonell of Glengarry.[42]

Highland Volunteer Rifles, 1850s-1860s

As a result of the Crimean War, the infantry's tailcoats trimmed with white buttonhole lace were replaced by tunics with relatively long skirts. This would not do for Highland uniforms that needed short skirts so that Highlanders were instead assigned a jacket with "Inverness" skirts (or "flaps") below the waist, each pointed and decorated with three buttons. Initially with two rows of buttons on the breast, it changed when Her Majesty was "pleased to approve a Pattern of a single-breasted Highland Jacket" on 13 October 1856 (WO 370/5). Early images of Canadian Highland volunteers, some of which are reproduced here, show various adaptations to the new Highland dress notably regarding the Inversness skirts on the jackets. Most are double-breasted indicating units raised before 1857. They were worn for several years and, up to 1866 in the case of the Toronto Highland Rifles. Full Highland dress for Canadian units, an expensive proposition for volunteers, thereafter took a pause

[40] LAC, RG8, C800, Clothing supplied for the service of the Williamstown [detachment of] Glengarry Highlanders in April 1838; LAC, RG9, IB1, vol. 69, Adjutant General to Lt. Governor, 29 May 1838; LAC, RG9, IB1, vol. 30, Col. C.C. Small, 4th North York Militia, to Adjutant General , Toronto, 21 November 1838 with note of approval by Adjutant General .

[41] Stuart L. Mason, *To Repel the Wicked: The Home Guard of Stormont, Dundas & Glengarry 1837-1843* (Cornwall: Stormont, Dundas & Glengarry Historical Society, 1999), p. 7.

[42] LAC, RG9, IC4, Vol. 2, MGOs, 16 April and 16 June 1852.

Clan Chief Archibald McNab, Lt. Col. of the 1st Renfrew Regiment, Upper Canada Militia, c. 1845. In this image, McNab appears to be wearing the same military style Highland costume shown in his portrait. Courtesy Library and Archives Canada, C2890.

Clan Chief Archibald McNab, Lt. Col., 1st Renfrew Regiment, Upper Canada Militia, c. 1845. McNab (1781-1860) was lieutenant-colonel of the regiment from 1836 to 1846. McNab (1781-1860) was lieutenant-colonel of the regiment from 1836 to 1846. This seemingly unsigned portrait might depict Laird McNab in his usual dress, the tartan jacket and kilt being in the red and black McNab tartan. His scarlet military style under jacket with silver buttons was likely the uniform he found suitable for the regiment. From an old print. Ph: RC.

Montreal Highland Volunteer Militia Rifle Company officers, c. 1857. On 1 July 1857, it was reported that "The company is now fully uniformed – the style of the uniform is as follows. Viz – Rifle green cloth coatee with scarlet facings trimmed with yellow tinsel braid, - trews and plaid of McKenzie tartan, - similar to the style of the 71st Highlanders. Blue cloth forage caps, with white check band – Highland bonnets red and white check trimmed with ostrich feather and scarlet hackle" as per Quarterly Returns of Volunteer Corps, LAC, RG9, IC2, Vol. 59. Courtesy Library and Archives Canada.

Lt. George Brown, Montreal Highland Volunteer Militia Rifle Company, c. 1857. He wears the rifle green uniform with scarlet collar and cuffs adopted by the company during the late 1850s. Original unsigned oil painting. Courtesy Black Watch of Canada Museum, Montreal. Ph: Earl Chapman.

Corporal of the Highland Rifle Company of Hamilton, c. 1862. In October 1862, the company reported being dressed in green jackets with "kilt, plaid & trousers of Hunting Stewart tartan" and bonnets. Note the large belt buckle and the shoe buckles in true "Highland Revival" style. One of its 49 Enfield rifled muskets is held by this corporal. Courtesy R.G. Harris.

(Left) Toronto Rifle Volunteer companies, c. 1862. At left, a private of the Toronto Highland Company; its uniform was the same as that of the 93rd Sutherland Highlanders with the exception of the feather bonnet — the Glengarry being worn — and the tunic, which was of green material with red collar and cuffs. Detail of a print after Roper. Courtesy Library and Archives Canada, C126600.

The Toronto Highland Rifles, No. 10 Company, 2nd (Queen's Own Rifles) Battalion, 1866. The company still wore its old uniform style consisting of a dark green jacket with red collar and cuffs, trews and sash of the the 93rd Sutherland Highlanders' tartan. Courtesy National Historic Sites, Parks Canada.

Scottish Rifles officers, Halifax Volunteers, c. 1862-1868. At left George R. Anderson and at right A.K. Mackinlay, both wearing the uniform "of a dark colour with plaid scarf…belts blackened." Private collection. Ph: RC

until the 1880s and 1890s when the 5th and 48th battalions assumed it.

The Toronto Highland Rifles Company was formed on 17 September 1856. It became No. 4 and later No. 10 company of the 2nd Battalion Queen's Own Rifles. Its uniform was "the same as that of the 93rd Sutherland Highlanders, with the exception of the feather bonnet — the Glengarry being worn — and the tunic which was of green material with red facings." It was initially "dressed in the kilt" and "wore the famous 42nd Black Watch tartan kilt and plaid.". It was deployed against the Fenians and fought at Ridgeway. Veterans recalled that "at Ridgeway" they "wore the tartan plaid of the 42nd Black Watch, albeit they had doffed the kilt and wore trews instead. These four survivors recall with pride the wearing of the tartan in that historic company. The tunic was of a dark green." Following the battle of Ridgeway, some members of No. 10 Highland Company applied for compensation to damaged uniforms including one "suit destroyed property of private John White wounded left arm since amputated." The damages were awarded. The company disbanded on 1 October 1868 due to the Militia Department's refusal to "grant an allowance in lieu of the ordinary uniforms" (Fraser, 25-26).[43]

The 2nd Queen's Own Rifles' No. 6 Company was originally the Whitby Highland Rifle Company that, in 1859, acquired a "rifle green Highland jacket scarlet facings, fine 42nd plaid pants – Glengarry cap & feather with plated ornament." It received compensation for clothing in lieu of the standard rifle uniform in 1863. Five years later, it gave up its Highland dress and "determined to put on pants and wear the shako so as to be uniform with the rest of the regiment."[44]

The Montreal Highland Volunteer Militia Rifle Company was authorised on 16 October 1856. It reported on 1 July 1857 that "The company is now fully uniformed – the style of the uniform is as follows. Viz – Rifle green cloth coatee with scarlet facings trimmed with yellow tinsel braid, - trews and plaid of McKenzie tartan, - similar to the style of the 71st Highlanders. Blue cloth forage caps, with white check band – Highland bonnets red and white check trimmed with ostrich feather and scarlet hackle." The officers wore the dirk and broadsword. In August 1858, the company went with other units to Portland, Maine (USA) on an exchange visit. Its uniform was noted with curiosity by the "Portland Transcript" as "green coated faced with red and gold, plain pants, tartan scarfs, Highland bonnet with ostrich plumes and red feather. The piper, in full Highland costume, with his kilt and bare knees, attracted much attention."

On 17 November 1859, the Highland Company, with eight other Montreal rifle companies, formed the 1st (Prince of Wales) Volunteer Militia Battalion along with eight other independent rifle companies in Montreal. But, on 9 October 1863, the company left the 1st Battalion over a dispute concerning its "Highland costume" and joined the 5th Battalion (Royal Light Infantry) where its men were permitted to retain some vestiges of their

[43]LAC, RG9, IC8, Vol. 4, Commanding Officers Reports…1862. There were many variations in the numbers of the Toronto companies. For instance, the Toronto Highland Company became the 4th and later the 10th company of the 2nd Rifle Battalion while its 6th Whitby Highland Rifle Company was later detached and, in 1866, joined to the 34th (Ontario) Infantry Battalion. Maj. C. Gilmor to Adjutant General, Toronto, 28 July 1866, LAC, RG9, IC3, Vol. 6; "Four Survivors of the Original Highland Co. Queen's Own Renew Old Memories", *Welland Tribune and Telegraph*, 19 May 1927.

[44]MF 1863; LAC, RG9, IC8, Vol. 4, Commanding Officers Reports…1862; *Volunteer Review*, 26 October 1868.

Highland uniform.[45]

The 2nd (Highland) Rifle Company of Kingston raised 1856 had the full Highland costume including a green jacket with black facings and had two pipers, but it existed only a year through lack of interest.[46]

Since 1856, the London Highland Rifle Company wore "rifle green with red facings – pants & kilts Hunting Stewart tartan – rifle green shako with Rob Roy band and black plume the same as Toronto Highland Volunteers." It seems to have given up its Highland dress in the early 1860s because it was not listed for compensation payments in lieu of uniforms in 1863.[47]

Several other Highland rifle companies were raised in the early 1860s. In 1862, the Brantford Highland Rifle Company wore a "Green Melton Inverness jacket with red facings; pants McKenzie clan tartan – dark blue Glengarry bonnet with small bronzed bugle in front fastened on left side." It did receive compensation in 1863, but not later. The Hamilton Highland Rifle Company had a "tunic green – kilt plaid & trousers of Hunting Stewart tartan – bonnet." It did not receive compensation in 1863.[48]

Tunic of the Halifax Scottish Rifles, Halifax Volunteers, c. 1862-1868. It is made entirely of black material. Only by substantially lightening the photo do the breast cords and sleeve Austrian knots become apparent. Courtesy The Army Museum, Halifax Citadel, Halifax, Nova Scotia.

In Montreal, the 5th Royal Light Infantry battalion had greeted the Highland Rifle Company in its ranks since a General Order of 9 October 1863 made it its No. 6 "Highland Company". It continued to wear its "expensive" uniform that would soon need to be renewed. Thus, from December 1863 or early 1864, it assumed the 5th Battalion's scarlet coat, but, because of the high proportion of Scots volunteers in the company, was granted "the exception that tartan trews and shako with check band, shall be allowed as a distinction." The seed of Highland dress was permanently planted in this unit and it went on thereafter. In a sense, it was thanks to the 5th Battalion that Highland dress survived and, eventually, prospered in the Canadian Volunteer Militia.

The Maritime provinces also had some Highland rifle companies before Confederation. In Halifax, Nova Scotia, the Scottish Rifles organised from 13 December 1859 had, an initial uniform of "grey cloth with royal blue piping on collar and sleeves, and belts of brown leather…a cap with diced band and horse-hair plume." From January 1862, the uniform became "of a dark colour with plaid scarf…belts blackened." The company was one of those that formed the volunteer battalion in 1860 and it appear to have worn their Highland dress until about 1868.

In St. John, New Brunswick, the Duke of Rothesay's Rifles (a Scottish Company), organised from 2 June 1860 wore "green with Scotch bonnet and plaid of pattern worn by 42nd Highlanders." It appears to have existed until 1863.

In Orwell, Prince Edward Island, a company of Highland Rifles was active during the 1860s. Its uniform was blue with red piping. While there likely were Highland features in their dress, they are presently unknown.

---

[45] LAC, RG9, IC2, Vol. 59, Quarterly Returns of Volunteer Corps.

[46] John Douglas MacKenzie-Naughton, *The Prince of Wales' Own Regiment* (Kingston: n.p., 1946), pp. 9-11.

[47] LAC, RG9, IC8, Vol. 4, Commanding Officers Reports…1862; MR 1863.

[48] LAC, RG9, IC8, Vol. 4, Commanding Officers Reports…1862; MR 1863. There was a Highland Rifle Company at Embro in 1862, but its dress is unknown as per DAG de Salaberry to Isaac Wallace at Embro, 8 January 1862, LAC, RG9, IC1, Vol. 285.

(Right) Officer, 79th Brome Battalion, c. 1890. Note the Cameron of Erracht tartan trews. Brome County Historical Society Museum, Knowlton, Quebec. Photo: RC.

(Bottom right) Sergeant, 78th Colchester, Hants and Pictou or Highlanders Battalion, 1880s. In spite of its "Highlanders" name, the battalion simply wore the standard militia infantry uniform. He holds the sword-bayonet of a Snider-Enfield short rifle that featured the ling "yatagan" blade. Courtesy Andrew Moss collection. Ph: CL.

Private, Halifax Scottish Rifles, c. 1862-1868. This image shows that all ranks had the same basic uniform, but the feather was seemingly shorter privates. Courtesy Andrew Moss collection. Ph: CL.

79th Brome Battalion officer's peaked forage cap, 1890s. It featured a red, white and dark blue diced border around the dark blue crown and a gold "79" in front, the black peak edged with gold lace. Brome County Historical Society Museum, Knowlton, Quebec. Photo: RC.

When formed in 1875, No. 1 Company of the 5th Royal Fusiliers in Montréal assumed a mixture of fusilier and Scottish dress. This 1877 image shows its members wearing doublets with trews, Fusilier's bearskin caps and Glengarry forage caps with red and white diced borders. From Chambers' history of the 5th Royal Scots.

Uniform of Ensign James Alexander McLennan, 5th Royal Fusiliers, 1875-1880. This remarkable and very rare assembly includes the sword with its white leather accoutrements and gilt fittings. Courtesy Black Watch of Canada Museum, Montreal. Ph: Earl Chapman.

Target shooting competition at the Pointe Saint-Charles range near Montreal, August 1881. Most of the figures shown in this print show personnel of the 5th Royal Scots Fusiliers wearing doublets and trews, most with white havelocks or white helmets. Contemporary print. Ph: RC.

Members of the 5th Royal Scots of Canada, c. 1885. A lone officer, second from right, is amongst a group of NCOs and corporals. All wear the scarlet doublet with dark blue collar, cuffs and shoulder straps, white piping, brass buttons and badges, Lorne tartan kilt with white sporran and two black tails, hose with white and black tops, dark blue Glengarry with diced border. Courtesy Black Watch of Canada archives, Montreal.

Manner of putting on and wearing the Highland plaid shoulder sash. Drawing by Douglas Anderson. Courtesy Halifax Citadel National Historic Site, Parks Canada.

Highland Battalions 1872-1901

The first battalion to have a Highland designation in the Canadian Militia was the 79th Shefford Battalion of Infantry (Highlanders) organised in 1872 in Waterloo, Quebec. However, it did not wear Highland dress with doublets and kilts. It has sometimes been assumed that only the officers had Cameron of Erracht trews and peaked forage caps with diced border during the later 1880s and 1890s. The Cameron of Erracht tartan was the one worn by the British army's 79th Queen's Own Cameron Highlanders. In August 1887, the 79th was paid compensation in lieu of the standard issue infantry "trousers and forage caps" and this strongly suggests that the NCOs and men of the battalion probably had trews and Glengarry caps with diced bands at that time (LAC, reel T4177/559).

Following Confederation in 1867, units in the Maritime provinces gradually took into wear the standard uniforms that were issued in central Canada since 1863, some units adding features of Highland dress. In Nova Scotia, the Colchester and Hants Provisional Infantry Battalion, formed on 6 April 1871 paraded at Truro, NS, on 23 June. According to the *Volunteer Review* of 17 July, the three "companies, Pictou Highlanders, from the neighbourhood of the West River, were also there, headed by their Pipers." This seven company battalion had a "greatly improved appearance" thanks to the importation of "Glengarry caps" by Col. Laurie, the commander. This unit was numbered and named 78th Colchester, Hants and Pictou Highlanders Infantry on 5 September 1879. Its Glengarry caps did not seem to have had a diced border, but the unit had pipers.

The volunteer militiamen of Cape Breton Island in Nova Scotia were organised into the Victoria Provisional Battalion of Infantry in October 1871. Cape Breton Island had been largely settled by immigrants from the Scottish highlands from the late 18th century and, because of a certain insular isolation, had retained their traditions. Following an inspection in 1874, Deputy Adjutant General J.W. Laurie was obviously surprised to report

Private, 5th Royal Scots of Canada, c. 1885. Scarlet doublet with dark blue collar, cuffs and shoulder straps, white piping, brass buttons and badges, Lorne tartan kilt with white sporran and two black tails, hose with white and black tops, white helmet with brass fittings, white accoutrements, dark blue Glengarry with diced border held by the waist belt, Snider-Enfield rifle. Courtesy Black Watch of Canada archives, Montreal.

that "as many of the men only speak Gaelic and do not understand English"; as late of 1941, it was said that the unit could easily triple the piper's band which had at least eight pipers who could "speak Gaelic as well as English; two compose marches, strathspeys and reel" for Highland dancing. On 12 June 1881, the unit's name was altered to 94th Victoria Infantry, Argyll Highlanders. For all their Highland heritage, the men of unit could ill afford full Highland regalia as some of their comrades in large cities, but some effort was made for a distinctive cap. The c. 1890 photo of a piper of the on Cape Breton Island shows a diced Glengarry worn with the standard infantry uniform and a c. 1899 image of this unit shows the men with diced Glengarry caps, its officers having white helmets.

In the mid-1870s, initially as a result of its 1875 reorganisation, the 5th battalion was designated "5th Battalion, Royal Fusiliers" and received permission to wear fusilier bearskin caps. Its "Flank" companies went a step further and provided themselves with scarlet Highland doublets having dark blue collar and cuffs, white piping and brass or gilt buttons, diced Glengarries and, for full dress, a fusilier's bearskin cap with its brass grenade badge and a white feather fixed at its left side and, most of all, Black Watch tartan trews that rounded out its Scots character. When General Selby-Smith inspected the battalion in 1876, he liked these companies so much that he commented that it was "a pity that the whole battalion was not uniformed like the Flank Companies." Encouraged by these comments, the 5th Battalion's officers decided to take steps to transform the unit into a Scots battalion.[49]

This was an expensive and time-consuming process that nevertheless went on; by 1878, the unit was about to order its tartan when John Campbell, 9th Duke of Argyll and Marquis of Lorne was appointed Governor-General of Canada. He and his wife, Princess Louise who was Queen Victoria's daughter, were the nineteenth century equivalents of international media stars today, and excitement ran high in Canada regarding this royal couple that had married seven years earlier. The officers of the 5th asked for and were granted permission to wear the Lorne tartan, which is one of the Campbell tartans. However, three different tartan samples of the Lorne tartan then came from Scotland and the officers of the 5th finally had to ask the Marquis as to which was the proper one. This was eventually settled and, in March 1879, the general officer commanding advised the deputy minister that the 5th was adopting "the Highland uniforms". On 24 May 1879, the Marquis of Lorne and Her Royal Highness the Princess Louise reviewed the battalion wearing its new Highland doublets and Lorne tartan trews. It was the first time that a full battalion of Canadian volunteers assumed doublets and trews. On 27 February 1880, it was renamed "5th Battalion, Royal Scots Fusiliers." That year, a visiting Scot journalist reported for the *Celtic Magazine* that the unit was "composed entirely of Scotsmen and Scottish Canadians, who wear the undress Highland uniform — Campbell tartan trews and plaid, with scarlet scalloped tunic, and Glengarry bonnet... The pipers wore the kilt..."

The battalion's ambition was to replace the trews with kilts and a "kilt fund" was set up in the unit's six companies. In June 1881, No. 1 Company paraded wearing kilts followed by No. 6 Company later that year; by 1883, all six companies had Lorne tartan kilts. That year, the "Regulations and Orders for the Active Militia of Canada" mentioned that "the 5th Battalion are permitted to wear the 'Kilt' with 'Bonnet' of established pattern

[49] On the 1863 trews and diced band shakos, see LAC, RG9, IC1, Vol. 199 kindly shared by Earl Chapman.

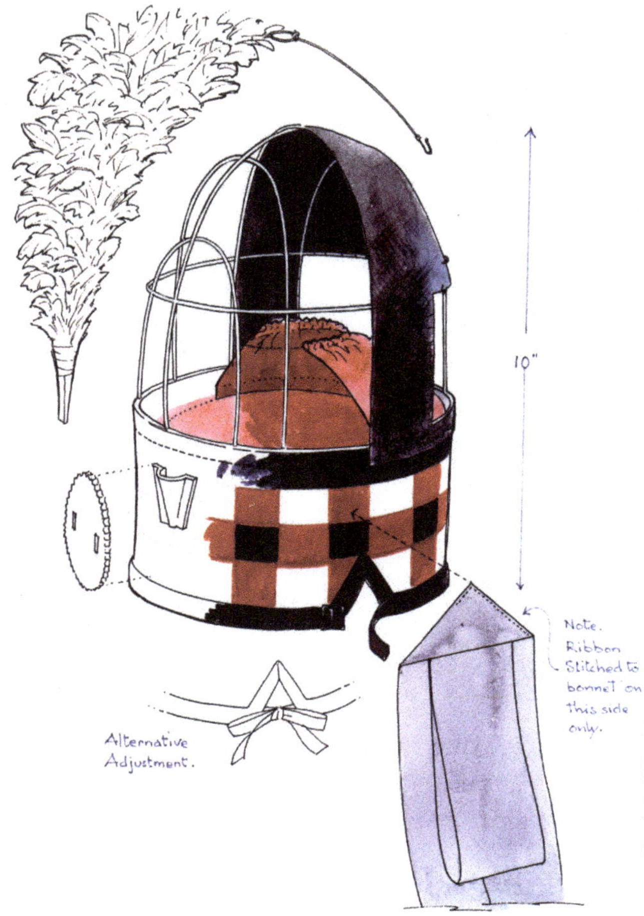

(Top left) Officer and privates, 5th Royal Scots of Canada, c. 1890. The dark hue of the Lorne tartan could be mistaken for that of the Black Watch, which was gradually adopted from 1891, and the hose tops are shown as black and red, which were not officially sanctioned before 1899. Print after H. Bunnett. Courtesy Earl Chapman.

(Top right) Highland bonnet construction. Drawing by Douglas Anderson. Courtesy Halifax Citadel National Historic Site, Parks Canada.

(Bottom left) Officer, 5th Royal Scots of Canada, 1898. Detail from a large plate by A.H. Hilder that was a supplement to 1898 "Christmas Globe" published in Toronto. Private collection. Ph: RC.

(Bottom right) Sergeant William Harp in full dress, 48th Highlanders of Toronto, 1894. Scarlet doublet with dark blue collar and cuffs, white piping and brass buttons, Davidson tartan kilt, black sporran with two white tails, red and black hose, Highland bonnet with black ostrich feathers and white plume. Courtesy Library and Archives Canada, C43488.

(Left) Sergeant William Harp in service dress, 48th Highanders of Toronto, 1894. The doublet is worn with a Glengarry and Davidson tartan trews. Courtesy Library and Archives Canada, C43486.

(Right) Sergeant, 48th Highlanders of Toronto, 1901. This image was the cover of a brochure highlighting the "complimentary diner" offered by the men of the 48th to their comrades who came home from service in South Africa. The unit's full dress is shown. Courtesy Toronto Reference Library.

(Left) Officer, 48th Highlanders of Toronto, 1898. A cadet of the Royal Military College of Canada at Kingston, Ontario, is in the background. Detail from a large plate by A.H. Hilder that was a supplement to 1898 "Christmas Globe" published in Toronto. Private collection. Ph: RC.

(Right) Officer, 48th Highlanders of Toronto, 20th century. Although published in 1951, this plate shows the officer's uniform as unchanged since the last decade of Queen Victoria's reign. It is still the regimental full dress. Print after "G.W.G." Private collection. Ph: RC.

of highland Regiments." These patterns were, of course, those of the British Highland regiments. Thus, Highlanders with bearskin caps would have been ridiculous and were replaced by Glengarry forage caps that year. On 29 February 1884, the designation "Fusiliers" was officially withdrawn and replaced with "5th Royal Scots of Canada". The "bonnets" with the large black ostrich plumes were not immediately taken into wear, due to cost. Only officers and pipers had them. The battalion instead adopted white helmets at about that time. In 1891, bonnet funds were started and, by 25 May 1895, when the 5th Royal Scots was officially allowed the red hackle, the whole battalion had its bonnets for full dress.

The 5th Battalion's hose tops seem to have been white and red from the time kilts were worn until formally changed by a battalion order of 18 May 1899, which echoed a GO of 1 February, that read "red and white checked hose changed to red and black" like the British imperial Black Watch. By then, the Marquis of Lorne had gone back to England since 1883 and, as time went on, the tartan's novelty had worn off while debates as to what really was the proper Lorne tartan went on. The decision was finally made in the late 1890s to revert to the Black Watch's tartan and the battalion was wearing it by September 1899. An interesting sidelight is that the adjutant of the 2nd Battalion of the imperial Black Watch was in Montreal at the time to brief the 5th on the details of all its orders of dress so as to conform exactly with the uniform of the British army's Royal Highland Regiment (Black Watch).[50]

In Ontario during the early 1880s, the commander of the 20th Halton Battalion Lorne Rifles wished his unit to have a closer connection with his Scots homeland. Thus, the battalion procured trews of the Black Watch tartan in about 1884. The following year, the *Canadian Militia Gazette* of 13 October reported that "the 20th Halton Lorne rifles turned out well accompanied by a brass and pipe and drum band and all the companies wore tartan pants and Scotch bonnets" at the 2nd District's training camp. The "bonnets" were Glengarries with a diced band. This novel appearance did not last for long because the tartan cloth of the trews turned out to be unsuitable for training. Thus, trews were not renewed the next year and the Highland experiment was expensive for all concerned as the battalion now had to procure the standard dark green trousers with red piping, which it had by early 1886. The diced Glengarry lasted a while longer and was still worn by officers in 1891 but, by the later 1890s, nothing remained of these Highland distinctions. At that time, the 20th purchased white helmets.[51]

During the 1880s and 1890s, the officers of Ontario's 15th Argyll Light Infantry and the 59th Stormont and Glengarry Infantry, although not then officially recognised as Highland units, were seen wearing dark blue peaked forage caps with diced bands.

Meanwhile, the rapidly expanding city of Toronto did not have a Highland unit from the late 1860s. The "Queen City's" large Scots community finally got together to raise no less than an entire battalion. Thus, 48th Highlanders of Toronto was organised from 16 October 1891 as a battalion of eight companies wearing a Highland uniform. This consisted of the scarlet Highland doublet with dark blue collar and cuffs, white piping, brass buttons, gold lace for officers, Highland bonnets with a white plume, kilts and sashes of the Davidson tartan selected in honour of the 48th's first Commanding Officer, John Irvine Davidson. The sporran was black with two white tails for NCOs and other ranks, three tails for officers. The hose tops were red and black. From 20 April 1894, white shell Jackets were worn by the 48th for drill and service dress. The unit also had white helmets, initially with white puggarees, tartan ones being adopted only in July 1905.

[50]Earl John Chapman, "The tartans worn by Canada's Black Watch 1863- present", *Canada's Red Hackle*, Spring 2013, pp. 64-65, 82. The Campbell, Marquis of Lorne tartan was elaborated in late 1870 for the Marquis of Lorne's wedding to Princess Louise. *Judy, or the London Serio-Comic Journal* of 7 December 1870 notes on the new Lorne tartan that "there are two patterns chosen, one of which is for the especial use of Lord Lorne's illustrious Consort ; and the difference is, that in the one there is a thin black line, which is red in the other." For examples of the Lorne tartan, one of which has a red line and a buff or white double line, see: http://www.tartanregister.gov.uk/tartanDetails.aspx?ref=533 and http://www.tartansauthority.com/tartan-ferret/search-results/?src=1&key=Lorne. The Royal Highland Regiment (Black Watch) sporran with five small tails was not taken into wear by Montreal's Black Watch until after the First World War.

[51] *The Lorne Scots* (Brampton, 1962), p. 10; *The Badges and Uniforms of the Lorne Scots* (Brampton, 2000), pp. 14-20.

Notes

Ernest J. Chambers, *The 5th regiment Royal Scots of Canada Highlanders : a regimental history* (Montreal, 1904); Alexander Fraser, *The 48th Highlanders of Toronto* (Toronto, 1900); Alex Morrison and Ted Slaney, *The Breed of Manly Men The History of the Cape Breton Highlanders* (Sydney, 1994); W. Boss, *The Stormont, Dundas and Glengarry Highlanders 1783-1951* (Ottawa, 1952); R.M. Barnes, *The Uniforms & History of the Scottish Regiments* (London, 1956); Frank Adams, *The Clans, Septs, and Regiments of the Scottish Highlands* (Edinburgh, 1908).

Chapter 5

Musicians

Musicians in European armies basically consisted of drummers, trumpeters and fifers whose duties were primarily to provide various calls during the day starting with reveille and ending with the tattoo, signal alarms, rolls of drums and trumpet calls at ceremonial events, as well as in battle. These musicians were on a unit's establishment and were under the authority of a drum-major or a trumpet-major.

The 18th century saw the advent of steadily increasing numbers of unit bands of musicians with various instruments under a bandmaster whose duties were much less martial, their musical contribution being nevertheless excellent for morale of the soldiers and also with the public. These musicians were initially privately paid for by the regimental officer and were not on the official establishment. Eventually, they were somewhat subsidised in the early 19th century British army which also established rules as to their numbers and payment.

Infantry drummer, c. 1828-1843. The drummer shown is of the Royal Marines, but could belong to a Canadian Incorporated Militia battalion or British line regiments. Apart from regimental details, their appearance was basically similar. Print after E. Hull. Fort Malden National Historic Site, Parks Canada. Ph: RC.

Canadian units followed British practices regarding drummers and trumpeters. However, bands were seemingly quite rare for militia units until the 1860s when many battalions were formed. Major L. T. Suzor's 1864 *Code Militaire* published in Quebec City intended for militia officers was largely based on the 1844 *Queen's Regulations*, which, for bands, stipulated that officers paid a yearly subscription of 20 days pay and an annual contribution not exceeding 12 day's pay "in support of band expenses". The size of the band was limited to a sergeant as bandmaster and 14 privates as musicians that were also to be fully trained as soldiers. The increasing popularity of having bands was eventually recognised by the Militia Department in 1879 when authority was granted for units having six companies to have 24 men in a band under the bandmaster that was henceforth ranked as a staff sergeant.

Drummers, buglers and trumpeters

Canadian regular soldiers and volunteers who were drummers, buglers and trumpeters were dressed in uniforms that followed the fashions and regulations of the British army during the 19th century. The dress of infantry drummers could be of a different colour than their comrades up to 1831. Thus, a unit with yellow facings at the cuffs and collar would have drummers wearing yellow coats with red collar and cuffs garnished with regimental lace and wings. The exception to this rule was the units that had scarlet faced with blue, which was the colours of the royal livery and worn only by "Royal" regiments. In these cases, the drummers, buglers and trumpeters were to dress in the same coat colours as their comrades, but with "royal" lace generously ornamenting their coats.

From 1831, all British infantry drummers and buglers henceforth had red coats trimmed with regimental lace in the line infantry and "royal" lace — blue fleurs de lis on white — in the Foot Guards and some "royal"

(Left) Back view of an infantry drummer in Canada, c. 1838-1843. This print after G. Tiret-Bognet shows the elaborate lace that was applied to the back of the coat at the seams and edges of pockets flaps and turnbacks. An artillery detachment is in the background. Private collection. Ph: RC.

Bugler and bandsman of the 60th Rifles, c. 1845-1855. Some Canadian volunteer rifle units, for instance the Montreal rifles, wore the same very dark green faced with scarlet uniform as the 60th. Its buglers jackets was further ornamented with mixed scarlet and black cords while musicians had plain red cords. Courtesy, National Army Museum, London. Photo 42222.

units. The line infantry's regimental lace was replaced in 1866 by the red crowns on white common to all regiment while the Foot Guards kept the fleurs de lis.

There is no known regimental pattern lace for Canadian units in the Victorian era except for plain white worn by the 1838-1843 incorporated militia battalions, which implies this was also the aspect of their drummer's lace. This was likely the situation for all Canadian infantry drummer up to 1866 when the white lace with red crowns was introduced. It edged the collar, cuffs, sleeve seams, wings and back seams. At least officially. From existing photos, sometimes, only wings were added to the basic private's tunic.

Up to the middle of the 19th century, buglers of rifle units might have wings and white lace or black (or black and red) hussar-style cords added to their dark green jackets. A good example of the latter is the bugler's c. 1830 jacket of the Montreal Rifles (see illustration). From the later 1850s, rifle buglers had the same uniform as privates except for dark green wings edged with scarlet and trimmed with black lace. Their bugles had dark green tasselled cords.

Although musical instruments used by the troops were generally somewhat alike, there might be differences in details. The wood cases of drums were painted in the colour of the unit's facing over which the royal and /or regimental designations were painted. Brass drums became popular towards the middle of the nineteenth century and a sealed Pattern for a brass drum was approved by the Queen on 12 March 1858, followed by one for a bass drum on 10 August and another for bugles, flutes and their brass cases on 29 November. These remained basically the same for the rest of the century. Band instruments were procured individually by the unit at its own cost although they were generally similar.[52]

Cavalry

By and large, cavalry trumpeters wore the same basic uniform as the troopers during the Victorian era. There was little to distinguish them from their comrades apart from their long trumpets with the cords used to sling them. Instruments came from England. For example, in 1862, 30 "Cavalry Trumpets with slings" and 150 "In-

[52] TNA, WO 370/5, Pattern warrants of 12 March, 10 August and 29 November 1858

fantry Bugles with slings" were bought; ten years later, the Militia Department ordered 75 "Cavalry Trumpets Complete" with mouthpieces and trumpet cords. To provide some distinction, the British dragoon and heavy cavalry trumpeters took to wearing aiguillettes, add piping to coat seams and have headdress plumes of a different colour from those of their regiment and this was in turn adopted by some Canadian units. Their pillbox caps might also have yellow cord on top and its crown might even be of a different colour. Some of these practices were followed by Canadian volunteers. From the 1890s, the trumpeters of the Montreal Hussars had white pillbox caps, the facing colour, trimmed with a yellow lace band, piping, cord and button.[53]

Montreal Volunteer Rifles, bugler's jacket, c. 1830. Very dark green including the collar, cuffs and wings, black or very green cords and fringes, white metal buttons marked to the unit. Photo by Christopher Bryant when the jacket was in a private collection. It is now in the Canadian War Museum, Ottawa.

Volunteer Cavalry Bands

Considering the expense associated with a mounted band, it is remarkable to find several references to such formations during the Victorian era.

On 23 September 1869, the "remarkably well mounted" Northumberland and Durham Squadrons of Cavalry entered the Militia Camp at Cobourg, Ontario, with "at the head of the squadron a splendid band, mounted, 15 in number, [that] played them into camp." On 9 July 1870, the squadron's band was reported to have a kettle drums, and said to be "only cavalry corps in the Dominion having a band." More details were given and it had originated in the late 1860s with the Durham Troop of Cavalry and was uniformed as hussars. Indeed, this "mounted band of 17, [had been] all instructed by the bandmaster of the 13th Hussars… Sometime ago [the] Municipality of Caven presented the band with instruments of the latest pattern, cost $500. On 1st July [the] town of Port Hope presented [the] band with [a] bass drums, Royal Arms highly emblazoned on it, [which] cost $80." Two years later, the Durham Troop was incorporated into the 3rd Dragoons, which eventually meant its uniform changed to scarlet faced with yellow. The mounted band seems to have vanished later on.[54]

The 2nd Cavalry appears to also have had, for a time, a sizeable mounted band complete with kettle drums and dressed as hussars, but with a brass helmet featuring a white horse hair plume. This is shown in the background of an 1886 watercolour of an officer by Richard Simkin, which appears to have been a private commission to this well-known British military artist (reproduced in the cavalry chapter in volume 1).

The Governor General's Body Guard had a band that, in the 1870s and early 1880s, wore tunics embellished with five rows of white hussar-style cords on the breast according to an image of musician Albert Browning. Later uniforms of the band were simply the standard dress of the unit. Trumpeters were mounted, but the band served on foot until 1934.

Furthermore, an 1880 print showing a escort party of the Ottawa's Princess Louise Dragoon Guards shows an amazing detail: a mounted State Trumpeter in dress uniform covered with gold lace is with the troopers next to the governor general's sleigh. State Trumpeters were attached to mounted units of the Royal Household in the United Kingdom. Such trumpeters were also found with the viceroy of India's escort. Following the 1878 appointment of the Marquess of Lorne, the office of governor general of Canada assumed a vice-regal quality because

[53] LAC, RG9, IIA6, Vol. 8, Order in Council, 12 September 1862 (the 180 instruments were $6 each for a total of $1080); LAC, RG9, IIB1, Vol. 518, DAG to [?] & Co. London, 27 June 1872. The trumpeters of the 10th British Hussars were seen in the 1880s with scarlet pillbox caps.

[54] *The Volunteer Review*, Ottawa, 11 October 1869 and 18 July 1870.

(Left) Bugler George Thomas Evans, 36th Peel Infantry Battalion, 1866. Evans was a 12 years old bugler when his photo was taken at Camp Niagara. He wears the Canadian Pattern 1863 scarlet tunic faced with dark blue and white piping to which is added a pair of scarlet wings trimmed with the white "drummer's lace" speckled with red crowns. Evans must have loved his youthful military experience; he went on to eventually be the 36th's lieutenant-colonel. Image in From Brock to Currie. Ph: RC.

(Right) Infantry drummer's lace, red crowns on white. This pattern of lace was worn from 1866 in the British infantry and appears to have been introduced at that time in the Canadian volunteer infantry. Private collection. Ph: RC.

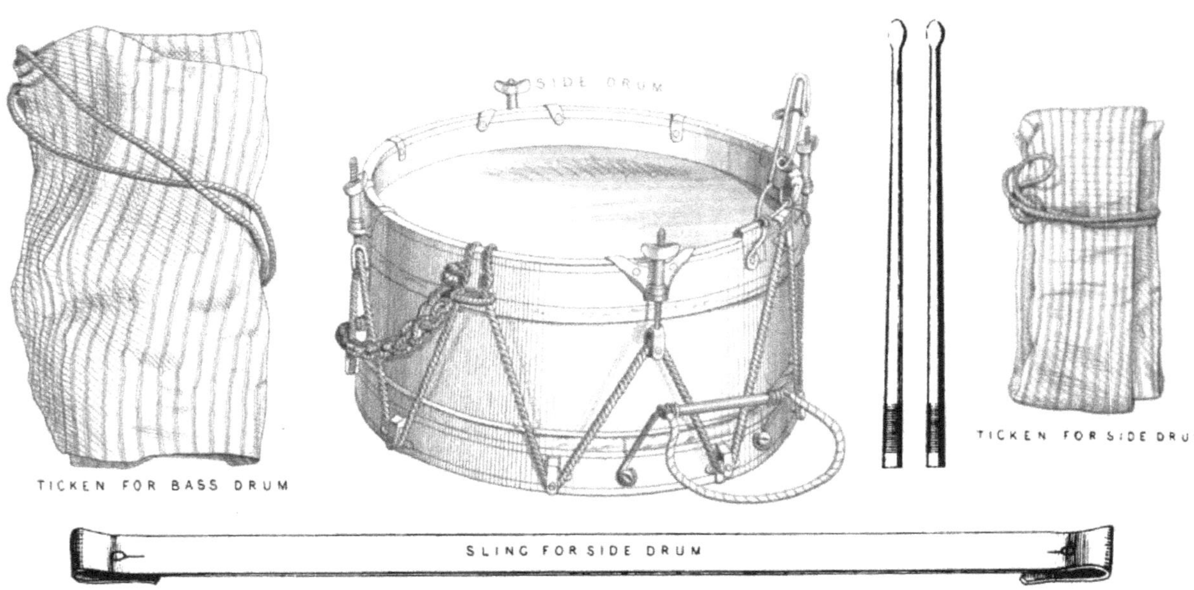

Brass side drum with sling, drumsticks and bag. Plate from Army Equipment: Infantry, 1865. Private collection. Ph: RC.

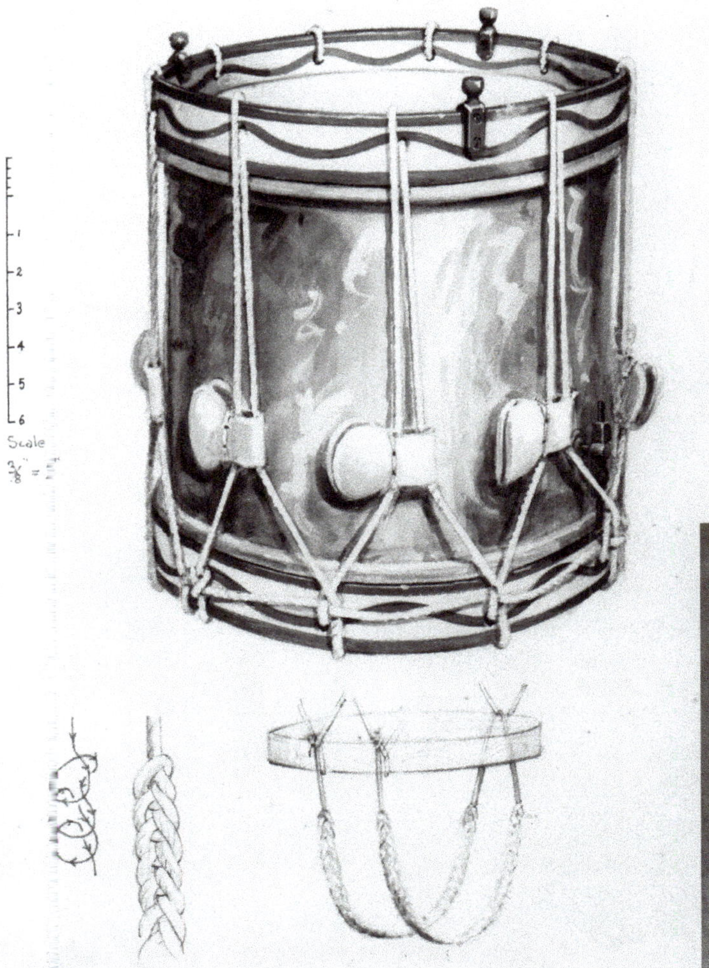

(Left) Brass drum with its cords. This was the type used by the British forces. Watercolour by Douglas Anderson. Halifax Citadel National Historic Site, Parks Canada.

(Right) Bugler, 20th Halton Infantry Battalion, c. 1866-1870. He wears the Canadian pattern 1866 infantry scarlet serge Norfolk frock with dark blue collar and cuffs, white piping in front and white metal buttons and numeral on his dark blue forage cap. There is no musician's distinction other than the brass bugle slung with a dark cord, possibly green. Courtesy Scott Duncan.

(Left) Drummer's tunic, 13th Infantry Battalion, c. 1900. Front view. Courtesy Hamilton Military Museum. Ph: Brenda Brownlea. (Center) Drummer's tunic, 13th Infantry Battalion, c. 1900. Back view. Courtesy Hamilton Military Museum. Ph: Brenda Brownlea. (Right) Drummer's tunic, 13th Infantry Battalion, c. 1900. Side view. Courtesy Hamilton Military Museum. Ph: Brenda Brownlea.

the Marchioness of Lorne was Queen Victoria's daughter, Princess Louise. Thus the addition of one or more trumpeters in the distinctive dress uniform of a crimson coat covered with strips of gold lace and embroidery. The chest and back had the crowned royal cipher. The headdress was the peaked black velvet cap. The practice of having such trumpeters attached to Princess Louise Dragoon Guards appears to have vanished after 1883 when the Marquess of Lorne and Princess Louise departed.

Volunteer infantry bands

Drums and fifes and bagpipes had always been an attraction in Canadian garrison towns during parades and reviews; many British regular regiments serving in Canada had bands of, usually, a dozen musicians or so and they sometimes gave concerts or provided music at plays.

The era Queen Victoria's reign saw a great expansion in music of all types; military music with its brass bands playing vigorous "marching music" became immensely popular. Before the 20th century, free and public access to concert music played by a fairly competent group of musicians was an exceptional and very popular treat. There was no radio, television or digital players to provide music. At that time, the state of band music in Canada, and in fact, all over America, was not very encouraging. The appearance in Canada of the British military bands during the 1860s did much to cause an improvement in this country, and also in the United States. Thus, as public literacy increased in the 19th century combined with an easier access to instruments and sheet music, bands organised by the local volunteer militia units became much more common in every city, town and even villages. They generally provided free concerts of popular music for various occasions that could be listened to by hundreds and even thousands of people. A round kiosk with its pointed roof appeared in almost every town park.[55]

The importance of military music to morale was huge. "On my way to school I had to pass the Grand Trunk [railway] station, recalled S.W. Perry. Upon this particular morning in May, 1866, the alluring sound of fife and drum led me to cross the railway tracks and join the crowd on the station platform. The volunteers of Peel had been called out to help repel the Fenian invasion… The fife and drum struck up the tune of Tramp, tramp, tramp, the boys are marching, the men began to sing, the crowd commenced to cheer and the train pulled out for Toronto…" Some 46 volunteer militia bands appear to have existed by 1869-1870. In 1869, the 45th West Durham Infantry at Bowmanville, Ontario, had "one of the best bands in the district" which numbered 21 performers while Montreal's newly raised 65th Carabiniers Mont-Royal were getting a 15-musician brass band organised. Other

[55] E.A. Cruikshank, *The Origin and Official History of the Thirteenth Battalion of Infantry* (Hamilton, 1899), p. 74.

units include the Grand Trunk Railway Brigade, the 28th, 31st, 55th, 63rd, 66th, 67th and 90th battalions.[56]

The white coat "with regimental facings" for British infantry bandsmen was introduced from 1830 and it was also applied to the undress jacket in 1837. Initially, Canadian militia bands procured the same types of uniforms as the British army bands: white tunics with collar, cuffs and wings of the facing colour — dark blue in the case of the Canadian Volunteer Militia infantry — and scarlet piping. According to issue returns, it was from the late 1860s that band uniforms were imported from England by the Milita Department and distributed to many units. According the 18 May 1870 *Volunteer Review*, the Chatham Volunteer Cornet Band attached to the 24th Kent Battalion received 20 band uniforms described as follows: the "tunic is made of white cassimere, with dark blue facings, trimmed with scarlet, blue and scarlet wings on the shoulders, silver gilt buttons, with the words 'Canada Militia' round a raised crown in the centre; the pants are of dark blue serge, with a broad scarlet stripe down the sides; and the cap has a scarlet top, dark blue band, a black patent leather peak, and in front a large silver gilt and chased ornament, surmounted by a Crown, and crossed band instruments in the centre… The uniforms were made by Tait, army contractor of London, England." St. John's 62nd Fusilier Battalion's band also had, in 1870, the standard "…white tunic with blue facings…until, by Imperial regulations, some few years later, musicians were uniformed the same as their regiments."[57]

Young dragoon trumpeter enjoying a huge pudding, c. 1870. Note the piping on the back seams and trefoil above the cuffs of his stable jacket. Plate in Seccombe's Army and Navy Drolleries. Private collection. Ph: R.C.

Since 1 May 1873, the tunic colour of British army infantry bands had been changed from white to scarlet, but this was not immediately applied for the Canadian forces. At that time, the regular Permanent Force had no bands and no band uniforms. As seen above, many volunteer battalions did have bands and they went on wearing white tunics for the remainder of the 1870s and into the 1880s. On 10 April 1879, a Canadian Militia General Order "applicable to all bands of the Dominion" finally stated that "Bands are to be dressed in clothing of the same colour as that worn by the respective regiment" to whom they belonged. Thereafter, the addition of wings trimmed with white lace on the standard scarlet private's tunic might be added to distinguish bandsmen. But there could be variety as can be seen by the illustrations. As the 1883 standing orders of the 7th Battalion Fusiliers in London, Ontario, put it: "Bandsmen will wear wings or shells on the shoulders, of alternate stripes of colours of the uniform and facings" while "Buglers wear twisted cord on the back and sleeve seams, and wings on the shoulder." By the end of the century, Col. Otter's *Guide* recommended, for infantry units, wings with "bars of half-inch white braid on blue ground" with the lyre badge on the right sleeve and white piping down the back seams of the tunic. This last feature does not seem to have been widely done.[58]

Grenadiers and Foot Guards

There might be exceptions. Bands of the British Brigade of Guards infantry regiments had scarlet coats garnished with lace and a few Canadian volunteer units preferred that colour for their band. In November 1867,

[56] S.W. Perry as quoted in William Perkins Bull, *From Brock to Currie* (Toronto, 1935) p. 237.

[57] LAC, RG9, IB2, Vol. 50, Return of clothing issued…in District No. 7 (and) No. 3, 1870-1872 - the 9th, 14th, 15th, 17th, 16th, 40th, 45th, 46th, 47th and 48th, 49th, 57th and 70th battalions received band uniforms or ornaments (for rifle units); E.T. Sturdee, *Historical record of the 62nd St. John Fusiliers (Canadian Militia)* (St. John, N.B., 1888), p. 115; MR 1869.

[58] LAC, RG9, IIJ2, Vol. 116, Militia General Order 10 April 1879; *Standing Orders of the 7th Battalion Fusiliers* (London, ON, 1885), p. 32.

Cavalry trumpeter, 1890s-1900s. During the 1890s, British dragoon trumpeters adopted an aiguillette worn at the left should for distinction and this practice was also followed by some units in Canada. A unit might also indulge in the cavalry version of the musician's sword. This figure by Robert Marrion shows a trumpeter in a regiment wearing scarlet with yellow facings, such as the 3rd Cavalry (Prince of Wales' Canadian Dragoons). Private collection. Ph: RC.

Trumpeter of the 2nd Cavalry, 1886. He wears the regimental hussar style uniform and holds a bugle with cords in the royal lace colours: red, yellow and blue. Watercolour by R. Simkin. Anne S. K. Brown Military Collection, Brown University Library, Providence, USA. Ph: RC.

Trumpeters, British 6th Dragoon Guards Regiment, 1890s. This unit's dress was the model for several Canadian volunteer units since 1871 and this possibly included trumpeters. This detail from a watercolour by Richard Simkin shows a trumpeter distinguished by a red helmet horse hair "plume" and a white aiguillette at the left shoulder. The tunic's piping is here shown as yellow although other period sources also shows white. He also holds a bugle. At right, a dismounted trumpeter in undress wearing a white pillbox cap with red band and button. Courtesy Anne S. K. Brown Military Collection, Brown University Library, Providence, USA.

Trumpet Major W.R. Glover, Montreal Hussars, 1898. Note the white pillbox cap trimmed with yellow on this somewhat damaged, but remarkable image. Royal Canadian Hussars Sergeant's Mess, Montreal. Ph: RC.

Mounted band of the 2nd Cavalry, 1886. The band's helmet plume was white with a red base while the rest of the unit had an all white plume. Detail of a watercolour by R. Simkin. Anne S. K. Brown Military Collection, Brown University Library, Providence, USA. Ph: RC.

Kettledrummer of the 13th Hussars Hussars, c. 1869. The North Durham Squadron of Cavalry formed a 17-member mounted band complete with a kettledrummer in the late 1860s. It was instructed by the bandmaster of the 13th Hussars. The Durham Light Cavalry wore the same uniform as the 13th Hussars and the dress shown this print by Capt. Seccombe could also apply to the band of the unit. Private collection. Ph: RC.

Governor General's Body Guard bandsman Albert Browning, c. 1879. The unit's uniform was made more colourful for bandsmen by adding white hussar-style cords. The undress uniform was the same as the rest of the unit. Courtesy Scott Duncan.

Governor General's Body Guard bandsman, 1899. This somewhat scratched photo was probably taken at Camp Niagara. He wears an elaborate white aiguillette, but this distinction was also worn by this regiment since about the mid-1890s. Dark blue uniform, white collar, piping, cords and trouser stripes, white metal helmet with brass fittings and white falling horsehair plume. Courtesy Niagara Falls Public Library, Niagara Falls, Ontario.

State trumpeter with a Princess Louise Dragoon Guards escort for Governor General the Marquess of Lorne and Princess Louise at the Ottawa Parliament, 1880. Detail from a print in Harpers Weekly, 14 February 1880.

the 13th Battalion's band now had brass instruments, drums and cymbals; its 12 "old members of the band" were "measured for the uniform which is similar to that of the rank and file, except with different facings in cross braid mountings on the breast" reported the review. This was a scarlet tunic with lace added. The 10th Royals Battalion is shown with a scarlet-coated band four years later with wings at the shoulders. Its drum-major is also shown wearing scarlet tunic with wings and a seemingly dark blue baldric edged with silver. By 1880, the 10th's drummers and buglers also wore fur caps. Seemingly from the 1890s, drummer's lace was added on the breast and on the sleeves, which gave them the appearance of British Brigade of Guards musicians. There was, however, another unit that wore this style since 1872.[59]

For drummers, of the Governor General's Foot Guard this consisted of a scarlet coat with dark blue collar, wings, cuffs and cuff flaps, lace with dark blue lilies at the buttonholes set in pairs and white fringes at the collar and wings; bearskin caps and trousers as the enlisted men The band of the Governor General's Foot Guard in Ottawa was organised in 1872 and obtained instruments in October at a cost of $263 (LAC, reel T4174/194). Four years later, it was expanded into an orchestra that played "almost without exception at all the functions at Government House" until the early 20th century. Its initial uniform was the same pattern as the enlisted men with shoulder wings added. The drum-major had gold lace and, from about 1887, so did the members of the band, which were henceforth dressed like the musicians of the Coldstream Guards. An 1899 order to obtain the Drum Major's tunic and belt from London mentions that the tunic was to be the Coldstream Guards pattern and that the belt was to have "the letters 'G.G.F.G' and the word 'Canada' are to be made; the embroidered raised letters are, however, to be on scarlet cloth instead of blue, the shape is to be the same…" with a length of "about 50 inches" (LAC, reel T387/285).

Rifles

The bands of volunteer rifle units were less colourful than those of infantry battalions. They wore the standard patterns rifle uniforms for enlisted men with wings added, like the rifle buglers. For further distinction, some city units might add scarlet cuffs and provide their bandsmen with shakos garnished with green-black drooping cocktail feathers. But the band uniform could vary from one unit to another. In 1867, the 17 bandsmen of the 39th Norfolk Rifles had the standard issue 1863 pattern tunic without musician's wings, but their shakos were distinguished by cock's feathers. In October 1875, the officers of the 2nd Queen's Own Rifles furnished fur Busbies that replaced shakos in British army rifle units from 1873. At least one band, that of the 90th Winnipeg Rifles, took part in the 1885 North-West campaign. "The brass band, particularly during the last few months of the campaign, for its playing improved wonderfully, was the pride and joy of the force, and there was almost being a row in the artillery lines at Battleford because some members of Colonel Otter's force ventured to speak disparagingly of the 90th band in the hearing of some of the gunners of "A" Battery. In the field, particularly during the fight at Fish Creek, the bandsmen performed invaluable service as an ambulance corps." Otter felt rifle bands should

[59] *Volunteer Review*, 11 November 1867; *Canadian Illustrated News*, 5 August 1871.

wear wings with "half-inch black braid" on scarlet, the lyre badge on the right sleeve and scarlet piping at the back of the tunic.

Apart from the buglers, some bands might have a distinct uniform. In 1880, the bandmaster of the 65th Carabiniers Mont-Royal wore the rifle officer's patrol jacket and a large grenadier bearskin cap. But its brass band of about 30 musicians, renowned for its concerts, wore a dark coat with fringed epaulettes and a French-style kepi. It has often been stated that rifles and light infantry units in the British forces had no drums, but this was certainly not the case with their bands in Victoria's Canada. Indeed, contrary to this opinion, the Duke of York himself condoned the issue of many drums to British regular corps during the Napoleonic Wars and they were seen in many Canadian Militia rifle units a hundred years later, including large bass drums.[60]

State trumpeter. This 1885 watercolour by Harry Payne shows the full dress of State Trumpeters attached to mounted units of the royal Household in the United Kingdom. Such trumpeters were also found attached to Princess Louise Dragoon Guards from about 1878 to 1883. Courtesy Anne S. K. Brown Military Collection, Brown University Library, Providence, USA.

Highlanders

Until the mid-19th century, British army Highland infantry pipers had basically the same type of coats as private soldiers in other infantry battalions, worn with bonnets, kilts and hose. Pipers are reported in the Glengarry Militia of Upper Canada in 1838 and in the 2nd King's Regiment of PEI two years later, but no details are known of their dress. From 1856 when doublets were issued to British army Highland units, pipers were henceforth served with distinctive all dark green doublets. British Highlander bands had white coats that were replaced by white doublets from 1856 until changed to scarlet in 1873.

In Canada, Highland volunteer units followed the British army's practices, but evidence regarding white doublets for bands before 1879 is presently lacking. The bands of Canadian Highland units consisted of "Pipes and Drums" with Montreal's 5th Fusiliers organising officially the first such group in 1875. Their uniforms were purchased from Hebbert and Company in London, England, and the purchase documents confirm that pipers wore dark green cloth tunics, "Highland" pattern, with yellow embroidered shoulder straps, trimmed with gold lace. They had the Royal Stewart tartan. Drummers wore fine scarlet cloth tunics with blue tweed trousers with fine scarlet cloth stripes down the sides.[61]

From 1891, Toronto's 48th Highlanders' pipers had the Davidson tartan worn with their dark green doublet. The 48th's band was organised in 1893 and wore the unit's uniform with wings on the scarlet doublet. Units not wearing Highland dress might also have pipers that wore their unit's standard uniform, but with a Highland cap as seen in the Carleton Place (Ontario) Volunteer Rifles during the 1860s or Nova Scotia's 78th Colchester, Hants and Pictou Highlanders 94th Victoria Infantry, Argyll Highlanders during the 1890s.[62]

---

[60] Bob Blackeley, *The Civilian Soldier* (Waterford, ON: Norfolk Mercantile, 2007), p. 48; E.J. Chambers, *The 90th Regiment* (1906) p. 53; sketches of the 65th Carabiniers's band by Henri Julien in *L'Opinion publique* (Montreal), 10 June 1880. See also illustrations in Chamber's histories of the 43rd, 65th and 90th. Issues of drums to British and light infantry units can be seen in WO 3/194, 196, 207.

[61] The 1875 purchase documents are in LAC, RG9, II A 1, dockets 02440-02452, kindly shared by Earl Chapman.

[62] S. MacKinnon, "The Bagpipe in Canada", *Canadian Geographical Journal*, IV, No. 4 (April 1932), pp. 233-241. A useful table of pipe

Band of the 36th Peel Infantry Battalion, 1866. The bandsmen wear white tunics with dark blue cuffs and cuff flaps, white forage caps with a red band. Other details are not distinct. Behind, presumably drummers and buglers followed by the rest of the battalion in scarlet tunics and mostly shakos. Some of the tunic cuffs have cuff flaps which indicates they were obtained before 1862. This image was taken at Orangeville, home of one of the 36th's companies, and later coloured by local artist Owen Staples. Courtesy Toronto Reference Library, pictures-r-1263.

Bandsmen of the 10th Toronto Infantry Battalion, 1868. Its caption mentions "Drum Major in Red, Bandsmen in White Tunics". Print after a photo by C.W. Jefferys. Ph: RC.

Bandsman and sergeant of the 24th Kent Infantry Battalion's band, 1870. They wear the uniform received in 1870 (described in text). Detail from plate by Robert Marrion. Ph: RC.

Band of the 55th Megantic Infantry Battalion, 1871. These bandsmen wear a dark blue Glengarry with red and white dicing. Courtesy Library and Archives Canada, C759.

Band of the 46th East Durham Infantry Battalion, c 1872. White tunic with blue collar, cuffs and wings piped red; dark blue trousers with red piping. The musician's headdress is an American style forage cap that was probably dark blue. Courtesy Ontario Archives, Toronto, 11747-15.

Band of the 13th Battalion, 1885. Detail of a print in the Illustrated War News, 2 May 1885 Courtesy Peel Library, University of Alberta.

The fife and drum band of Montreal's 6th Fusilier Battalion, 1880s. This was one of the two bands maintained by this battalion, the other being a brass band. Courtesy Canadian Grenadier Guards Museum, Montreal.

Band of the 32nd Bruce Infantry Battalion, 1886. Wings trimmed with white lace have been added to the pattern 1878 tunics to denote the men as musicians. They are also wearing American pattern white tropical helmets. Canadian War Museum, Ottawa. Ph: RC.

Band of the 66th Princess Louise Fusiliers, c. 1895. This detail from a large image shows all wearing wings added to the standard infantry tunic with some drummers having the "drummer's lace" applied on the seams. Courtesy Library and Archives Canada, C650.

(Top left) Musician's tunic, 79th Shefford Infantry of Waterloo, Quebec, c. 1895. Scarlet with dark blue collar, cuffs, shoulder straps and wings, white piping and lace, brass buttons and numerals. The tunic has five buttons and was made for a boy. Courtesy Lorne Waid.

(Top right) Back view of a musician's tunic, 79th Shefford Infantry of Waterloo, Quebec, c. 1895. Courtesy Lorne Waid.

(Top left) The Mahone Bay (Nova Scotia) detachment of the 69th Annapolis County Infantry Battalion's band, c. 1898-1905. Only the bugler and drummer have the distinctive musician's wings on their pattern 1896 infantry frocks. Settlers Museum, Mahone Bay, NS. Ph: RC.

(Top right) Drummers and buglers of the 77th Wentworth Infantry Battalion, c. 1899-1905. All wear the Canadian pattern infantry uniform with no wings or other distinctions for musicians. Only the sergeant (standing in the back row) has crossed bugles over his chevrons indicating he is a bugle-major. The volunteer sitting at the centre appears to have shoulder straps for a bass drum. Royal Hamilton Light Infantry Museum. Ph: CL.

Band of the 10th Royal Grenadiers of Toronto, 1880. Courtesy Toronto Reference Library, pictures-r-5294.

Regular Permanent Units

The few regular Canadian units formed from 1871 had drummers, buglers and trumpeters that followed the regulations, patterns and practices of the British army. Indeed, instruments and much of the uniforms and equipment came from Great Britain as can be seen by surviving images, some of which we reproduce in this book.

The 1871 Manitoba battalion would have had rifle uniforms until 1872 when it was converted to the scarlet-coated infantry. The 1884 Infantry School Corps (Royal Canadian Regiment from 1892/1893) had drummers and buglers wearing the standard infantry uniform with the distinctions as for volunteer infantry battalions, scarlet tunic with dark blue collar and cuffs, white piping, brass buttons, white lace with crowns. The unit also had a five breast buttons service dress frock in the same colours from 1884 until the adoption of the Canadian Pattern serge frock in 1896. In service dress, the buglers usually wore the ordinary uniform, the bugle and its cords being their distinction.

The data is vague concerning bands of early Canadian regular units. In 1870-1871 Manitoba, a band of the two rifle expeditionary battalions was formed and, when the reinforcements arrived at Fort Garry on 18 November 1871 to form a permanent battalion, they were greeted "to the tune of British Grenadiers, a good crowd of people cheering us heartily" recalled Private Adshead. On 30 November, clothing was requested for the band, but it did not obtain it since a report of 13 June 1872 stated there was "No Rifle Band clothing in store." Nevertheless, it was later declared to be "the only efficient band in the province" and thus was "permitted to play at [the] Batchelor's Ball" in Winnipeg's Grand Hotel in November 1872. Since the battalion's strength was gradually reduced, the band had probably vanished by the time this unit was disbanded in 1877.[63]

There are also occasional references of bands for the regular artillery batteries created in 1871 (Royal Canadian Artillery from 1893) and, from about 1884 the Infantry School Corps (Royal Canadian Regiment from 1892/1893) had such a band. The Cavalry School Corps raised in 1884 (Royal Canadian Dragoons from 1892/1893) did not have a mounted band, but did have a "Musical Ride" that first performed in 1894. This was akin to the Royal Canadian Mounted Police famous musical ride and was made up of 40 men in dress uniforms armed with lances bearing red over white pennons manoeuvring in "drill pattern formation riding to music at three and a bit paces" finishing with the spectacular final cavalry charge.[64]

In February 1894, an Order-in-Council authorised "a Band to form part of the Permanent Militia Force at each of the Station of the Force throughout the Dominion" and the Royal Canadian Artillery had such a band since at least 1874, but it had become "disorganised" by late 1897 (LAC, reels T4174/567, T4179/209, T278/195). In his 1898 Militia report, Maj. Gen. Edward Hutton stated that a "good permanent military band is much required,

bands mentions that the pipe bands of the later Pictou Highlanders had the Seaforth tartan, the Cape Breton Highlanders had the Black Watch tartan, the Montreal Highland Cadets had Mackenzie, etc. A c. 1905 photo of the personnel of the 94th has pipers wearing the scarlet frock without wings but one with a plain Glengarry with an eagle feather, another with a diced Glengarry, and three with diced balmorals. The other enlisted men have white helmets as do some officers while other officers have peaked caps as per Courtesy Robert MacLellan Collection (Donated by Brenda Beaton, Upper Leitches Creek, NS, 2015) Cape Breton Military History web site.

[63] MA, MG6, B3, George Adshead Journal, 21-22 October, 18 November 1871 entries; MG6, B1, Military District No. 10, letter book 1, pp. 3, 112; ibid, letter book 2, 30 November 1872, 16 August 1873, 5 November 1874.

[64] Larry Worthington, *The Spur and the Sprocket* (CFB Gagetown, 1968), pp. 39-41.

Drummer, 10th Royal Grenadiers of Toronto, 1899. Watercolour by P. Holmes. Anne S. K. Brown Military Collection, Brown University Library, Providence, USA. Ph: RC.

Drum Major of the Governor General's Foot Guard, 1888. Watercolour based on a period photograph by Derek FitzJames. Anne S. K. Brown Military Collection, Brown University Library, Providence, USA. Ph: RC.

Drummers of the Governor General's Foot Guard, 1888. The bugler and the drummer wear the royal lace, the bandsman behind has a gold-laced coat. Courtesy, Governor General's Foot Guard, Ottawa.

which should form a Militia School of Music for improving the existing Militia bands…" His wish was soon blessed because, the following year, he reported that a "permanent band for the Royal Canadian Artillery has been organised at Quebec under a competent bandmaster."

Notes

J.R. Madden, *The History of Bands in the Canadian Army*, Historical Report No. 47, Historical Section, Army HQ (Ottawa, 1952); R.G. Harris, "Cavalry Mounted Bands…", *Military Modelling*, January to December 1983; N.G.G. Beach, "Of drums and drummers" and "Of Bands and Bandsmen", *Military Band Journal,* February 1966; W.D. Otter, *The Guide: a manual for the Canadian Militia (infantry)*, (Toronto, six editions from 1880 to 1906).

Presentation of a Silver Mace to the 2nd Queen's Own Rifles "by their Lady Relatives and Friends" at Toronto, 25 May 1863. The mace is show at the bottom. At left, the band's shakos have flowing cock-tail plumes as a distinction. Courtesy Toronto Reference Library, pictures-r-5751.

Band of the 14th Queen's County Rifle Battalion photographed in Kingston during April 1872. This rare and revealing image shows that the band opted for the shako with a black "cocktail" drooping feather rather than a busby for dress. The lighter hue of many cuffs almost certainly indicates that their original dark green was altered to scarlet; the same appears to be true for some of the wings. The bandsmen with swords are sergeants and the bugle-major might be the senior NCO at left who wears the officer's pattern dolman. Fort Henry, Kingston. Ph: RC.

(Right) Rifle unit musician, c. 1863-1867. This trombone player was photographed by R.J. Clench at Niagara, Canada West. He wears the Canadian 1863 pattern rifle tunic with barely visible corporal's stripes on the right sleeve. There are no musicians insignias visible. His forage cap has a bugle badge and a black patent leather chin strap. The bayonet is for a long infantry Enfield rifled musket as is the blackened belt with the round black buckle. Courtesy Don Blyth Collection, Guelph, Ontario.

(Left) Bandmaster of the 65th Carabiniers Mont-Royal, 1880. The uniform is not described, but a bearskin cap was the obvious distinction of the bandmaster. The jacket appears to be similar to the patrol jacket worn by regimental officers. The trousers have a wide stripe. Print after Henri Julien in L'Opinion Publique, 10 June 1880.

Band of the 65th Carabiniers Mont-Royal, 1880. This fine brass and percussion band was very popular in the city and is shown here giving a concert in Montreal's Viger Park. Its dress was a double-breasted coat with fringed epaulettes, trousers with a wide stripe and a French-style kepi. It was a social as well as a military band and is led by an orchestra master for this concert. A rifleman wearing the 65th's dress shako can be seen below the podium. Print after Henri Julien in the Canadian Illustrated News, 18 September 1880. Ph: RC.

Band of the 2nd Queen's Own Rifles, c. 1898. This remarkably large band had some 39 musicians of which 14 are drummers including two bass drummers. It was organized in 1860 and is the oldest continually serving volunteers militia band in Canada. Except for the officer at the centre, all wear the basic 1870 Canadian pattern dark green tunic and trousers with added distinctions for rifle musicians on the tunic: dark green wings laced with scarlet and scarlet piping at the seams, here visible on the sleeve seams. Army and Navy Illustrated, 11 June 1898. Ph: RC.

(Top left) Band tunic of the 6th Duke of Connaught's Own Rifles, 1890s. This tunic is smaller and has six buttons indicating that it was tailored for a lad by altering a standard private's rifleman's tunic. Scarlet shoulder wings trimmed with black lace and cord and scarlet and white cord at the sleeves ending in a trefoil at the top of the cuffs. Black on scarlet function badge on left sleeve. British Columbia Regiment Museum, Vancouver. Ph: CL. (Top right) Back view of the 6th Duke of Connaught's Own Rifles, 1890s. The scarlet and white cord is applied to the back seams. British Columbia Regiment Museum, Vancouver. Ph: CL.

Band of the 6th Duke of Connaught's Own Rifles, 1901. They wear the dark green rifle frock assigned to rifle units, most having wings at the shoulders. Courtesy Vancouver City Archives, AM54-54Mil P250.

The Carleton Place Rifle Company, c. 1866. The surprising presence of a piper shows that Highland traditions could be present in units that were not officially Scottish. In March 1866, the company paraded "in full uniform guns and bayonets in first best style… led by the good old Scotch bag pipes and drum, well played" (Brockville Recorder). He wears a Glengarry cap while the others have a forage cap with a bugle horn badge. All wear the 1863 pattern rifle tunic with blackened infantry belts (round buckles). The weapon is the model 1853 Enfield rifled musket. Courtesy Carleton Place & Beckwith County Historical Museum, Carleton Place (Ontario). Photo National Historic Sites, Parks Canada.

Pipers of Montreal's 5th Royal Scots, 1877. The pipers wore kilts and plaids of the Royal Stewart tartan. Detail from a large composite photo of No. 1 Company of the 5th made between 1875 and 1877 reproduced in Chambers' 1904 history of the 5th Royal Scots. Ph: RC.

Piper Gillis, 94th Victoria Infantry, Argyll Highlanders, c. 1890. This somewhat blurry image is nevertheless a rare view of a piper in this unit based in Sydney and Baddeck on Cape Breton Island in Nova Scotia. The unit obviously could not afford to import Highland musician's dress so it wore the standard issue Canadian infantry uniform, but with a diced border around the Glengarry and marched to the sounds of pipes. Private collection. Ph: William O'Shea.

(Right) Piper's or musician's doublet and plaid, c. 1890. Dark green doublet including cuffs and collar with scarlet piping edging, brass "Canada Militia" buttons with beaver and crown, plaid of the Royal Stewart tartan. The outfit is somewhat small and may have been meant for a young musician attached to a Highland unit. Or to a battalion with a Highland pipe band. Fort Henry, Kingston. Ph: RC.

A group of artillerymen at a training camp, c. 1890-1896. They appear to belong to one of the regular batteries of the Royal Canadian Artillery. Some are bandsmen as seen from the instruments in the foreground, but no musician's distinctions are visible. Most wear the regulation tunic, the absence of grenades at the collar revealing that they probably belong to a volunteer militia field battery. Some have the five button serge jacket with its lace trefoil on the cuffs introduced in 1882, a rare view of this garment. The three men wearing the frock jackets with pockets at the chest (the one second from right appears to be homemade) and grenades at the collar may be officers. All wear the pillbox cap and boots except the gunner at right. The guns are 9-pdr Rifles Muzzle Loading cannons. Fort Henry, Kingston. Ph: RC.

Artillery bugler or musician, c. 1900. This hazy and damaged image shows a young artilleryman with his regulation uniform adorned by cords (possibly Busby cords) that are probably yellow and meant to hold an unseen musical instrument and its case. The forage cap is also decorated with this cord. Courtesy National Historic Sites, Parks Canada.

Band of the Infantry School Corps, c. 1890. The scarlet tunics are ornamented with white-laced wings and seams. Courtesy The New Brunswick Museum, St. John.

Trumpeter Bishop, 3rd Montreal Field Battery, Royal Canadian Artillery, 1890s. The 3rd Field Battery gathered many of Montreal's volunteer gunners and, by the 1890s, was the only battery wearing the distinctive fur busby. The trumpet was actually a bugle in all artillery and cavalry units. Image in the Army and Navy Illustrated of 14 May 1898. Ph: RC.

Band of the Yukon Field Force, Dawson City, May 1900. Nearly all bandsmen belong to the Royal Canadian Regiment and most wear the 1896 pattern infantry serge frock except for a few gunners. Officers wearing field service caps are at the back. Courtesy Vancouver City Archives, AM54-S4-Mil P269.

Chapter 6

Engineers and Communications

Quebec Royal Engineer Rifles, 1st and 2nd Companies, 1837-1838. Dark blue cap with red band and brass bugle horn badge, cream white capot with blue collar, cuffs, piping, buttons and dark blue bar on the skirt, dark blue trousers with wide red stripe, black accoutrements with a brass belt-plate bearing a bugle horn design. He is armed with what is probably a Baker rifle. Plate by Peregrine Pouchbelt and Rodrick Ramrod. Courtesy Library and Archives Canada, Ottawa, C40797.

Canadian engineer units during Queen Victoria's reign were few. Up until the evacuation of the British garrison from central Canada in 1871, the Royal Engineers maintained a few officers and detachments of enlisted men based mainly in Quebec City to provide specialist engineering services. A similar establishment was posted in Halifax until 1906. There was also some 230 officers and men of the Royal-Engineers who laid the foundations for much of British Columbia between 1858 and 1863 whose important peculiar story and status is outlined below. Later, detachments served in Esquimalt from 1893 to 1906.

Early Units

The 1837-1838 Quebec Volunteers included two companies of "Royal Engineer Rifles" and one of artificers, wearing a mixture of Canadian winter costume and uniform items such as peaked caps; please see illustrations. These units were disbanded following the end of the rebellions.

An African-Canadian "sapper corps" existed in Fredericton (NB) from the 1830s to 1849. There appears to have been no engineer units until the Halifax and Dartmouth companies were raised in 1860, but these only lasted a few years. In 1861-1862, the West Military District Engineer Company was formed at St. John (NB) and it wore scarlet with blue velvet facings, most likely an identical uniform to that of the Royal Engineers. For the dress of engineers, the New Brunswick dress regulations of 7 April 1863 stated that NCOs and men's jacket was to be mostly the same as that of the infantry, scarlet with blue collar, cuffs and edging, scarlet shoulder straps with blue edging, yellow cord "sleeve ornament knot…seven inches deep" that appears to have been an Austrian knot, dark blue trousers with 1 1/4 inch wide scarlet stripe, same cap as the infantry, but with a bronze grenade (please see Vol. 1, p. 238). Officer wore much the same, but the collar edged with yellow silk cord, no shoulder cords and yellow silk sleeve knots "traced in and out with small silk braid eight inches deep and figured for captains, seven inches deep and plain for lieutenants." On the whole, this regulation made New Brunswick engineer's uniforms almost similar to those of the British Royal Engineers except for the fur busby.

Royal Engineers in British Columbia 1858-1863

During 1858, about 30,000 mostly American gold-seekers poured into the new colony of British Columbia. The few hundred British residents were overwhelmed. To maintain order and the British flag on the North-West. A 160 Royal Engineers under Col. Richard Moody were detached to the colony to carry out a program of public works, laying out townsites, aiding in road construction and serve as a defence force. A further detachment of 65 men would assist with the work of the British Boundary Commission, which included Royal Engineers officers, to establish the 49th parallel border with the United States. These works went well beyond strictly military matters since they actually created the colony's infrastructure and the men who built it were more of skilled artisans that could turn into the trained soldiers they also were if need be. They were encouraged to settle with their families in the colony and, in a way, this pioneering work made them volunteers. In 1863, when the detachment was called back, all but 15 opted to remain as settlers. Insofar as the dress worn by the members of these detachments they undoubtedly arrived with their uniforms. Lt. Wilson's photo showing him wearing a buckskin suit and other images show other civilian clothing used by members of the Boundary Commission.

Sergeant, Quebec Loyal Artificers, or Faugh a Ballagh, 1837-1838. "[Paid] 2/6...each day when on duty. 1 Company of 80 Men." Dark brown fur cap with green bag, cream white capot with green collar, cuffs, wings, piping, bar on the skirt and buttons, yellow chevrons on red ground, dark blue trousers with wide red stripe, moccasins with red tops, crimson sash, white gloves. Plate by Peregrine Pouchbelt and Rodrick Ramrod. Anne S.K. Brown Military Collection, Brown University Library, Providence, USA. Ph: RC.

From 1856, the British Royal Engineers wore a scarlet tunic with blue velvet collar, cuffs and piping, Oxford mixture dark grey trousers with wide scarlet stripe. Officers had gold buttons and lace, dark red morocco belts trimmed with gold lace. The shoulder belt featured a wavy gold line at its centre that was a unique distinction reserved to engineer officers. In 1857, a fur Busby with a blue bag, gold cord and badge holding a white plume on the left side was introduced. Enlisted men had bass buttons, badge and yellow cord. This dress uniform was widely adopted by engineer units throughout the British Empire including in Canada.

Lieutenant Charles William Wilson, Royal Engineers, c. 1858-1861. His most unusual dress his explained in that he was secretary to the British Boundary Commission, whose duty it was to map the 49th parallel between the Rocky Mountains and the Pacific Ocean so as to determine the Western border of Canada with the United States between 1858 and 1862. He thus wears a buckskin suit decorated with short fringes and a cloth cap with a long cord and tassel. He is armed with what appears to be a trade musket. This dress may well have been akin to that of several other engineers attached to the commission. Wilson (1836-1905) went on to conducting Ordnance surveys in the Sinai, Palestine, Scotland, Ireland and campaigned in Egypt and the Sudan. He was knighted and became Major-General. Courtesy British Columbia Provincial Archives, Victoria.

Canadian Volunteer Militia Engineers

In late 1861, thought was given to include engineer units as part of the Canadian Volunteer Militia, but, as the Adjutant General wrote on 7 January 1862, "Engineer Corps are limited, one to each military district." Instead a second company was raised in Montreal. By July, companies were being organised in Ottawa, Quebec City, Lévis, Port Hope, Sarnia and St. Catharines. As the following years and decades

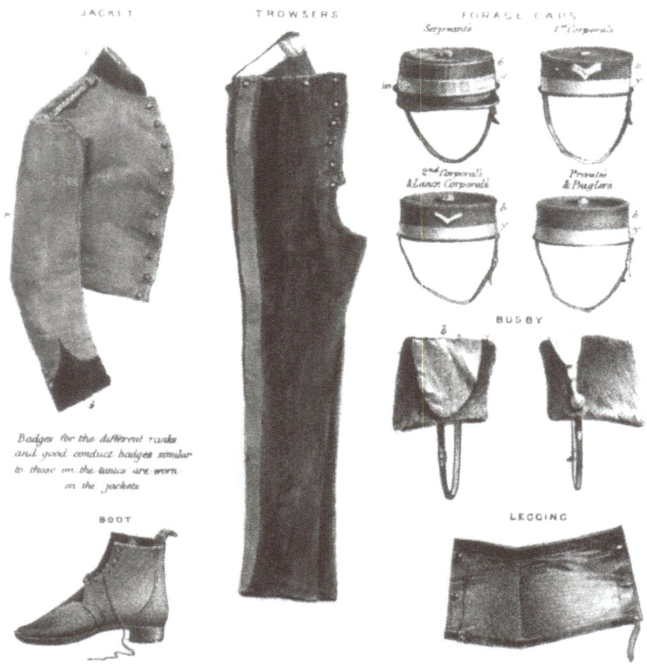

Royal Engineers NCOs and enlisted men's jacket, trousers, caps, busbies and shoes. The Canadian engineer units had similar garments. Print after J. Ferguson in Equipment of a Company of Engineers, London, 1866.

(Right) Royal Engineers officer, full dress, c. 1865-1866. Canadian volunteer militia units always strived to have similar uniforms to those of the British army. There were few Canadian engineer units, but its seems many of their uniforms were imported from Britain. In the background, an officer in full dress with a bicorn worn at court and ceremonial occasions. The officer with a cloak and a scarlet stripe on the trousers in Dress Order for such occasions as church parades, funerals, etc. Print after J. Ferguson in Equipment of a Company of Engineers, London, 1866.

(Right) Royal Engineers, sappers in dress and service uniforms, 1855-56 to 1870. Canadian volunteer engineer units generally wore the same uniforms as the British corps, usually putting on the yellow metal/gold buttons and lace of the regulars rather than the silver worn by British volunteers. Unsigned watercolour. Courtesy Anne S.K. Brown Military Collection, Brown University Library, Providence, USA.

Royal Engineers officer, undress order, c. 1865-1866. All dark blue frock coat with black flat lace, dark grey trousers with red stripe, dark blue cap with gold lace. In the background, an officer in drill order wearing the peaked cap and another officer in staff order with a bicorn hat. Print after J. Ferguson in Equipment of a Company of Engineers, London, 1866.

(Right) A Canadian volunteer sapper between 1870 and at least 1878 would have looked much like this sapper of the Royal Engineers. From about 1878, the Home pattern helmet was worn by the Montreal and Charlottetown, but the St. John, Brighton and Toronto units may have kept the Busby. Watercolour by Richard Simkin. Courtesy Anne S.K. Brown Military Collection, Brown University Library, Providence, USA.

(Above left) Royal Engineer officer's uniform, c. 1870-1878. Following the departure of the British Royal Engineers from Canada, the country had hardly any military engineering capacity except for a handful of volunteers. Their uniforms were similar to those of the Royal Engineers. This plate is an advertisement from Jones & Co., a military outfitter in London, intended for Canadian officers. At left, scarlet tunic with blue velvet collar, cuffs and piping, gold buttons and lace, Oxford mixture dark grey trousers with wide scarlet stripe, dark red morocco belts trimmed with gold lace, Busby with gold cord and badge holding a white plume. At centre, scarlet vest with blue collar and cuffs, small gold ball buttons and lace, dark blue peaked cap trimmed with gold. At right, blue-black patrol jacket with black lace and cords. ¬Private collection. Ph: RC.

Officer, Toronto Engineer Company, 1876-1881. This plate, based on an image of Lt. Col. T.C. Scoble, a professional engineer who led the unit, shows that its uniform was similar to the British Royal Engineers. Watercolour by R.J. Marrion. Courtesy Canadian War Museum, Ottawa.

Officers and NCOs of the Charlottetown (Prince Edward Island) Engineer Corps, 1880. All have the regulation engineer's uniform complete with the dark blue "Home Pattern" spiked helmet. *Canadian Illustrated News*, 31 July 1880. Private collection. Ph: RC.

Second Lieutenant James W. Moffat, Montreal Engineer Company, 1882. He wears the regulation engineer's uniform with a dark blue Home pattern spiked helmet with the dress uniform. Ph: CL .

would show, engineer companies were much more difficult to maintain than units of the other arms of the service.[65]

The two companies of Montreal Engineers were raised in December 1861 and May 1862 respectively. By October 1862, they had no uniforms, except for a "sample made: scarlet coat, blue facings trimmed with white cording – pants black 1 3/4 red stripe – Busby & forage cap." These uniforms were similar to those of the Royal Engineers, except that the cords were white and the buttons were, presumably, of white metal. The 10 March 1863 *Rules and Regulations* of the first company called for "a single-breasted scarlet tunic with blue facings embroidered with white braid; pantaloons of Oxford mixture, with a scarlet stripe one and three quarters inches wide, a busby with blue fall [bag] and white grenade; forage cap of blue cloth, white piping, and white band." The 1866 *Montreal Directory*, which noted that engineer officers had "silver lace" on their scarlet uniforms with "blue facings." The Montreal Engineers thus took the same dress as the British volunteer engineer units.

The Ottawa Engineer Company organised during 1862 obtained: "Scarlet cloth with blue facings – seal skin Busby with blue ball and horsehair plume. – acquired last July. Note: the cloth is of admirable texture, scarlet with blue facings, trousers Oxford grey with scarlet stripe – the Busby is of piled sealskin with blue bag and white horsehair plume – a forage cap of blue Kilmarnock worsted with yellow band…and the company is clothed in strict conformity to the Army regulations for corresponding corps." This uniform was the same as the British regular Royal Engineers. In 1863, infantry uniforms were issued to companies at Lévis and Port Hope while allowances in lieu of clothing were granted to companies at Montreal, Ottawa and St. Catherines to have engineer uniforms.

In 1866, only the Montreal Engineers remained, but the Grand Trunk Railway then raised its own engineer unit as No. 7 Company of its 1st Battalion in Montreal and new uniforms would be needed for other units also. Thus, in August and September, the Militia Department ordered "55 Pvts tunics same as Royal Engineers" and another "110 tunics same as Royal Engineers adding Canadian Militia buttons" with "110 trousers same as Royal Engineers" and 165 "Busbies same as Royal Engineers." Since the tunics were the "same as Royal Engineers", the cords would have been yellow and the buttons presumably of brass (or gold) like the Royal Engineers. The unit also obtained its own knapsacks and haversacks.[66]

---

[65] LAC, RG9, IC1, Vol. 114, Adjutant General to Charles Legge, 7 January 1862, Adjutant General to George Brown, 8 January 1862, Adjutant General to Samuel Keefer, 8 January 1862; Vol. 285, Return showing the number of volunteer corps in Canada, 10 July 1862; TNA, WO 370/5 (Royal warrants), 2 September 1856 and 26 December 1856.

[66] LAC, RG9, IC8, Vols. 3 and 4. Commanding Officers Reports…1862; LAC, RG9, IC1, Vol. 109, W. Powell to P. Tait & Co. (Limerick), 6 August and 13 September 1866, LAC, RG9, IC8, Vol. 4. Commanding Officers Reports…1862. The Chateau de Ramezay Museum in Montreal has a scarlet mess jacket with blue cuffs and collar, silver lace and shoulder cord, small silver ball buttons in front and silver shoulder cord button bearing a crowned VR and "Montreal Engineers." It also has a c. 1870-1880 officer's tunic with gilt buttons having the crowned VR and "Montreal Engineers."

By 1872, there were only the Montreal, St. John (NB) and the Grand Trunk Railroad companies left and the latter unit soon vanished. The New Brunswick Engineer Company at St. John existed since 1861 and was issued 55 engineer uniforms with Busbies in May 1870. The Toronto Engineer Company was raised from January 1876, the Charlottetown (PEI) Engineer Company began in 1878 and the Brighton (NB) Engineers formed from January 1880. But the Toronto Company was disbanded in 1881 and the one in St. John (NB) in 1882. By 1890, only the Montreal, Brighton and Charlottetown companies remained. Part of the problem was a lack of support for engineering units because, in the then tiny Canadian forces HQ, there was no engineering component until 1884 when a Engineer Branch was created. As engineer officers increasingly graduated from Kingston's Royal Military College, this created more interest in engineering, which led to the creation of a regular Canadian engineer unit in 1903.

Royal Canadian Engineers officer's tunic, c. 1890. This garment is associated with Donald McLeod Vince, officer in the Brighton (New Brunswick) company of engineers. Scarlet with dark blue collar, cuffs and piping, gold lace and shoulder knots, silver grenades at collar, gilt Royal Engineers buttons with crowned VR. Carleton County Historical Society Museum, Woodstock, NB. Ph: RC.

The 4 March 1876 Canadian Dress Regulations prescribed the same uniform for Canadian engineer officers as worn by British engineer officers since 1870. The main novelty was the smaller Busby whose plume was now in front. For officers, this plume had a blue base and their Busby was ornamented with a gold cockade and gold cords. In Britain, the Busby was replaced by the dark blue Home pattern spiked helmet from 1878. However, the Busby remained the regulations headdress in Canada for another couple of years.

By Militia General Order of 30 July 1880, the Canadian Engineers were allowed a white helmet rather than the dark blue Home pattern helmet and this was confirmed by the 1 May 1886 Canadian Dress Regulations, which described the white helmet "as for Artillery with spike except [the] badge, which is to be wreath of maple leaves enclosing grenade."

Nevertheless, the Montreal and Charlottetown (PEI) engineer companies procured dark blue Home pattern helmets instead from about 1878. Indeed, all units imported their uniforms from England. On the whole, this could be an expensive proposition since, from the early 1870s, the government stores did not have engineer uniform before 1897. Thus, in 1876, the Toronto Engineer Company paid over $2,000 for uniforms and equipment, which was far more than the compensation offered by the Militia Department. The Montreal Engineers obtained imported uniforms in May 1878 that almost surely included Home pattern helmets. According to the November 1885 *Canadian Militia Gazette*, the Montreal Engineer Company again obtained new uniforms in October that "looked particularly neat" and, three years later, more uniforms came for them from England free of duty.

The officer's serge patrol jacket with breast pockets was introduced by Militia General Order of 24 March 1892. It was scarlet with Garter blue collar, cuffs and shoulder straps. From October 1897, enlisted men's scarlet serge frocks with "collar, shoulder straps and cuffs to be made of blue cotton velvet" and trimmed with yellow "Artillery braid" put on the "bottom of collar and on cuffs only" along with "small Engineers buttons" became available from the Militia Department. Surviving scarlet Canadian engineers frocks have pleated breast pockets, dark blue velvet collar, cuffs and shoulder straps, yellow braid below the collar and forming a crow's foot at the cuffs, but no yellow braid edging the shoulder straps. The buttons have a crowned VR with MILITIA ENGINEERS CANADA on the garter. In June 1898, 100 such frocks were sent to the Brighton (NB) Engineers.[67]

67 LAC, RG9, IID1, Vol. 11, Memo for Engineers Jacket, Scarlet serge (31 October 1897); LAC, T384/1124, A. Benoit to Sanford, 8 June

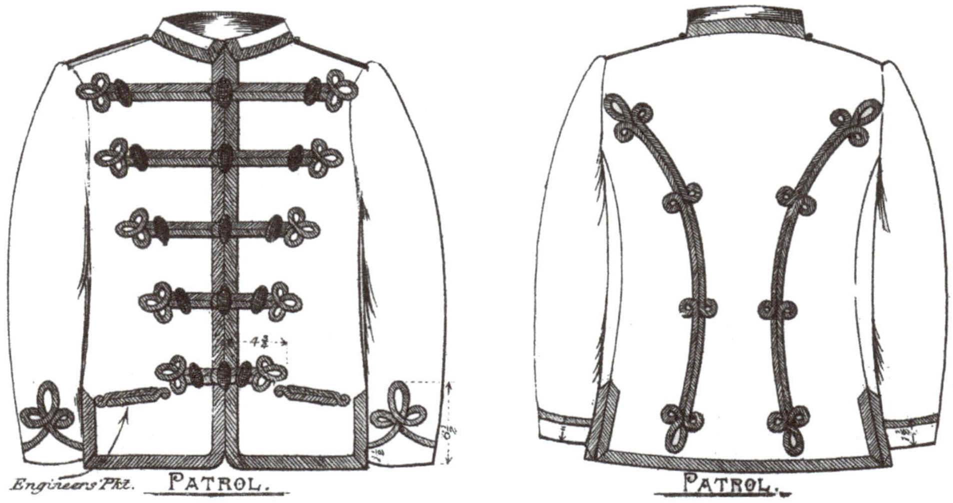

Officer's patrol jackets pattern, Royal Engineers, 1880s-1890s. Canadian engineer's patrol jacket would have been similar. This pattern hardly changed since the late 1860s. The most notable addition was shoulder straps from 1880 in Britain, late 1885 in Canada. Dark blue with black lace. Plate in T. H. Holding's 1894 Uniforms…Ph: RC.

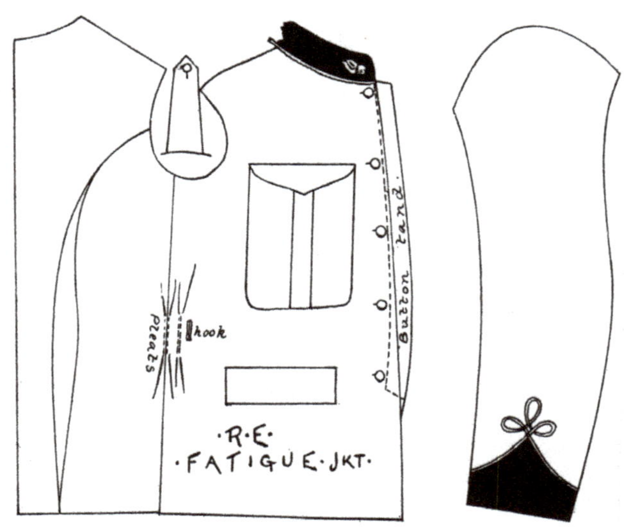

Officer's fatigue jacket, Royal Engineers, 1890s. Canadian engineers mentioned in the Canadian 1892 regulations would have been the same as the British corps. Scarlet with dark blue collar and cuffs, gold buttons and lace. Plate in T. H. Holding's 1894 Uniforms…Ph: RC.

Volunteer militia engineer's sergeant-major's tunic, 1880s. Scarlet tunic, dark blue collar and cuffs, silver buttons and lace. Sheldon Kasman collection. Ph: RC.

Submarine Miners Volunteers, Royal Engineers, 1890s. From left: sergeant in working dress other ranks in seagoing working dress; other ranks in marching order; officer in undress uniform; officer in full dress. Plate by R.J. Marrion. Ph: RC.

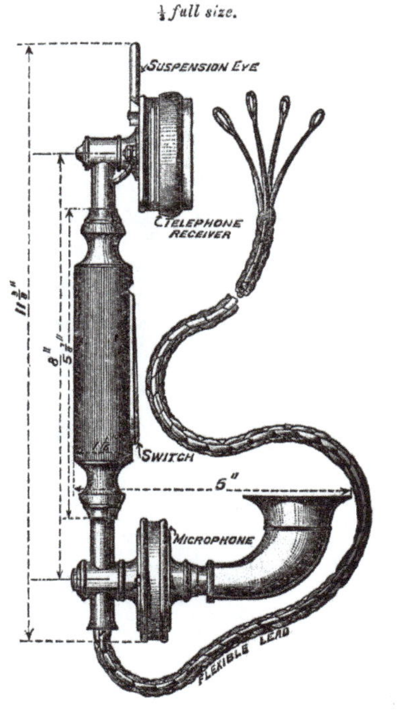

Approved plan of a model telephone for switchboards as per the List of changes in War Materiel of 30 March 1900. Ph: RC.

The signal squad of a British volunteer unit, 1880s. This scene would be exactly the same in a Canadian volunteer's summer training camp. The flag at left is white with a blue central stripe, the flag on the ground is blue. Print after G. Giles. Ph: RC.

A heliograph with five inch mirrors and its carrying case, 1880s-1900s. This is the most common type and is based on the British army's signal corps's Henry Christopher Mance's patent taken at Kurrachee, India, on 23 January 1877. Military Communications and Electronics Museum, Kingston. Ph: RC.

The Canadian engineer units retained the same uniform as the British Royal Engineers until, by a GO of August 1904 decreed that Canadian military engineers would henceforth have distinctive badges and buttons marked "Royal Canadian Engineers".

Halifax Volunteer Submarine Miners 1888-1907

The Submarine Mining Service, of the Royal Engineers was formed at Chatham (UK) in 1871. The Service installed, maintained and operated underwater defences in the approaches to harbours in Britain and the Empire's main naval bases. Submarine miners were to position and fire a series of mines placed around defended ports. Its vessels were crewed by uniformed volunteers. The uniform of this service was the same as the rest of the corps except that the men serving on vessels were also issued with sailor's clothing. From 1888, there was also in Halifax an "Auxiliary" detachment of 30 "Submarine miners" that would have been Canadian volunteers (WO 33/32). They served with the regular Royal Engineers detachment (part of 40th Company) posted in Halifax. From May 1900, another Royal Engineers submarine miner detachment (48th Company) with half of 44th Fortress Company was at Esquimalt, British Columbia. Because no auxiliary detachment could be formed there, one officer and 14 sappers were added "to the strength of the Sub-Marine Mining Company" during 1901 (LAC, reel T279/462).

The volunteer submarine miners wore the uniform of the Royal Engineers, but trimmed with white metal buttons and badges, white lace and piping, the officers having silver. Service being often on boats, sailor's trousers, dark blue double-breasted pea jackets, jerseys, trousers, sea boots and sou'wester caps were also issued. The service head dress was the dark blue sailor's cap with "Royal Engineers" on the ribbon. Officers on service also had a dark blue frock and, from about the 1890s, a scarlet frock with outside pockets and with blue collar and cuffs with a dark blue field service cap. The British garrisons at Halifax and Esquimalt were replaced by Canadian troops in 1906 and the Halifax volunteer submarine miner detachment was obviously disbanded between then and 1907 when defence measures to counter submarine mines were no longer deemed necessary.[68]

Signals & Communications

Early Signalling

Until the second half of the 19th century, signals and communications by land armies were generally carried out by musical instruments such as drums and trumpets as well as by flag waving and dispatch riders.

1898. Engineer frock c. 1900, Canadian War Museum, Ottawa, No. 3-4-336; Canadian engineer frock, c 1900-1914, Glenbow-Alberta Institute, Calgary; these frocks have seven buttons in front, one to each pocket and shoulder strap, one at the side of the cuff and one above.

[68]Don Thomas "Sailor Sapper", *The Mountaineer*, Canadian Military Engineer Museum, CFB Chilliwack, http://mypage.uniserve.ca/~echo2/Don.htm#sailor; Ray Westlake, "Submarine Miners", *Military Modelling*, March 1985, pp. 192-194. In 1905, it was decided to take away the duties in connection with the submarine defence of the military and commercial ports from the Royal Engineers, and transfer them to the Royal Navy. In 1906, a proportion of the N.C.O.s and men of the Royal Engineers accepted permanent transfers and some of the officers were appointed to the Canadian Army.

Two signallers of the 48th Highlanders, c. 1898. They are using a Five Inch Heliograph. Vignette in Fraser's 1900 history of the 48th Highlanders. Ph: RC.

Amongst other means, carrier pigeons might also be pressed into service at times. Optical telegraph lines were also used in Europe from the end of the 18th century using a code of semaphore signals. But these were still rather slow so their spread was limited. Early telegraphs, be they visual or electric, were usually manned or under the responsibility of engineers. The earliest dedicated unit appears to have been the Portuguese army's telegraph corps that manned lines of towers with signals that could send messages from the Spanish border to Lisbon in a few hours during the Peninsular War. The extraordinary technical inventions that became prevalent from the middle of the 19th century revolutionised communications in all fields including in armies and navies.

Semaphore flags

A much more common way of communication for units in the field was by a soldier or sailor using two flags to make signals by holding them in different positions to signal letters of the alphabet and numbers. This became widespread in field armies from about the 1870s. There was, at least in Canada, no specialised signal corps until 1903, but it was a common practice in the British forces that a squad or a few individuals be selected amongst interested soldiers to be signallers and this was also done in Canadian volunteer units. In early 1876, three sets each of "Flags, hand, large…for gun practice" and "for signalling orders to detached forts &c" were ordered from England (LAC, reel T270/201). Early semaphore flags were usually one blue and one white with a blue central band (that were shown as such on the arm badge worn by signallers) and were fairly large so as to be visible afar; they could be two or three feet (61 and 91 cm) square (LAC, reel T395/997). One flag at a time was waived using both arms.

Electrical Telegraph

The advent of the electrical telegraph, which transmitted by wires on poles usually built next to railway lines, was first commercially successful in England during the 1840s and revolutionised communications. Implications in all fields were obvious including military strategy now that the outcome of a military engagement could be telegraphed and published in a newspaper in a matter of hours. Telegraphs lines were built across Europe and North America. The next step was linking both continents and this was achieved in 1864 by the cable linking Britain to Newfoundland. This also linked senior government officials as well as naval and military HQs on both sides of the Atlantic. The impact on field armies was not as important insofar as the transmission of the more rou-

The Signal Squad of the 77th Wentworth Battalion, c. 1900. A telescope used by the squad is at left. Private collection. Ph: CL.

tine instructions by dispatch riders and bugles were still the primary signalling means on battlefields or manoeuvres.

Nevertheless, military applications of specially trained soldiers were contemplated and, in 1870, a telegraph company (later battalion) was created in the British army within the Royal Engineers. The signallers would make use of any telegraph line in the area where they were deployed. Secrecy of messages eventually became a concern. In Canada, a simple telegraph cipher code was used during the 1885 North-West Campaign, although its efficiency was questioned. Secrecy for Department of Militia correspondence was again raised in January 1896, but considered "superfluous" except with regards communications regarding arms purchases via the Canadian High Commission in London; these were transmitted in cipher code. In May 1900, "Cipher Code words selected for names of Canadian Corps in South Africa for Telegraph messages" were: unblamable: regiment infantry; unbleach: mounted infantry; unlight: medical staff corps; uncheck: C Battery; un-cheerful: D Battery; E Battery; unchurch: First Battalion mounted rifles; uncial: Second Battalion mounted rifles; uncialp: Strathcona Horse (LAC, reels T277/535 and 572; T4175/718; T4180/531; T279/171). Since telegraphs provided newspapers, press censorship could be used to prevent information from reaching the enemy.[69]

Telephone

A speaking device, the telephone, was invented in Canada by Alexander Graham Bell in 1876 and was used early on by the Canadian military forces. In August of that year, Bell made the world's first long distance telephone call, about 9.6 km (6 miles) between Brantford and Paris, Ontario. However, it was only in 1878 that it was offered commercially in North America, first in New Haven, Connecticut, for a few subscribers. Canadian military officers were obviously watching developments because, in April 1880, the ledger of the Deputy Minister of Defence records approval to instal "a Telephone to connect Militia Officers, Quebec, with Citadel, Military Stores, &c." for $30 per connection. The same source mentions that by February 1886, there are also "Bell Telephone Company" instruments already installed in Fort Henry (Kingston, Ontario) and probably in certain other bases (LAC, reels T4175/718; T4177/275). The early telephones were certainly the models sold by their makers. The British army also acquired telephones and, by 1900, approved an official model for switchboards that was soon communicated to Canada via Halifax.

Heliograph

The heliograph was a solar telegraph, usually using the Morse code, by flashes of sunlight reflected by a mirror mechanism. Light reflected off its "collector mirror" was transferred to the "signal mirror" that had been pointed in the direction of a receiver. The flashes were produced by momentarily pivoting the mirror, or by masking the sun's beam with a shutter. The British army's heliograph was developed in India during the 1870s, found useful during the Great Indian Survey and, during 1877, in the Jowaki Afridi expedition. In Afghanistan, it "proved of great value" as it enabled communications when telegraph lines had been cut by the ennemy. Word soon spread of this signalling system and the US Army used it from 1880.

[69]W.H. Goodenough and J.C. Dalton, *The Army Book of the British Empire* (London: HMSO, 1893), pp. 564-566.

In Canada, individual signallers in some units might be equipped with "Five Inch" heliographs. In August 1883, there is reference to a heliograph for Quebec City's 8th Royal Rifle Battalion (LAC, reel T4176/476) and later the 48th Highlanders had this instrument as, no doubt, many other units.

Cannon signals

The sound of cannons shooting blank rounds as signals were used from the time they were invented, not only as military alarm signals, but also as a way to tell certain times of day in towns when most individuals did not have watches or other time-telling instruments. Thus, like church bells, cannon shots would be fired at specific times of day, notably the "boom" of the noonday gun still heard in a few Canadian cities such as Halifax and Québec. This was done by the imperial troops that garrisoned Canada and was eventually taken over by Canadian regular troops. In at least one case, volunteer gunners also performed this duty. In March 1869, the "Ottawa Volunteer Artillery" was requested to establish and fire "a 12 O'Clock gun" on Parliament Hill that is still heard today (LAC, RG9, IIB1, vol. 515).

Another aspect of gunfire signals involved older guns that were turned over to the Canadian Marine Department. For instance, a 12-pdr cannon was requested in 1870 "to be used as a Signal Gun" at St. Paul Island, off the northeast end of Cape Breton Island, where a lighthouse was first installed in 1839. Some 32-pdr carronades were sought "for signal purposes" in 1881, and again in 1882 "for Light House Stations in the Lower St. Lawrence" so keepers could also to warn ships in fogs. Some of these guns are still seen in coastal villages. (LAC, reels T4173/379, T4176/169 and 290).

Notes

P.H. Kealy, "The Uniforms of the Corps of Royal Engineers up to 1914", *Royal Engineers Journal*, June 1934, December 1934 and September 1935; J. Ferguson, *Equipment of a Company of Engineers*, (London, 1866); "The Royal Engineers in British Columbia 1858-1863 and 1893-1906", *Royal Engineers Journal*, July 1906; Jack L. Summers and René Chartrand, *Military Uniforms in Canada 1665-1970* (Ottawa: Canadian War Museum, 1981), pp. 83-85.

An old carronade used for signals to ships during foggy weather. These were placed near lighthouses. This one was at the lighthouse of L'Islet on the south shore of the eastern St. Lawrence River. Ph: RC.

Chapter 7

Non-Combatant Corps

Medical Officers and Chaplains

Officer of the Army Medical Staff, 1865. Dress uniform: scarlet tunic with black velvet collar and cuffs, white piping, gold lace and buttons; dark blue trousers with a gold stripe; bicorn hat with black feather; black leather belts with gilt badge and buckle. The cuffs were pointed from 1868. Plate from Army Equipment. Ph: RC.

During Canada's colonial periods, medical services to the armed forces were provided by the mother countries. They maintained medical officers permanently in Canada to provide expertise and care in two distinct types of health care. The first type consisted of the medical officers attached to units that consisted usually of surgeons and assistants serving as part of a regiment's staff. They were not initially full-fledged university-trained doctors although this changed during the second half of the 19th century in when local civilian doctors joined their area's volunteer militia regiment. They did not have elaborate facilities and a tent with some medical supplies acted as a clinic at camps.

The second type of medical care consisted of specialised and long-term care that required supervision by professional doctors, who ranked as non-combatant staff officers serving at military hospitals in garrison cites and sizeable clinics in larger outlying forts. Such facilities, which were also available to Canadian volunteers, mostly vanished when the British garrison left Canada in 1870-1871, except at Halifax and Esquimalt. However, since the French Regime, governments also subsidised civilian hospitals, often kept by religious orders, for the care of military patients. Indeed, some of the surgeons, physicians and pharmacists at those institution could also be part of their local militia unit.

British military medical services initially came under the authority of the Army Medical Department officers, reorganised into the Army Medical Staff in 1855 for medical officers only and the Army Hospital Corps in 1857 for enlisted men, re-designated Medical Staff Corps in 1884, both organisations being merged as the Royal Army Medical Corps in 1898. Dentistry was not then a separate specialty and army surgeons extracted teeth although, from 1857, they were encouraged to conserve them and instruments were provided for this purpose. Following the 1870-1871 withdrawal of the British garrison from Central Canada, the Department of Militia and Defence continued the practice (abandoned from 1873 in the British Army) of appointing civilian physicians as medical officers to permanent and volunteer militia units; they also made up the Medical Staff at district headquarters. According to the 1885 *Militia List*, there were some 221 medical officers, including 18 veterinary surgeons, in all parts of Canada.[70]

The outbreak of the North-West Rebellion in the spring of 1885 showed that there was actually hardly any regular Canadian military medical service and authorities in Ottawa now rushed to appoint medical professionals to establish field hospitals and ambulance services for the force going west. The medical officers appointed were nearly all civilian physicians assisted by medical students in Toronto and Montreal who volunteered as dressers for hospital accommodation provided at Saskatoon, Battleford and elsewhere. At Fort Otter, the "casualties [were]

[70] Of nine surgeon-majors, three were medical doctors (MD); of 123 surgeons, 73 were MDs; of 69 assistant-surgeons, 42 were MDs for a total of 118 MDs out of 203 medical officers. The 18 veterinary surgeons could not be considered MDs.

in a Field hospital, temporarily arranged and managed by a party of some 20 to 30 Medical students from Toronto University…all members of 'K' Company, Queen's Own Rifles, and had been given the status of 'Dressers' in a Temporary Field Dressing Station Unit. They hastily collected some [white] cotton sheets and red flannel and made up a Red Cross flag which was flown over their 'Station,' the first Red Cross emblem to fly in Canada on active service over Canadians actually wounded in the field."[71]

Thereafter, there was still no regular Canadian medical corps since the permanent force was too small to justify such an establishment. However, as years passed, medical needs grew. In Canada, medical officers continued to belong to regiments, but, from 1896 when Col. Hubert Neilson, MD, was appointed Director General of a new permanent force medical sub-department, the situation changed. Theraising of a Canadian Bearer Company (later called ambulance) in Halifax fostered the organisation of the regular Canadian Army Medical Staff in February 1898. The following year, there were six bearer companies and six field hospitals and their existence led to the creation of the Canadian volunteer militia's Army Medical Corps on 20 May 1899 (which became, in 1904, the permanent and regular Canadian Army Medical Corps) as well as recognising regimental surgeons and incorporating them as the Regimental Medical Service. Thus, when the Boer War broke out, Canada could contribute medical services and a complete unit, the 10th Canadian Field Hospital, reached South Africa in March 1902.

Officer of the Army Medical Staff, 1865. Service dress uniform: dark blue double-breasted frock coat, gold buttons and collar rank badges, dark blue trousers with red stripe, dark blue cap with black lace and visor, gilt badge. Plate from Army Equipment. Ph: RC.

Nursing Services

During nearly all the Victorian era, there was no military corps of nurses in the regular force nor in the volunteer militia. Nevertheless, nurses were deployed with troops in several instances. During the 1885 North-West Campaign, a dozen volunteer nurses, including four sisters of the Order of St. John in Toronto, went west with the troops and soon took care of some 80 wounded volunteers in improvised field hospitals.

In 1898, four nurses of the Victorian Order of Nurses (founded by Lady Aberdeen in Ottawa the previous year) were joined to the Yukon Field Force. They were lauded by its officers for their outstanding work during a typhoid fever epidemic that would have been much more severe had it not been for these nurses. This Order was also requested by the military authorities in Halifax to provide medical care for the "wives and children of the [British] soldiers in barracks" at that time.

Late the following year, four nurses were selected to sail with the first contingent to South Africa followed by four more with the second contingent. By Militia General Order of 25 January 1900, they were given officer status with the rank and pay of lieutenants, a progressive outlook as to the status of women in the military structure of armies of that time, which was eventually followed in most major armies during the 20th century. Eight more landed in South Africa with the third contingent in March 1902. By then, the regular Canadian army's nursing service had been created since 1 August 1901.

[71] C.F. Winters, "The North-West Rebellion, 1885: A Memoir by [Governor General's Foot Guard] Colour Sergeant (later General) C.F. Winters," Daniel S.C. Mackay, ed., *Saskatchewan History*, XXXV: 1 (Winter 1982), p. 12. A distinct Canadian Red Cross Society was established in 1896.

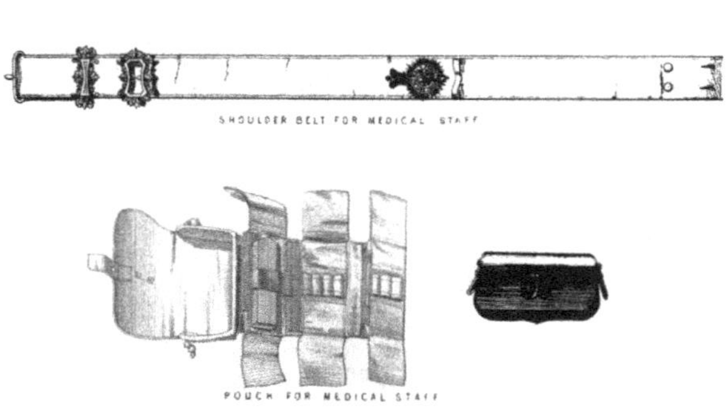

Shoulder belt and pouch for medical staff, 1865. Plate from Army Equipment: Infantry, 1865.

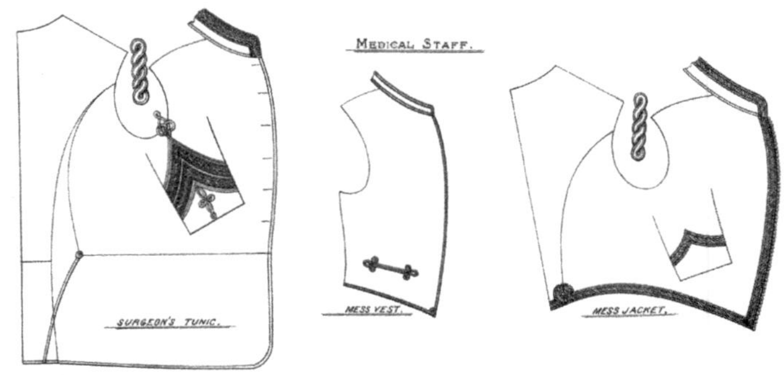

Patterns for Medical Staff tunic, mess vest and mess jacket. The tunic's cuff and collar denotes the rank of surgeon of the Medical Staff. From T. H. Holding's 1894 Uniform...Ph: RC.

Veterinary Services

Care for the health of military horses also became gradually more important due to the rise in numbers of militia cavalry units as well as military transportation that increased as new roads, farm fields and towns appeared where wilderness had been. Veterinary services were thus increasingly sought and, especially from the second half of the 19th century, veterinary officers were increasingly found in attached to cavalry and field artillery units. For instance, in April 1880, a "Veterinary Surgeon for [the] horses of B Battery" was approved (LAC, reel T4175/715). The Canadian Army Veterinary Corps was raised in 1910. The British Army Veterinary Department, that later became Canada's model, was formed in 1881. The Canadian Army Veterinary Corps was created in 1910.

Uniforms

Medical Officers

During the long reign of Queen Victoria, British army medical and veterinary officers that were part of regimental staffs generally wore the same uniform as other officers in their unit except that they wore cocked hats rather than shakos or helmets. Officers of the British Army Medical Department wore a scarlet coat with black velvet collars and cuffs and gold buttons and lace. The officers of the Army Medical Staff, as reorganised from 1884, henceforth wore dark blue coats with black velvet collar and cuffs and gold buttons and lace, dark blue trousers with a wide scarlet stripe, dark blue helmet with a ball. They commanded the NCOs and men of the Army Hospital Corps (Medical Staff Corps from 1884) who, from 1873-1874 wore dark blue with red piping. From 1874, British veterinary officers were assigned a dark blue uniform with maroon velvet facings and gold buttons and lace, which became their only uniform after 1881 when veterinary officers were no longer part of regiments, but henceforth members of the Army Veterinary Department.[72]

The uniforms for Canadian medical and veterinary officers were essentially identical to their British counterparts. Following the 1867 Confederation, the 1876 and 1886 Canadian dress regulations stated that "medical officers wear the uniforms of their regiments, except in the following particulars": cocked hat with a four inch long black cock's tail feather "drooping outwards", shoulder belt and instrument case in brown un-tanned leather. It would appear that senior medical volunteer officers in a Militia district would also wear the British Army Medical Staff uniform as seen in a c. 1886 photo of Dr. G.E. Fenwick. Since August 1879, he had the rank of surgeon major, the most senior medical officer in the 5th Military District as well as being attached to the Mon-

[72]"Original and copy records, pamphlets, correspondence etc., including re criticisms of the Army Medical Service 1804 & 1897" ledger at the Wellcome Library in London, UK, has many notes, notably on uniforms. Also accessible at: http://search.wellcomelibrary.org/iii/encore/record/C__Rb1840055

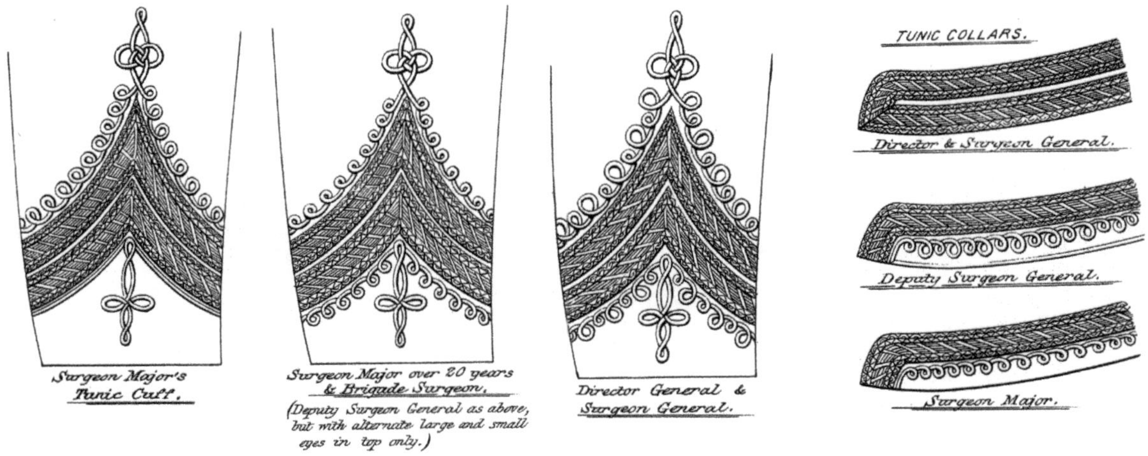

Sleeve rank and collar distinctions for the Medical Staff. From T. H. Holding's 1894 Uniforms…Ph: RC.

treal Field Battery. By the Militia General Order of 13 May 1892, veterinary officers were to have sword belts, pouch belts and instrument cases of the same pattern as those prescribed by the British army dress regulations of 1891.

This was somewhat refined in April 1894 for "the Medical and Veterinary Officers of the Permanent Force. "Any new uniforms obtained" were to be of the pattern "for officers of similar rank in the Imperial service", but with the following exceptions: shoulder straps for tunic, mess jacket, patrol jacket and serge jacket of same colour, and material as facings of tunic. The tunic and mess jacket the shoulder-straps were to be edged "with gold lace of the same width as worn on top of collar" and have badges of rank embroidered in silver. For patrol Jacket, the badges of rank were embroidered in gold. For serge jacket, greatcoat and winter coat, the badges of rank were in gilt metal. The black leather sabretache bore the "Royal and Imperial cypher and crown, in gilt metal." The pouch ornament and the badge on field service forage cap had the royal and Imperial cipher and crown in gilt metal.

Surgeon of the 2nd Queen's Own Rifles, c. 1876. He wears the distinctive bicorn for medical officers with its black plume, silver (gold from 1886) cockade loop and end tassels. The crown on his collar indicates the rank of lieutenant. Private collection. Ph: CL.

At that time, the basic guide for the uniforms of the officers of the Canadian Militia Army Medical Staff was the British "Dress Regulations for the Officers of the Army, 1894" as reaffirmed in Canadian orders of May 1900, but "with the exception of the following alterations and additions" for Canada: the 1st Honorary Surgeons to the Governor General had a "special gold and black sash, as for Honorary Surgeons or Physicians to the Queen" and "an aiguillette as for personal A.D.C. to the Queen, worn on the right shoulder." The Director General Medical Services, and Principal Medical Officers had the frock coat with departmental buttons and aiguillettes.

Furthermore, Canadian medical officers were to have the tunic as described in the "Dress Regulations for Officers of the [British] Army, 1894, with this addition: the skirt square in front, open behind, with a blue cloth flap on back of each flap. Flaps edged with round gold cord, traced inside with gold Russia braid, three buttons on each flap behind and two at the waist behind" while cocked hat were no longer worn, "except by the Director General and Principal Medical Officers of Divisions." Canadian medical officers henceforth wore helmets of the home pattern with its plate bearing the "Royal Arms in gilt metal." The peaked forage cap was "similar to the staff pattern

14th Prince of Wales' Own Rifles Battalion medical detachment, c. 1885. The sergeant (centre, with sword) and enlisted riflemen wear the unit's rifle green uniform with the short-lived dark green "home pattern" spiked helmet. Note also the red cross insignia on white, which was adopted during the 1880s. On each side are the medical officers wearing their distinctive bicorn hats. The small company camp colours are behind. Fort Henry, Kingston. Ph: RC.

Medical staff treating wounded men at Fish Creek, 1885. Print after T. de Thulstrup. Harpers Weekly, 9 May 1885. ph: RC.

(Top) 7th Fusiliers of London, Ontario, medical detachment at Swift Current, Saskatchewan, late April-early May 1885. The men are holding the rolled up stretchers. The medical detachments wore the unit's uniform with the red cross on a white armband. Pencil drawing by Alexander Campbell in his diary of the North-West Campaign. Courtesy of the University of Saskatchewan Library, Saskatoon.

(Right) 2nd Dragoons' surgeon in dress uniform, c. 1890. The medical officer shown is probably G.M. Warren who served with the 2nd from 1884 to his retirement in 1903. Dark blue uniform, white collar, cuffs (invisible due to white gloves), piping and trousers stripes, gold buttons and lace, black staff pattern bicorn hat with black plume. Courtesy Ontario Archives, Toronto, S16109-1.

A first aid field dressing station during the North-West Campaign, May 1885. Nearly all the wounded men are infantrymen as can be seen by the trefoil at their cuffs. Standing at the centre are a sergeant-major of artillery and a scout. At lower right, a medical officer wearing an officer's undress frock with an armband bearing the Geneva Red Cross. *Illustrated War News,* 30 May 1885. Courtesy Bruce Peel Special Collections, University of Alberta, Edmonton.

Wounded men at a first aid field dressing station during the North-West Campaign, May 1885. All wear the infantry tunic with its trefoil cuff lace and the Glengarry forage cap. At centre, a soldier has the Geneva Red Cross armband, but the illustrator has mistakenly put the number 90 on his cap to indicate the 90th Rifles. In fact its badge was different and rifle tunics did not have cuff trefoils. Such images nevertheless give a good sense of the conditions and appearance of such stations. *Illustrated War News*, 20 June 1885. Courtesy Bruce Peel Special Collections, University of Alberta, Edmonton.

of the British Army, 1899" with the following changes in Canada: "Black Cloth band around the crown instead of scarlet" and "Gold embroidered enamelled peak for Officers of field rank, black enamelled leather peak for other ranks. Above the peak the A.M.S. [Army Medical Staff], special badge will be worn. The field service cap was the infantry pattern, with A.M.S. badge on left side." In effect, Canadian medical officers would now wear this uniform rather than regimentals.

Officers also wore the dark blue serge frock of infantry pattern with departmental buttons. Officers of Head-Quarters and Divisional Medical Staffs had "scarlet gorget patches on the collar." The corps' gilt buttons had "a star; on the star, a garter surmounted by the Crown. The garter inscribed 'Army Medical Staff'. Within the garter, the Royal Cypher." This was the same as the British army's and, as late as 1906, there was "no button especially designed for the Canadian Army Medical Corps" (LAC, reel T402/430). The badge, approved in 1899 was a "maple leaf of white metal one and one-fifth inch from stem to point, in the centre a Geneva cross of red enamel for officers A.M.S., pierced through the metal, and with scarlet cloth backing for N.C.O. and privates of the A.M.C. [Army Medical Corps]. This badge will be worn on both sides of the collar of all orders of dress, except on the great-coat, point upwards, also on the front of forage and winter caps and on the left side of fatigue caps."

Enlisted Men

Enlisted men in Canadian volunteer units with a medical detachment wore their unit's uniform with the red cross on a white armband following Britain's 1870 adherence to the Geneva Convention. With the 1898 organisation of the Canadian Bearer (later called Ambulance) Company, enlisted men now part of the medical services. Their uniforms were ordered from England in December 1898 "for clothing for the Halifax Bearing Company… of the pattern used by the [British] Royal Medical Corps", but with "a distinctive Canadian badge, a design for which is being prepared".[73] From the 1899 photo of Corporal Roue, the British style dress uniform does not seem to have been ordered for this company. He wears an all dark blue frock with five brass buttons, without breast pockets, a dark blue pillbox cap and dark blue trousers that, however, seems to have the dull cherry piping of the Royal Army Medical Corps. The cap and collar badges identify his Canadian status. On his arm, the red cross in a roundel badge that identified all medical personnel. In February 1899, some 336 "armlets" [armbands] bawling the letter "S.B." were ordered for the bearer companies (LAC reel T386/424).

[73] LAC RG9, IID1, Vol. 15 p. 5109, A. Benoit to Secretary of High Commissioner, London, 16 December 1898.

Field hospital staff and patients at Moose Jaw, 1885. No. 1, Dr. Sullivan in command. No. 2, Nurse Cottrell of Toronto. No. 3, Nurse Mackenzie of Toronto. No. 4, four nursing sisters of the Order of St. John, Toronto. Note that several patients have had an arm amputated. Civilian medical staff have no uniforms except for those attached to a nursing or religious order. The Army Museum, Halifax Citadel. Courtesy Parks Canada.

Colour-Sergeant Curzon, 10th Royal Grenadier, under a "shower of bullets" rescues a badly wounded private of the Midland Battalion at Batoche in May 1885. All wear the scarlet tunic; Curzon's sergeant's sash is shown. Illustrated War News, 13 June 1885. Courtesy Bruce Peel Special Collections, University of Alberta, Edmonton.

Uniform of Lt. Col. Thomas Walker, medical officer of the 62nd St. John Fusiliers, 1895. He wears the regimental scarlet tunic with dark blue collar and cuffs, gold buttons and lace and gold grenade at the collar tunic, dark blue trousers with scarlet piping, black hat with gold tassels and black feather plume. Courtesy of the New Brunswick Museum, St. John.

Corporal William Roue, Canadian Medical Corps, 1899. He went to South Africa and served there in the 10th Canadian Field Hospital. Canadian War Museum, Ottawa. Ph: RC.

The four Victorian Order of Nurses attached to the Yukon Field Force, 1898. From the left: Margaret Peyson, Rachel Hanna, Georgiana Powell and Mary Scott. Courtesy Victorian Order of Nurses, Ottawa.

Nurse Minnie Affleck, first Canadian contingent to South Africa, with some of her patients, 1900. This somewhat damaged image shows a radiant Nurse Affleck wearing the white apron over her khaki uniform with the detachable white collar and cuffs and the red cross armband, straw hat with wide white band and a red aigret.. Her patients (except for one with a forage cap), who may be mostly British, wear the British army's grey hospital outfit. Private collection/Wikimedia.

"Canadian Sisters and Miss L. Trotter", a British nurse, standing outside tents in South Africa, c. 1900-1901. This images gives a good side and front view of the Canadian nurse's khaki jacket and long skirt. The detachable white collar (but not the cuffs) is worn and one nurse holds what is probably an apron. Straw hat with wide white band and a red aigret. Courtesy Royal Army Medical Corps Archives and Manuscripts/557 Box n.o/s 31.

Head Nurse Deborah Hurcomb, Canadian Medical Corps, second contingent to South Africa, 1900. Miss Hurcomb wears the nurse's contingent khaki uniform with brass buttons and a white armband bearing the red cross. Courtesy Library and Archives Canada, PA57339.

A note might be added concerning the equipment of the 10th Canadian Field Hospital in South Africa. According to Adami, its "equipment differed from that of the regulation British field hospital in that it was a combination of British and American usages. The Hubert tent (so-called after Colonel Hubert Neilson, who devised it) replaced the British hospital tent, and the wagons were lighter than the British model, being built upon the lines of a Canadian express wagon. It carried with it another innovation of Canadian origin – namely, a mobile acetylene gas plant." It was present at the Battle of Hart's River, brought along with the 2nd Canadian Mounted Rifles and a "bullet-marked Red Cross pendant from one of its ambulance wagons remains as a relic of this engagement."

Reverend J. Edgar Hill, Chaplain of the 5th Royal Scots of Canada, c. 1890. His all black frock has no shoulder straps and the two gold stars on the collar denote the 4th class or captain according to the British 1883 dress regulations that obviously also applied in Canada. From Chambers history of the 5th. Ph: RC.

Nurses

Nurses that accompanied the 1885 North-West field Force did not have uniforms although the sisters of the Order of St. John wore black and white habits. The 1898 Victorian Order of Nurses contingent with the Yukon Field Force was provided with a special uniform consisting of a "neat brown duck [linen] suit with a short skirt with bloomers and gaiters in the style of a natty bicycle suit" with a tarpaulin hat, boots with hobnail on the soles and, for winter, a heavy blue blanket, with a big hood to match, lined with quilted silk." A photo shows that the four nurses also had long skirts, white blouses and caps. The Order's regular uniform was a pale blue-grey dress with detachable white collar and cuffs, the Order's badge (approved by Queen Victoria) pinned on the left breast of the dress, a white apron and a white cap. A dark blue coat and hat were worn in cold weather. Nursing supplies were carried in a black bag marked with the letters "V.O.N." on the outside.[74]

Nurses serving in South Africa from 1900 to 1902 were provided with a khaki uniform featuring a "Russian" style jacket with two rows of gilt buttons that buttoned on the left side with a breast pocket with pointed flap, a standing collar with two buttons at the side, short "bicycle" skirt, a straw hat with a small and seemingly red aigret, a white nurse's cap and a white apron. From photos, the nurses also had a khaki longer skirt and detachable white collar and cuffs for the jacket. These were described by the *Quebec Saturday Budget* of 4 November 1899 as "pretty khaki nurse uniforms, with the distinguishing red cross on their arms…" Canadian army nurses continued to be uniformed in khaki until 1907 when a dark blue uniform was finally allowed.

Chaplains

There have been military chaplains in Canada since the 17th century. Anglican chaplains were the only official churchmen with the regular British army until 1836 when Presbyterian and Roman Catholics were admitted. In the Canadian Militia, many units already had Catholic chaplains, notably in Lower Canada, but wherever they served, chaplains were rarely mentioned in published Militia lists. In 1885 for instance, the 9th Voltigeurs de Québec is the only unit that lists its chaplain, Father Raymond Casgrain, but it is certain that just about every other unit had a chaplain too. The history of the 90th Winnipeg Rifles mentions the excellent services of the Reverend D.M. Gordon during the 1885 North-West Campaign. All this was certainly a matter of course at the Department of Defence that, seemingly as an afterthought 1897, announced that authority was given for the appointment of "Honorary Chaplains" as long as they would incur no expense. Chaplains were considered to be of officer status,

[74]Susanna McLeod, "Lady Aberdeen's legacy", *Kingston Whig-Standard,* 24 February 2011.

Chaplain of the 4th Class, c. 1885. His rank is shown by the two gold stars on his collar. His uniforms is otherwise completely black. The narrow black lace on the frock coat, rather than twist cord, may be for a hussar regiment as it is part of a set of undated sketches signed "Macdonald" that concern mostly British hussars and the medical corps. Courtesy Anne S.K. Brown Military Collection, Brown University Library, Providence, USA.

but it was only in 1903 that their rank was defined as captain. Chaplains were also present with regular Permanent Militia units since, from 1875, "Clergymen or Priests" were compensated for the divine services they held as well as for visits in hospitals (LAC, reel T272/549). The British Army's Chaplains Service was created in 1858 as a corps. The regular Canadian Chaplains Service was only established in 1921.

Chaplains traditionally wore the robes of their denomination and, in the British forces, did not have an assigned uniform until the Dress Regulations of 1864 that specified a black cloth frock coat, single-breasted, with "turn-down collar, six buttons down the front and two on hips", lined with black silk. Ranks was defined as 1st class (colonel) having a "1/2 inch of black mohair braid on the collar with a black crown and star on each end, three buttons and notched holes on cuffs"; 2nd class (lieutenant-colonel): same, "but with crowns only on collar"; 3rd class (major): same, "but with stars only on collar"; 4th class (captain): "No lace on collar, but both crowns and stars. Plain round cuffs without buttons or notched holes." The trousers were black and so was the peaked forage cap.

Nearly twenty years later, the British 1883 Dress Regulations hardly changed the coat except that the collar was now standing and that black twist cord false button holes was mentioned. Buttons were defined as being of black silk. The collar rank badges were now for 1st class: a crown and two stars; 2nd class: a crown and one star; 3rd class: one crown and 4th class: two stars at each end. Trousers had "black braid an inch wide", the gloves were of black leather, the greatcoat and cape also black and the black forage cap had 7/8 inch black embroidery, a black lace "staff pattern" band 1 3/4 inches wide, "black netted button and braided figure on crown."

By the early 1890s, this had evolved slightly. Rank badges "in black and gold" were now on shoulder straps of twisted black cord and the collar had a Maltese cross badge "worked in black and gold" and an all black patrol jacket decorated with flat black mohair braid was now added for serving in the field with gold rank badges on the shoulder straps according to Holding's 1894 tailor's pattern book.

In Canada, there were no instructions regarding the dress of chaplains until 1897 when the newly defined honorary chaplains were instructed by the Defence Department to be dressed in uniform. This dress was not defined, which indicated that many chaplains already had uniforms; these obviously followed the British army's patterns described above. It would appear it was the protestant chaplains of city units in anglophone Canada that went to the expense of procuring uniforms for their military duties. Photos of Canadian protestant chaplains in city volunteer units confirm that they wore the British patterns. They obviously depended on British tailors or Canadian outfitters that had copies of the British regulations and such pattern books. Those attached to rural units would likely be less eager. Insofar as French Canadian units wherever they were, there seems to be no evidence the Roman Catholic clergy wore such uniforms.

Storekeepers, Paymasters

The HQ staff at Ottawa and the Military District had a few officers that were responsible for supplies. Before 1871, these services were essentially provided by the Commissariat and Ordnance corps of the British army. After the British garrison left Canada, the small size of the Canadian regular force and the modest requirements of the volunteer militia did not justify the creation of a large staff devoted to logistics, but a few officers were appointed. The 1874 dress regulations assigned them a dark blue uniform with dark blue velvet collar and cuffs, the trousers with two white stripes, silver buttons and lace. This was, except for silver instead of gold buttons and lace, the dress of the British Commissariat (see Vol. 1, pp. 55-60). Not mentioned in the 1874 regulations, but briefly in those 1886 stating that they were to have regimental Canadian uniforms except for the cocked hat, sword, pouch and belts "as per Imperial Pay Department." These services were formed into the Canadian Army Service Corps in December 1901, the Canadian Ordnance Corps in 1903 and the Canadian Army Pay Corps in 1907

Notes

J. George Adami, *The War Story of the Canadian Army Medical Corps* (Westminster, UK, 1919). On the British Medical Corps, see the amply illustrated: "Royal Army Medical Corps 1660-1998", *Regiment*, No. 29 (1998); Barbara Dundas, *A History of Women in the Canadian Military* (Montreal: Art Global, 2000); Countess of Aberdeen, *What is the use of the Victorian Order of Nurses for Canada* (Ottawa, 1900); Sheila Penney, *A Century of Caring: The History of the Victorian Order of Nurses for Canada* (Ottawa: Victorian Order of Nurses, 1996). On the uniforms of the British Army Veterinary Corps, see the appendix 1 of Frederick Smith, *A History of the Royal Army Veterinary Corps 1796-1919* (London, 1927).

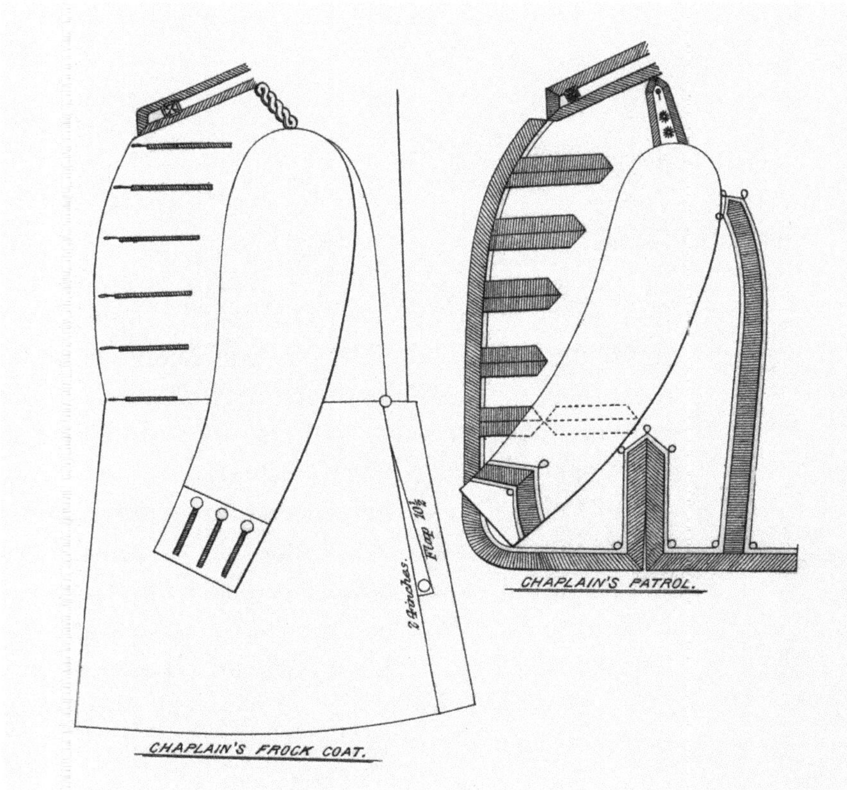

Patterns for chaplain's frock and patrol jacket, 1890s. These garment were completely black except for rank badges. From Holding's 1894 uniform pattern book. Ph: RC.

Officers at a camp in Montreal, c. 1890. This remarkable image show two officers of the medical services. At left, Surgeon-Major G.E. Fenwick, 3rd Field Battery of Montreal, but wearing the Medical Corps uniform. At right, Veterinary-Captain Charles McEachren of the 3rd Field Battery in the uniform of the 3rd Field Battery wearing a bicorn. The other officers are Lt. Col. Montizambert, Regiment of Canadian Artillery, with three officers of the 3rd Field Battery recognizable by their Busbies and an infantry officer (Lt. Percy Girouard) with the white helmet. This image reproduced in Chambers' history of the 3rd Field Battery is dated 1890, but some rank badges are visible on the collar, which dates this photo as being at about 1886 when rank badges were officially authorised on the cord shoulder straps. Ph: RC.

Chapter 8

Winter Uniforms

Soldiers at a bivouac, c. 1837-1839. They wear the grey greatcoat with its shoulder cape worn over the coat with the trousers, shoes and the shako. In Canada, this order of dress was worn, depending on the weather, until the late fall and from the early spring. Anne S.K. Brown Military Collection, Brown University Library, Providence, USA. Ph: RC.

In a northern country such as Canada, keeping warm during the colder part of the year was a constant preoccupation of the pre-Columbian First Nations as well as early French settlers and soldiers. There soon evolved a practical and comfortable costume that was a mixture of European and indigenous dress, worn from the 17th to the later part of the 19th century. Soldiers, volunteers and militiamen were possibly more exposed to the rigours of winter. They had to mount guard in the coldest winter nights and might be campaigning during the middle of winter, in what can only be described as an alien environment.

British Forces 1830s-1871

In the 1830s, enlisted men in the British infantry wore the grey greatcoat introduced since 1801. The greatcoats worn by Canadian militiamen called on active service were similar, and indeed often came from British army supplies. Initially double-breasted, the greatcoat had a single row of buttons in front by the 1820s and possibly earlier. A detachable grey cape covered the shoulders. The cuffs and collars of the sergeant's greatcoats were of the facing colour in infantry units. The buttons were of the colour of the tunic's buttons. For British soldiers, this mean white metal until about 1856-1857 when tunic buttons were changed to brass. The basic appearance of the greatcoat changed little between the 1820s and the 1850s when a few buttons were added to the cape so as to fasten it in front. A "new pattern" greatcoat of kersey with detached cape appeared in 1864-1865, externally quite similar to the older model.

The British infantry officer's winter dress in Canada usually consisted of a fairly long coat of thick grey, grey-blue or dark blue material. This was worn with a scarf over the collar and fur gauntlet gloves. A fur cap with ear flaps and peak, also covered with fur, and a hussar-like red bag would complete the outfit of the fashionable officer. The 1837 Rebellions eventually brought about official changes to the winter dress of officers. It was probably during that period that fur trimming added to the greatcoats became very popular. Many British regiments were sent to Canada at that time to contain the rebels and the following "hints" were recommended to officers departing from England face the winter:

> "1. A large, not dandily-tight, P[ea] jacket, lined with woollen.
> 2. Fine blue pilot-cloth regimental frock-coat, double breasted, regimental buttons, sleeves lined with flannel, black Astracan collar - not cuffs of the same, for it is difficult to draw the P[ea] over them.

Soldier at "22 degrees below zero" in 1838. To the regular grey cloth greatcoat, a peaked fur cap with flaps to cover the ears, fur mitts and high beef leather boots are added. The accoutrements, knapsack and blanket roll are also worn. Watercolour by Sir Archibald Cope. Courtesy Library and Archives Canada, TC731.

Trooper, Quebec Volunteer Cavalry, 1837-1838. "Paid by Government. 1 Troop of 80 Men." Dark brown fur cap with a dark or black bag with tassel, dark blue coat with white fur collar and cuffs, white metal buttons and shoulder scales, sky blue trousers with wide white stripe, buff gauntlet gloves, black boots with buttons at sides, black sabretache with small brass badge, dark blue valise with white lace, red schabraque edged with white lace. Plate by Peregrine Pouchbelt and Rodrick Ramrod. Anne S.K. Brown Military Collection, Brown University Library, Providence, USA. Ph: RC.

"Infantry, Winter Costume", c. 1838-1840. This watercolour by Henry James Warre gives a rare side-rear view of a soldier wearing a winter uniform. It could be a Canadian volunteer or a British army regular. Grey greatcoat, brown beef leather boots, brown mitts and dark brown peaked fur cap. Courtesy Library and Archives Canada, C31249k.

Troopers of the 1st King's Dragoon Guards escorting prisoners, winter of 1838-1839. Three squadrons of this regiment served in Canada from 1838 to 1843. Dark blue pilot coat with brass shoulder scales, long black boots, black gauntlet gloves, black cap with red bag. Print after M.A. Hayes. Private collection. Ph: RC.

Trooper of the Queen's Light Dragoons of Toronto, 1837-1838. Dark blue cloak with buff collar, fur cap with red bag. Plate after Barry Rich. Private collection. Ph: RC.

Winter dress of the 60th, or King's Royal Rifle Corps, 1846 and 1867. The 60th's 2nd Battalion served in Canada from 1844 to 1847 and the 1st Battalion from 1867 to 1877. The winter dress of this regiment was widely followed by Canadian volunteer units. At left, a private in 1846, at centre, an officer in 1867 and at right, an officer in 1846. Print in the 1879 Regimental Chronicle. Ph: RC.

Infantry officer in winter dress, 1848-1849. Print in the Illustrated London News of 17 Februay 1849, after a sketch sent by a correspondent in Montreal. Courtesy Library and Archives Canada, C2729.

"Quarters of the Army. (Quebec). 20 below zero 6 A.M." This print of a private of the 2nd Battalion of the Rifle Brigade which served at Quebec during 1847-1848 gives an idea of the kind of exposure to the cold sentries had to endure. Courtesy Library and Archives Canada,

Infantry officer, New Brunswick Militia, c. 1863-1868. He wears the dark blue double-breasted frock coat with gilt buttons underneath a grey greatcoat with a wide cape; a black glazed cover is worn over his cap. Courtesy Andrew Moss collection. Ph: CL.

The Montreal Volunteer Cavalry, most likely the 1st Troop, in winter dress, 1866. This humorous rendering by British army Capt. Francis G. Coleridge show the "command" for the troopers to blow their noses in their handkerchiefs on a winter day. They wear fur caps with a red bag, dark blue coats trimmed with black fur, cords and buttons with dark blue trousers with two white stripes. The 1st Troop obtained such fur caps and coats in the fall of 1861. Courtesy Library and Archives Canada, PA102551.

3. Pantaloons of fine pilot-cloth — regimental colour, if to be got, if not blue.
4. The regimental forage-cap, with band and ear-covers of fur, is the best and cheapest winter piece. The adjuncts are to be had in any Canadian town.
5. Blue Indian-rubber cloak, with sleeves.
6. Do. do. haversack.
7. Waterproof Wellington boots (Campbell and Page, Pall Mall East) — ditto goloshes, well up to instep. N.B. — Winter boots and shoes to be large, so as to wear the thickest sort, or even two pair, of socks. Connemara socks the best."

The "hints" also added that the Canadian blanket-coat (or "capot") was a "most confortable vestment in quiet cold weather." The officers of the 43rd Light Infantry, who sported fox fur "as boas", had already taken the hint: they wore white blanket coats trimmed with "deep blue borders, scarlet wings, seams, etc.," in 1838. In 1842, officers of the 52nd Light Infantry indulged in a "new uniform for Canada, from all accounts, should be very handsome, and most useful. A fur cap, a good pea-coat, with waistcoat underneath, and a great coat — all both comfortable and ornamental." Eventually, the leading favourite style was a thick grey or blue cloth double-breasted coat with astrakhan fur collar.[75]

It was no doubt bearing this in mind that, finally, during the summer of 1845, the Commander of the Forces in British North America appointed a Board of officers to determine a common winter uniform for officers. The approved conclusions of the Board were issued as general orders at Montreal on 30 August of the same year. Infantry officers were ordered to wear:

> "Cap — To be Wedge shaped, eight inches high, to close flat when off the head, to be made of dark Otter Skin, Peak of the same material, with ear covers, and turn down over the neck.
> Coat — Grey double breasted Frock Coat as nearly as possible the same colour as the Great Coat worn by the Soldier, — the Collar to be turn'd down, lined with Grey Astrackan Lamb Skin, — and made sufficiently deep to turn up over the face and ears, if required. Button — Black Horn, with Crown and Regimental Number, or as may be Established Regimentally. — Black Sling sword Waistbelt. — Sash of the Regular Pattern.-
> Winter Boots — Of unblacked Leather of the Country to come up as high as the bind of the knee for company officers, for Mounted officers to come over the knee, and to be worn at the discretion of officers Commanding Regiments.

[75] "Hints to officers embarking for Canada. By Bugle.", *United Services Journal*, April 1839, pp. 518-519, 521. I am indebted to Philip Haythornthwaite for this reference; R.G.A. Levinge, *Echoes from the Backwoods* (London, 1846), p. 243. The 43rd marched from Fredericton, New Brunswick, to Quebec City leaving on 11 December 1837 and arriving 18 days later. Levinge reports, on page 114, wearing "pairs of socks...then mocassins, over which I had large worsted sort of long boots, P trowsers, Thick PP coat, over that a coat made of seal-skin, rather outré but very warm and comfortable; then a buffalo skin muff, fur cap, ear-covers, and lots of gloves, mits, boas, etc." I am indebted to Francis Back for this reference. Captain Jarvis, 52nd Light Infantry, "Regimental Soliering 1840-1843", *The Oxfordshire Light Infantry Chronicle*, 1905, p. 228.

Gauntlets [gloves] of dark otter skin which will prevent the necessity of having fur round the Cuff of the Frock Coat."

Greatcoat for NCOs and sappers of the Royal Engineers, 1866. The cut of the greatcoat was the same for all foot troops of the British and Canadian forces. The colour could vary. For engineers and the artillery, the greatcoat was dark blue with red piping at the shoulder and waist straps. For infantry, the greatcoat was grey without piping. Plate from the 1866 Equipment of Royal Engineers. Ph: RC.

On 26 September, officers serving on the General Staff were ordered to wear black Astrackan caps with ear covers and peak of the same, five and a half inches high and the crown nine inches in diameter; blue cloth double breasted greatcoats with five black cord loops on the breast "and one row of olivet buttons on each breast," black lamb skin lappels and stand and fall collar 5 inches deep; black lamb skin gauntlet gloves.[76]

The above 1845 regulations remained the standard dress for infantry and staff officers into the 20th century with usually fairly minor variations such as cord added on the breast and fur at the cuffs of all infantry officers' greatcoats from the late 1850s.

Occasionally, a rifle regiment would be posted to Canada. At first, there was little to distinguish their winter dress from that of the line infantry; officers of the 2nd Battalion King's Royal Rifle Corps, in Canada from 1844 to 1847, wore a frock coat of dark grey or black cloth trimmed with black Astrackan fur, matching fur cap and mitts, and long black boots. The enlisted men wore a dark grey greatcoat and black cap with peak and ear flaps. The men of the Rifle Brigade's 2nd Battalion, in Canada from 1843 to 1852, wore the same type of greatcoat, but with a wedge shaped fur cap. When the 4th Battalion of the King's Royal Rifle Corps arrived in 1861, the officers took to wearing black patrol jackets trimmed with black Astrackan fur as well as a round black fur Busby, a dress also adopted by the officers of the 1st Battalion in 1867 except for a black fur wedge cap. The 1st Battalion of the Rifle Brigade marched overland from Saint John, New Brunswick, to Rivière-du-Loup, Québec, during January 1862 wearing "great-coats, fur caps and moccasins...the cape of the great-coat being turned up and tied with a woollen conforter outside. Over all, a blanket with a hole cut for the head as a 'poncho'...".[77]

The Royal Canadian Rifle Regiment, a regular regiment formed from 1840 in Canada from veterans of the British Army that served mostly along the border with the United States until disbanded in 1870, was a model to many Canadian volunteer rifle units of the 1850s and 1860s. Its officers initially adopted a "grey great coat made double-breasted, with grey fur collar and cuffs" defined in 1848 as the type "directed generally to be worn by all Officers of Infantry." By the 1860s, the winter uniform for rifle officers consisted of a Busby of "astrakhan or other fur with covers for the ears", a double breasted grey greatcoat with horn buttons and trimmed with grey astrakhan fur, "Canadian" boots and fur gloves. The rifle's enlisted men had the same type of grey greatcoat as did the line infantrymen and were also issued with fur cap, flannel waistcoat, drawers, and a pair of long boots. The cuffs and collars of the sergeant's greatcoats were of dark green cloth in rifle regiments.[78]

Until the middle of the 19th century, artillery officers generally wore the same winter dress as infantrymen: grey greatcoats, but with brass buttons along with fur caps and mitts. Thereafter, Royal Artillery officers opted for dark blue greatcoats, trimmed with black fur from about the late 1840s. This winter dress was sanctioned officially

---

[76] LAC, RG8, C1194B, General Orders, Head Quarters, Montreal, 30 August 1845.

[77]Lewis W. G. Butler, *The Annals of the King's Royal Rifle Corps* (London, 1913), vol. 6, pp. 34-35; Sir William H. Cope, *The History of the Rifle Brigade (The Prince Consort's Own) formerly the 95th* (London, 1877), p. 458.

[78]The *Regulations for the Dress of General, Staff, and Regimental Officers of the Army, 1st April 1846* (London, 1846) footnote on p. 154, mentioned that the officers of this unit already had this winter dress; WO 3/107, Adjutant General,Horse Guards, 26 September 1848; War Office, *Army Equipment, Part V, Infantry, June 1865* (London, 1865), p. 32; *Standing Orders of the Royal Canadian Rifles, issued May, 1861* (Montreal, 1861), pp. 30-31.

(Left) Private, 21st Richelieu Light Infantry Battalion, winter dress, c. 1865. This French print after Sinnet shows the typical dress of a volunteer soldier of the "Troupes du Canada" on active service in winter. He wears a grey greatcoat with is cape, mitts, winter boots and fur cap. The artist in Paris may have received the information to make the illustration from a correspondent in St. Jean, Québec. Anne S.K. Brown Military Collection, Brown University Library, Providence, USA. Ph: RC.

(Above) Lt. J. Peter, 10th Battalion, c. 1866. Grey greatcoat with grey cords and olivets, trimmed with light grey Astrackan fur and fur cap. Courtesy The Royal Regiment of Canada Museum, Toronto.

(Left) Officer, 4th Troop, Montreal Volunteer Cavalry (Royal Guides), winter dress, c. 1862-1866. Dark blue jacket with black cords, lace and fur trim, black fur cap with cap lines and chinstrap, dark blue pantaloons strapped with black leather and double white stripes, black boots. The distinctive pouch belt worn by officers of the Royal Guides was of red or crimson leather laced with silver including a central wavy line. Courtesy Department of National Defence, Ottawa.

(Right) Lt. Joseph McKay, 41st Brockville Rifles, c. 1866-1870. Very dark grey winter coat with black cords and Astrackan fur. Black belts with silvered badge, whistle and chain. McKay went on to become lieutenant-colonel of the 41st. Courtesy Library and Archives Canada, C29316

(Left) Infantry officer, winter uniform, c. 1866. Grey greatcoat with grey cords and olivets, trimmed with light grey Astrackan fur and fur cap. Courtesy National Historic Sites, Parks Canada.

(Right) Infantry officer in winter dress, c. 1870-1873. He wears the 1870 officer's pattern greatcoat with the trefoil ended cords. Note that the sash is worn over the greatcoat. Print in Capt. Seccombe's Army and Navy Drolleries published in London, c 1873. Ph: RC.

(Below) Lt. Col. Charles T. Gilmor, 2nd Queen's Own Rifles, c. 1880. Very dark grey winter coat with black cords and grey Astrackan fur. Black belts with silvered badge, whistle and chain. He wears the pre-1870 pattern winter coat with grey instead of black fur specified from 1870 for rifle regiments. Painting by J.W.L. Forster. Courtesy Queen's Own Rifles, Toronto. Photo: C. Lawson.

(Below) Gunners of B Battery, Royal Canadian Artillery, 1873. The event is a shooting competion on the iced up St. Charles River at Quebec City. All wear greatcoats that would be dark blue and dark fur caps. The field guns are mounted on sleds. Canadian Illustrated News, 19 April 1873. Courtesy Library and Archives Canada, C59098.

Snowshoes were not a general issue, but the permanent regular artillerymen had them in the early 1870s. Detail from a print in the Canadian Illustrated News, 21 February 1874. Ph: RC.

on 15 March 1862:

"Clothing of [Royal Artillery] Officers serving in North America to be:— Frock-Coat, dark blue cloth, double breasted with rolling collar, black astrachan fur on collar, breast, and cuff, braid same as regimental frock-coat. To be worn as an overcoat with scarlet scarf or muffler. Cap: black astrachan fur, 4 inches deep in front and 5 inches behind. To be worn with ear-flaps in cold weather. Boots: Canadian. To be worn over the trousers, and to reach the knee. *Gauntlets*: black astrachan."[79]

Initially, enlisted artillerymen usually wore the same winter dress as infantrymen: grey greatcoats, but with brass buttons along with fur caps and mitts. In 1851, their greatcoats became dark blue, trimmed with scarlet collars for those having the rank of Sergeant. By 1860, the privates had the scarlet collar as well. There appears to have been little change thereafter for artillery officers or enlisted men until the turn of the century.[80]

For the mounted field batteries, the 1860s Royal Artillery regulations and the 1876 Canadian ones specified a dark blue cloak with sleeves and fall collar whose skirt was "to reach within 8 inches of the ground." In 1883 for British artillery officers and 1886 for Canadians, "Shoulder straps of same material as garment; a small button of regimental pattern on top. Badges of rank in gilt metal" were added. There was also a dark blue cloth shoulder cape that could be attached to the cloak.

For officers of the Royal Engineers, and their enlisted men, called Royal Sappers and Miners until 1855, the winter uniform was, by and large, the same as that of the artillery. The greatcoats of the men were grey with brass corps buttons up until 1853 when they became dark blue, with scarlet collars for Sergeants. Privates began to wear the scarlet collar in 1861. This winter uniform remained practically the same until 1902.[81]

Few British cavalry regiments served in Canada. From 1838 to 1843, the 1st (King's) Dragoon Guards and the 7th Hussars served in Canada. The very ornate dress of cavalry at that time proved utterly unsuitable for winter and a warm "uniform" was adopted. This consisted, for the King's Dragoon Guards, of a fur cap with ear flaps, a dark blue double breasted pea jacket, fur mitts and long boots lined with sheepskin. The 7th Hussars had a similar dress and also used their fur-edged pelisses. Cavalry units also had the long cavalry cloak of the unit's coat colour with its collar of the facing colour that protected the horse as well as the rider. From 1840, cavalry cloaks sent to Canada were made with sleeves, which were eventually added to all British army cloaks. The next, and

[79]Quoted in R.J. Macdonald, *The History of the Dress of the Royal Regiment of Artillery, 1625-1897* (London, 1899), p. 105.

[80]D. Alastair Campbell, *The Dress of the Royal Artillery* (London, 1971), p. 49; watercolour dated "February 1850 Montreal" in the David M. Stewart Library, Montreal. Gunners on a forced march to Quebec in March 1863 were given "abundant and excellent" clothing: "Each man had a fur cap with ear-lappets, a woollen comforter, a chamois waistcoat, and a flannel shirt; warm gloves, thick woollen stockings, and moccasins instead of boots. They always wore great-coats, and packed their knapsacks and carbines in their sleighs." Francis Duncan, *Our Garrison in the West or Sketches in British North America* (London, 1864), p. 233. I am indebted to Francis Back for this reference.

[81]P.H. Kealy, "The uniforms of the Corps of Royal Engineers up to 1914", *Royal Engineers Journal*, June 1934, pp. 186-208; December 1934, pp. 504-513; September 1935, pp. 403-418. We have not seen a specific reference to the Canadian winter dress of the officers but it was in all likelihood similar to that of the officers of the Royal Artillery.

last, British cavalry unit to serve in Canada was the 13th Hussars from 1866 to 1869. The officers had a dark blue pea jacket braided and trimmed with black Astrackan fur, dark blue thick cloth pantaloons with two white stripes, long boots, black astrakhan fur gauntlet gloves and fur cap with white crown embroidered with gold "and the fur cap made loose so as to be well pulled over the ears." The enlisted men probably had the same type of dress but without the astrakhan fur and gold lace.[82]

Royal Canadian Artillery, winter 1885. The gunners are shown on a gun with its limber mounted on sleds, wearing dark blue greatcoats, red scarves covering the red collars, buff winter moccasins and mitts, fur caps with red bag and brass grenade badge. Canadian Army Journal, April 1955. Ph: RC.

Canadian Militia and Volunteers 1830s-1870

During the French Regime and the early British colonial period, militiamen usually wore their own civilian winter clothes. Following the adoption of the grey greatcoat by the regular army in 1801, the authorities would issue these to units of Embodied Militia on regular service. Men in Sedentary Militia units, which were called up only for short periods, would wear their own civilian winter clothes over which were worn the cross-belts (and musket) issued in emergencies.

During the Rebellions of 1837-1838, numerous corps of Loyal Volunteers were raised and kept in service for varying periods of time that often occurred during the winters of 1837-1838 and 1838-1839. In Quebec City, most of the volunteer units adopted blanket coats without hoods and trimmed in various colours for distinction, which made for a very handsome appearance as seen in contemporary artwork. In Montreal, the volunteers usually opted for the dark blue pea jacket and this garment was seen as well in Kingston and many other communities. These types of coats represented *both* the "regimental" distinctive uniform for the men who wore them as well as the protective winter clothing. At the same time, thousands of volunteers on duty were also issued the standard British army winter dress consisting of grey greatcoats and cavalry capes rounded out with fur caps, mitts, beef boots, scarves, etc. Some units in Lower Canada were issued grey greatcoats that doubled as their "regimentals" during the winter of 1837-1838, notably Col. Heriot's thousand men strong Eastern Townships Volunteers. In Upper Canada, the military grey greatcoat was issued to a number of infantry units embodied in late 1837 such as the 1st and 2nd Frontenac and the 1st and 2nd Addington regiments at Kingston. On the warmer Niagara Frontier, the Queen's Niagara Fencibles had the usual fur caps, mitts, etc., but no greatcoats. The battalions of Incorporated Militia, which served from 1838 to 1843, had the standard British infantry grey greatcoat.[83]

Blue pilot cloth jackets were also worn as winter clothing in addition to standard uniforms, notably by volunteer cavalry troops. For instance, the Quebec Light Cavalry was seen wearing fur caps and dark blue pi-

[82]John Luard, *A History of the Dress of the British Soldier* (London, 1852), p. ix; LAC, RG8, C1036, p. 168; Michael Mann, *A Particular Duty* (Salisbury, UK, 1996), p. 164. A. Shirley to Military Secretary, Montreal, 29 August 1842, complains that the pea jackets sent to the 7th Hussars were unsuitable in the previous three winters; *The Volunteer Review and Military and Naval Gazette*, Ottawa, 4 March 1872.

[83] McCord Museum, Montreal, J.S. McCord Militia & Rebellion Papers, Box 3. Lt. Col. Griffin to Col. McCord, 20 November 1838; LAC, Picture Division, Powell's Scrapbooks, vol. 1; Sir Richard Bonnycastle, *Canada, as it was, is, and May Be* (London, 1852) Vol. 1: pp. 310-311; LAC, RG9, IB1, vol. 25. Board of Officers to Governor General [May 1838]. The Queen's Niagara Fencibles were to have red coats faced blue but obtained grey because of a shortage of red cloth. WO 1/538, Militia General Order, Toronto 31 October 1838.

Royal Canadian Artillery winter gunnery practice, 1888. The gunner wear dark blue greatcoats. It must have been a mild day as the officers wear patrol jackets complete with swords and sabretaches. Illustrated London News, 14 April 1888. Ph: RC.

Lt. Col A. McKeand, 90th Winnipeg Rifles, c. 1885-1887. He wears a non-regulation but quite attractive Astrackan fur coat with gauntlet gloves and cap of the same fur; an example that some commanding officers took some liberties to better the dress regulations. He is outfitted for mounted duty as seen by his sabretache and boots. Lt. Col. McKeand commanded the battalion during the 1885 North-West Campaign. From Chamber's history of the 90th Winnipeg Rifles. Ph: RC.

Cavalry officer's winter dress, 1890. From left: Lt. W.P. Stull, 2nd Cavalry; Capt. E.H.T. Heward, Royal School of Cavalry; Capt. C. McArthur, 6th Cavalry; Lt. Col. J.F. Turnbull, Royal School of Cavalry; Lt. A. Binnington, Prescott Troop of Cavalry; Capt. E.F. Lessard, Royal School of Cavalry; Lt. A.D.K. Servos, 2nd Cavalry; Lt. A.F. Clerk, 6th Cavalry. Note the cords on the coat and different cap for the 6th Cavalry. Officer's Mess, Royal Canadian Hussars, Montreal. Ph: RC.

Officer's winter jacket, Royal Canadian Artillery, c. 1895. Dark blue with scarlet shoulder straps bearing a crown (for the rank of major), black Astrackan fur covering the collar and edging the front and the skirts. The lace on the chest and at the cuffs was originally black, but has discoloured to a beige hue. This pattern of jacket was approved for officers of the permanent batteries on 26 December 1889. The scarlet shoulder straps were approved for this jacket on 17 November 1893. Fort Henry, Kingston. Ph: RC.

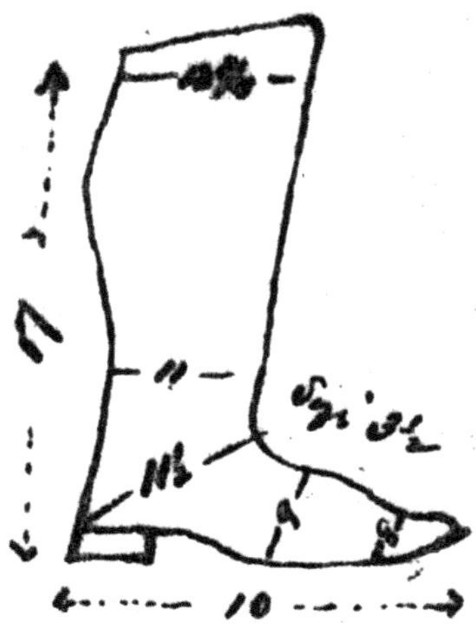

(Left) Sketch for a contract for winter (top) and ankle (bottom) boots for, in this instance, Canadian Artillery gunners in June 1897. These items were likely the same for other non-mounted units. Courtesy Library and Archives Canada, RG9, IID1, vol. 10.

(Right) Officer's winter grey greatcoat trimmed with fur, 1890s. The fur on the collar and open lapels has fallen off and the black cords and olivets have faded, but this garment remains in remarkable condition. Royal Canadian Regiment Museum. Ph: CL.

Winter dress, early 1890s. At left, a Royal Military College cadet wearing the infantry officer's pattern grey greatcoat trimmed with pale grey fur with pale grey fur mitts and cap. At centre, a Royal Canadian Artillery detachment aiming a field piece mounted on a sledge carriage; they wear the greatcoat, fur mitts and black fur cap with red bag and brass grenade badge. At right, a Royal Canadian Infantry Regiment private wearing the grey greatcoat with black fur cap and gloves. Illustrations by A.H. Heming published in Harper's Weekly on 3 March 1894. Ph: RC.

Lt. Col. William Otter, Royal Canadian Regiment, c. 1895-1900. He wears the officer's grey greatcoat with grey Astrackan fur on the collar and open lapels, black cords and olivets. The gold aiguillette indicates a staff appointment. Private collection. Ph: RC.

lot coats with shoulder scales. Troop orders of the Frontenac Light Dragoons given at Kingston on 14 December 1837 called for "Winter Clothing" consisting of "a plain Blue Pilot Cloth or Flushing round about Jacket, single breasted, without pockets, and one row of Ball buttons with Prussian Collar with hooks and eyes. Pantaloons of the material to button in front with welt to cover the buttons, without pockets, buttons for straps at bottom - A fur Scull Cap with leather Peak." Toronto's York Dragoons had become the Queen's Light Dragoons in 1837 and placed on active service. It was issued with "good serviceable [dark] blue cloaks…with a fur cap" that was "a sort of imitation dog skin; they looked as if made from an oblong piece of fur, doubled at its centre and stitched up each side with a bag of red cloth with a tassel on one side, somewhat like the present busby bag" recalled Capt. Denison.[84]

In September 1839, Montreal's Queen's Light Dragoons contracted for making a cold weather uniform in the following terms: "...Forty blue cloth coats and a like number of trowsers of similar cloth to the exact fit of the men....the whole according to pattern and similar to the coat and trowsers made for trooper Michael Reilly, of said Queen's Light Dragoons... coat and trowsers will be fur trimmed...This agreed that should the fur-trimmings...exceed ten shillings, the surplus will be paid..."[85]

Later on, in 1862, the 1st and 4th Montreal Cavalry troops reported a winter dress consisting of dark blue jacket trimmed with black fur and cords and black fur caps. The 4 March 1872 issue of the *Volunteer Review* featured a description of the "winter costume for officers of Cavalry in Canada" that consisted of a "dark blue pea patrol jacket braided and trimmed with black astracan fur — black astracan gauntlets, thick blue cloth pantaloons, with two white stripes…, long boots and jack spurs with steel understep, embroidered gold lace scarlet waistcoat (or white do., as worn by the 13th Hussars when in this country) a black astracan fur cap…with scarlet or white crown embroidered with gold and gold buttons — no cap lines, and the fur cap made loose so as to be well pulled over the ears in cold weather, undress snow belt and black patent leather Sabretache undress pouch and belt when mounted. The long cavalry cloak with removable cape, can, if required, be worn over all, the cloak protects the horse and rider. This is the uniform worn by the Quebec and Montreal V[olunteer] Hussars." As seen in an 1890 image, although the above colours were used, officers of most units chose to have the Hussar's dark blue patrol jacket trimmed with black fur and its black flat and wide lace on the breast, although the 6th Cavalry had black cords instead. The 1890s NCOs and seemingly the enlisted men of the Montreal Hussars had dark blue coats trimmed with black fur down the front and black fur caps with white bags. Otherwise, the standard greatcoats and cloaks generally followed British army patterns, but there could be some differences. Cloaks made from September 1896 for the Royal Canadian Dragoons generally followed the British army cloaks, but there were some differences in the Canadian pattern. The Canadian cloaks were of thick wool cloth with serge lining, 45 to 51 inches long in front, brass buttons six inches distant from each other, 28 to 32 inches for length of back opening, collar five inches deep closed with one set of hooks and eyes (three sets on the British), pockets cut diagonally (horizontally on the British), cape 18 to 20 inches long in front, two and a half inch wide belt by 47 to 51 inches.[86]

[84]LAC, RG8, C1203 1/2, Troop orders of the Frontenac Light Dragoons, Kingston, 14 December 1837; Frederick C. Denison, *Historical Record of the Governor-General's Body Guard* (Toronto, 1876), pp. 10-11.

[85]Archives nationales du Québec à Montréal, greffe Etienne Guy, 2530, Agreement between Lieutenant John Shaz, Queen's Light Dragoons, and William Adams, 25 September 1839.

[86] LAC, RG9, IC4, Vol. 2, Militia General Order, Toronto, 11 October 1856, offered British army greatcoats at half price but few units reported this item in 1862 although a number mostly in Canada East (Québec) mentioned wedge caps. The 1st and 4th Montreal Cavalry

A trooper and an officer of the Royal Canadian Dragoons, 1896. The trooper standing wears the grey long-skirted greatcoat, fur cap with yellow bag and holds a Martini-Henry rifle. The mounted officer wear a shorter grey coat tried with fur at the collar, cuffs and bottom of the skirt. Cutting from an unknown period publication. Vinkhuizen collection, Draper fund. Courtesy New York Public Library/Wikimedia.

A Militia General Order given at Toronto on 11 October 1856 offered 4,000 British army greatcoats at half price sent to Canada for the "service of the Active Force of the Province", but this was not taken up by most volunteers. Few volunteer units reported this item in 1862, although some, mostly in Canada East (Quebec), mentioned winter wedge caps. From 1863 on, the basic uniform was issued to enlisted volunteers and a Militia General Order of 10 April mentioned that greatcoats could be purchased for $4 each. Again, no doubt due to cost, most volunteers don't seem to have had military greatcoats. These supplies were similar to those used in the British army described above. Fur caps, mitts, scarves and winter boots were worn with the greatcoat. The British army made some changes to its greatcoats that were noted in Canada by 1867, but those issued to Canadian volunteers continued to be the older version in spite of the apprehended "chagrin this will cause among the men." These were grey except for the volunteer artillerymen who, like the Royal Artillery, were to have dark blue greatcoats. In February 1866 however, there were not enough in the Montreal stores and 750 grey greatcoats were issued to the gunners instead along with 460 dark blue ones. More were later ordered, no doubt in both colours. In 1868 for instance, 5000 "Old pattern greatcoats of present quality of imperial government cloth" were purchased in England at 18 shillings each. Another 5,000 grey greatcoats were urgently ordered from England "by telegraph" on 12 April 1870 that were partly intended for the troops of the Red River Expedition. Until 1871, stock on hand from the British army could be turned over to Canada as well purchase of the Militia Department. Thereafter, convicts at the Kingston Penitentiary made greatcoats, seemingly of the old pattern.[87]

Additional items such as snowshoes or buffalo fur coats could be seen in certain instances. The officers of the Canadian Volunteer Militia and the Permanent force obtained winter clothing generally similar to that worn by officers in the British regular service described above. For instance, the 2nd Queen's Own Rifles' regimental orders of 18 December 1862 specified that its officers wore "a dark grey coat with grey fur facings; forage cap with a band of grey fur and black leather peak; knickerbockers reaching to the knee and long boots." The unit changed it in October 1867 for "a black cloth patrol jacket, trimmed with black fur, black fur cap, dress trousers and short leggings." That year, the Toronto Military School cadets appeared in "neat winter uniform" and "fur hats instead of the ordinary forage caps. The new hats are made of bear skin, and although not very neat in appearance, are yet reasonable" commented a journalist.[88]

According to the 1 May 1866 *Standing Orders of the Quebec Garrison Artillery*, the officer's winter dress for its four batteries was a "dark blue cloth cap, with band of Astracan fur, 4 inches high in front, 5 inches be-

troops also reported a winter dress of dark blue jacket trimmed with black fur and cords and black fur caps as per Commanding officer's reports, 1862, LAC, RG9, IC8, vol. 3; LAC, reel T382/809.

[87] LAC, RG9, IC8, Vol. 5, unknown to McDougall, 13 March 1866 adding that the negative feeling would be "strengthened by the fact that many of the lowest class – the very 'roughs' of Toronto are wearing the old pattern military overcoat which has been largely obtained by them either by thieving, charity or purchase from among the condemned military stores…"; Military Secretary's Office, issue notes for January and February 1866; LAC, RG9, IC1, Vol. 110, Walker Powell to Peter Tait & Co, 24 August 1868; LAC, reel T4173/300, Deputy Minister's ledgers, April 1870. In Canada, grey or blue greatcoats were regulation until 1903 when future issues were specified to be drab.

[88] Ernest J. Chambers, *The Queen's Own Rifles of Canada* (Toronto, 1901), pp. 48, 70; *Volunteer Review*, 14 January 1867.

A field detachment of "B" Battery, Royal Canadian Artillery, in Quebec City during March 1897. All are in mounted order about to move out with their field pieces mounted on the sledge carriages used by Canadian gunners. They wear the winter uniform, but without the scarves and ear muffs as the weather gets milder in March. Note the mounted officer dressed with the 1889 pattern winter jacket. Fort Henry, Kingston. Ph: RC.

hind, 23 inches in circumference on top, with gold button and braid on crown, same pattern as forage cap. Pea jacket of blue pilot cloth, trimmed with Astracan fur, frogs and braid. Trousers, undress. Crimean boots. Sword worn under jacket; Gloves, White." The men's overcoat was "as issued by Government. Cap, Blue cloth with Astracan fur band. Gloves, White woollen."

Canadian Militia and Volunteers 1870-1901

The gunners of the School of Gunnery regular batteries organised from 1871 had black fur caps with a brass grenade in front. The fur caps initially had a round shape, but this evolved into a wedge shape with red bag and brass grenade in front by the 1880s. The enlisted gunners had dark blue greatcoats, long black winter boots, mitts as well as moccasins, snowshoes and red scarves. All Canadian artillery officers were assigned similar winter uniforms to Britain's gunners consisting of a dark blue, double-breasted greatcoat trimmed with black astrakhan fur, black Astrackan fur cap and gauntlet gloves, long boots reaching the knees and a red scarf. The 1876 and 1886 Canadian dress regulations also prescribed the Royal Artillery's dark blue cloak with its cape. For the permanent force artillery batteries, a new officer's winter jacket was approved on 26 December 1889. It was to be "exactly the same as the patrol jacket described in Dress Regulations, but will in addition be trimmed all round with black Persian lamb or Astrachan fur — the cuff excepted — stand and fall collar of fur lined with cloth five inches deep. A false collar of the same pattern as the stable jacket will be worn with this jacket."[89]

Since volunteer militia units often had to obtain winter dress at their own expense, some variety was predictable from the British and Canadian regulation patterns although everyone insured that the clothing would have a military look. From the 1850s, as seen above, the officer's winter coat was trimmed with black Astrackan in the case of dark blue coats (artillery, etc.) and grey Astrackan for grey coats (infantry, rifles, staff, etc.). A Militia General Order of 25 October 1870 specified that the "Winter Great Coat for Officers of Infantry, to be made double breasted of cloth of the same colour as that worn by the men, skirts four inches below knee, cuffs and collar of Grey Astrachan; buttons regulation pattern. Officers of Rifle Corps to wear the same description of Coat, substituting Black Astrachan for Grey, on cuffs and collar." From 1870, winter greatcoats with fur collar and cuffs, had a slight alteration in the placement of the cord decorating the breast, which were to have trefoil ends. This was also adopted in the British army for the fur-trimmed greatcoats of British officers serving in Canada. Large gauntlet fur mitts were also very popular at that time.

The 1 May 1886 Canadian dress regulations described the "optional" winter greatcoat (which had already been assigned to the Infantry School Corps by a Militia General Order of 12 December 1884) as follows:

> "Milled cloth ([dark] blue for General and Staff Officers' and Officers of Artillery, [dark] grey for other officers); double-breasted, with lapels, 4 1/2 inches wide at the top, and 2 1/2 inches at the bottom; the

[89] In winter staff sergeants and sergeants were allowed to wear dark blue double-breasted pea jackets with regimental buttons and chevrons on both sleeves as per regimental order, 6 December 1873 as per LAC, RG9, IIJ1, Vol. 21. For instance, capes could be worn over the greatcoat when the temperature was very low (Regimental Order , 19 January 1882); ibid, 26 December 1889.

Members of the Yukon Field Force, 1898. From left, an officer and a sergeant of the Royal Canadian Artillery in winter dress, an officer of the Royal Canadian Dragoons in a cool weather dress wearing a fur cap with the scarlet serge frock, and a private of the Royal Canadian Regiment in special winter dress suited to the Yukon inspired from Inuit clothing. Plate by R. Marrion. With the kind permission of the Company of Military Historians. MUIA 346. Ph: RC.

Trooper, Royal Canadian Dragoons, c. 1898. In winter dress wearing a buffalo coat and a fur cap. Navy and Army Illustrated, 2 April 1898.

Sergeant, Duke of York's Canadian Hussars, 1899. Winter dress. Royal Canadian Hussars Sergeant's Mess, Montreal. Ph: RC.

Officer and private, Royal Canadian Regiment, c. 1898-1900. Dark grey greatcoat trimmed with dark Astrackan fur and black cords for the officer who also has his crimson sash at the waist, plain greatcoat for the private soldier with a scarf around his neck. Both have the fur cap with the regimental badge. Private collection. Ph: CL.

Private, Royal Canadian Regiment, 1899. He wears the greatcoat also issued to infantry sent to South Africa. Courtesy The Army Museum, Halifax Citadel.

shape as for frock-coat; to reach within a foot of the ground. For mounted officers the coat will reach to the ankles. The body, skirt, and sleeves lined with real or artificial fur, optional; the collar and cuffs on the outside and the lapels on the inside lined with black Astrachan fur for General and staff Officers, Artillery and Rifles, grey Astrachan for other officers. The Astrachan forms an edging 1 inch wide on the inside of the collar and down the lapels, on the outside, to the waist. Four loops of 1/4 inch black flat cord down the front on each side. Each loop forms an eye in the centre at the top and bottom and a crow's foot at the end; a black netted olivet on each loop near the crow's feet, two olivets at the waist behind. Pocket inside the left breast. Pockets in the plaits of the skirts behind. Three hooks and eyes on the collar. Shoulder straps of the same material as the garment; a black netted button at the top. Badges of rank in gold for winter wear."

However, while regulations mentioned astrakhan fur, photographs of infantry and mounted infantry officers show greatcoats trimmed with what appear to be bear's or otter's fur, which was easily procured in Canada. Its hue was usually darker than astrakhan and would be in various hues of brown. The fur caps matched the coat's fur trim. Indeed, a photo of the Mounted Infantry Corps shows it had a peculiar type of greatcoat heavily trimmed with fur and without cord for officers, all wearing a large fur cap and gauntlet mitts, the whole outfit obviously meant to provide protection from the extreme cold in some parts of Canada. In such cases, fur coats (often of Buffalo) were resorted to and also Inuit-style clothing.[90]

By a Militia General Order of 12 December 1884, the Infantry School Corps' fur cap (otter's fur for officers) had a "blue bag and badge at the end of bag." A Militia General Order of 7 October 1887, the specified for permanent corps "winter caps for Staff Sergeants and Sergeants of all arms, except corps stationed in Manitoba" that were to be "of Coon skin according to pattern approved…wedge shaped of Russian lamb, so made as to permit their being drawn down over the ears, no ear flaps; cloth bag on right side, for Cavalry buff, for Artillery scarlet, and for Infantry blue. No gold lace or other trimming on the bag." This would have been followed in those volunteer units that bought their winter dress. Badges were more or less officially added. During the 1890s, the colours of the fur cap bags continued to be as above with, from 1892, the addition of yellow for Royal Canadian Dragoons.[91]

Some volunteer units procured winter dress at their own cost. This was obviously fairly similar to what the regular troops wore as described above, but some units seeking distinct features could take some liberties on details. For instance, Toronto's 10th Royal Grenadiers "black wedge" winter caps had their blue bags "edged with scarlet" according to the *Canadian Militia Gazette* of 28 January 1892.

The gentlemen cadets of the Royal Military College in Kingston had a winter uniform that was generally inspired from the 1870 and 1886 dress regulations for infantry officers, but the double-breasted dark grey greatcoat trimmed with grey astrakhan had gilt buttons without cords and olivets and the astrakhan fur cap was not as high as the wedge type and had a fold-down astrakhan flap for the ears and back of the head.

---

90 Some of the more prosperous city regiments specified their own winter dress items. The 10th Royal Grenadiers adopted a winter cap for its NCOs and men in early 1892 consisting of "the black wedge with bag edged with scarlet." The bag appears to have been dark blue, the facing colour. CMG, 28 January 1892.

91 LAC RG9, IID1, Vol. 6, Purchase order for Russia lamb fur cap with yellow bag for Sgt. Harris "B" squadron, RCD, 2 March 1896. Simkin's plate of an RCD trooper wearing with a fur coat (taken from a B&W photo) and cap mistakenly coloured its bag red.

Notes

René Chartrand, "Winter Uniforms in Canada, 1665-1871", *Military Collector & Historian, Journal of the Company of Military Historians*, XLIII, 1991; Peregrine Pouchbelt and Rodrick Ramrod, *The Quebec Volunteers*, Quebec, 1839 [prints of the volunteers in uniform by Sir James Archibald Hope].

Prime Minister Sir Wilfrid Laurier biding farewell to the officers and men of Strathcona's Horse on Parliament Hill in Ottawa, 1900. The men wear forage caps and cavalry cloaks (with the five inch collar and without the cloaks on their shoulders), whose rarely seen backs are shown in this illustration by A.S. Boyd after a photograph. Anne S.K. Brown Military Collection, Brown University Library, Providence, USA. Ph: RC.

Chapter 9

Military Education

Four members of a School of Military Instruction, c. 1865. From their age and their batons, they appear to be instructors but no rank badges are visible. Each wears a scarlet serge frock edged white, collar of a darker shade possibly meant to be blue, brass buttons, dark blue trousers with scarlet piping, dark blue cap with a brass badge. Courtesy Scott Duncan.

Military Schools of Instruction 1865-1871

Most Canadians today believe that education and training for an officer's commission in the regular army (or the volunteers) dates from 1876 when the Royal Military College opened in Kingston. It actually goes back much further. From 1685 to 1760, some 280 gentlemen cadets were educated in Canada to become officer for the regular Compagnies franches de la Marine who garrisoned New France. After 1760, because they were Roman Catholic, French Canadians could not serve in the regular British army since commissions were reserved for Protestants although there were a few exceptions to this rule. In any event, there was no huge need for officer training since British colonies were garrisoned by a system that rotated a regular regiments from colony to colony. However, from time to time, informal schools were held to train provincial and militia officers.

The considerable rise in the numbers of volunteer from the early 1860s as well as the increased literacy of the population brought forth the need for having structured courses for officers. During the American Civil War, military schools staffed by British Regulars were established for adult male students to undergo a three month long military course in Toronto and Quebec City in 1864, and at Montreal, Quebec City, Kingston, London and Hamilton in 1865. Although the military colleges in London and Hamilton disbanded in 1865, the schools at Quebec City, Montreal, Kingston and Toronto were retained at Confederation, in 1867. In 1868, schools of cavalry and artillery were formed in Toronto and a school of artillery was formed in Montreal. Since these were not Boarding schools, students lived in the communities.

Camps of instruction were formed in September 1865 at Laprairie (south of Montreal), Quebec City, Kingston, Toronto and London. The cadets were divided into the Right battalion from Upper Canada 397 men, the Centre battalion made up of English-speaking men of Central Canada, 310 men, and the Left or French Canadian battalion with 420 men, for a total strength of 1,127 cadets. Col. Garnet Wolseley, who commanded the camp at Laprairie, reported that he had "no hesitation in saying that they compared most favourably with regular troops" and that they identified "themselves with the army whose uniform they wore, and in my opinion they were most successful in doing so."

(Left) A.H. Gilmour, School of Military Instruction cadet, c. 1865. Scarlet serge frock with dark blue collar edged white, dark blue trousers with scarlet piping, dark blue cap with a badge. Private collection. Ph: RC. (Center) Robert Scirving, School of Military Instruction cadet, c. 1865. Scarlet serge frock with dark blue collar edged white, dark blue trousers with scarlet piping, dark blue cap with a badge. Courtesy Ontario Archives, Toronto. S14947. (Right) Volunteer Militia officer cadet, School of Military Instruction, c. 1865. Although damaged and somewhat faded, this image gives a rare view of what these officer cadets might have looked like in colour. Scarlet serge frock edged white, collar of a darker shade possibly meant to be blue, brass buttons, dark blue trousers with scarlet piping, dark blue cap with a brass badge. Fort Henry, Kingston. Ph: RC.

The 1865 and 1866 cadets were issued a uniform consisting of "a loose serge red frock and…dark trousers" which gave the cadets a "decidedly military appearance" according to a reporter of London's *Macmillan's Magazine*. The red serge frock was a fairly loose-fitting garment with a dark blue collar edged with white, white shoulder cords, four buttons in front and an outside pocket with a pocket flap on the left breast. Trousers were of dark blue serge with red piping. Each cadet also wore a dark blue standard issue Kilmarnock forage cap. The was an "ornament" for the cap and also a fur cap. From various images, the collar appears to have been dark blue, the buttons possibly both of brass and/or of white metal. Many images also show the forage cap bearing a small round badge with a small dart or point at its bottom.

In 1870-1871, from December to May, six schools conducted officer training for cavalry, infantry and artillery. Instructors from the British garrison operated the schools at Halifax, Saint John (New Brunswick) and Quebec while Canadian militia staff and former British army drill sergeants operated the schools at Montreal, Kingston and Toronto. Thereafter, Canadian militia staff replaced the British regulars when the imperial garrison was withdrawn from Canada (except for Halifax). The first regular Canadian units, A and B Batteries at Kingston and Quebec, organised gunnery schools on a year round basis in which artillery courses lasted from three to 12 months with the possibility of extension. After 1866, there appears to have been no special clothing for the cadets as previously and all would have worn their own unit's uniform.

The instructors at these various camps wore, as a rule, the uniforms of the corps to which they were attached. The courses were usually given in service dress that basically consisted, for officers, of the double-breasted dark blue frock coat until1867 when it was replaced by the patrol jacket. There is a possibility that British regular army schoolmasters may have also participated in aspects of the training. According to the July 1865 issue of the *Journal of Education for Upper Canada*, the dress of the schoolmasters was "under strict regulation":

> "Army schoolmasters, being enlisted soldiers, with the rank of non-commissioned officers, are required to appear in uniform. The dress first established viz., a blue frock-coat with braid, silk sash, sword and waist belt, and forage cap with red cloth band, was found to resemble too much the undress uniform of a commissioned officer, and gave rise to inconveniences which were complained of by commanding officers, and even by

(Right) School of Military Instruction officer cadets, instructor and friends at Laprairie, c. 1870. The three officer cadets wear scarlet frocks and forage caps, the instructor has a dark blue patrol jacket and laced cap. Courtesy Library and Archives Canada, PA99976.

(Center) Military officer as instructor, c. 1863-1867. He wears the service dress dark blue double-breasted frock and, if an officer, could add a crimson shoulder sash. From 1863, instructors below commissioned officer's rank did not wear sashes but had rank badges (rifle pattern) on the frock coat to avoid confusion. Print of the 1860s in Punch. Private collection. Ph: RC.

(Bottom left) Military officer as instructor, c. 1867-1890s. From 1867, the instructors usually wore the service dress, which consisted mainly of the patrol jacket with cords until the 1890s when gradually replaced by the serge jacket with pockets. Print of the 1870s in Punch. Private collection. Ph: RC.

Gentlemen cadet's uniform, Royal Military College, Kingston, 1876. This print after Henri Julien was widely published in Canadian newspapers and magazines. The uniform chosen was a scarlet tunic with dark blue collar and cuffs, white piping, gilt buttons and lace, dark blue trousers with scarlet piping, black shako with red and black lace bands, brass chin strap and plate, white plume. Royal Military College of Canada Museum, Kingston, Ontario. Ph: RC.

"The Old Eighteen" in about 1878-1880. These were the 18 first gentlemen cadets at the Royal Military College in Kingston seen here wearing the pillbox cap. They are with the captain of cadets Capt. J.B. Ridout and Sergeant Major John Mortimer, both wearing peaked caps. Royal Military College of Canada Museum, Kingston, Ontario. Ph: RC.

Plate of the Royal Military College cadet's shako, 1876-1879. Royal Military College of Canada Museum, Kingston, Ontario. Ph: RC.

Gentlemen cadet's shako, Royal Military College, Kingston, 1876-1879. Black with red worsted top band, black edged with red worsted bottom band, brass plate and chin strap, white horsehair drooping plume. The shako was replaced by the white helmet at about 1879. Royal Military College of Canada Museum, Kingston, Ontario. Ph: RC.

Presentation of awards to gentlemen cadets by Her Royal Highness Princess Louise on 31 May 1879. The gentlemen cadets are at left (except one at right and another, at the centre background, bare-headed, going to receive his prize) in full dress including their shakos with the drooping white horsehair plumes. The faculty is either in civilian dress or in uniform including, at right of Princess Louise, the commandant, Colonel E.O. Hewett, wearing his uniform of the Royal Engineers and holding a prize. Behind him is a staff officer and the college's chaplain. The other attendants would be mainly relatives and friends. Private collection. Ph: RC.

some of the schoolmasters themselves. In 1863, His Royal Highness the Field-Marshall Commanding-in-Chief, with the concurrence of the Secretary of State for War, was pleased to approve of a dress more in accordance with the military rank of the schoolmaster1 viz., a blue frock coat as before, with chevrons (on both arms) of the colour and pattern of the chevrons worn by rifle regiments. The sash was suppressed."

Gentlemen cadet Battalion sergeant-major's tunic, 1878-1880. This tunic was worn by Gentleman cadet Sergeant-Major L. Homfray Irving at the Royal Military College. It was of the 1868 officer's pattern and is the oldest known Royal Military College tunic known to exist. It is scarlet with dark blue collar and cuffs, white piping, gold lace and buttons marked "Cadet Company / Military College / Canada". The sergeant-major's gold chevrons and crown are placed above the cuffs, three years before this was officially sanctioned in Canada. He was appointed to this rank on 1 September 1878. Royal Military College of Canada Museum, Kingston, Ontario. Ph: RC.

Royal Military College 1876-1901

On 1 June 1876, the Royal Military College of Canada in Kingston, Ontario, opened its doors to its first 18 officer-cadets for a four-year course. It was a success as a military and academic institution, which continues into our 21st century. Thanks to "RMC", competent military leadership for Canada's small regular force and its volunteer militia was assured. It was also the nation's first true engineering school. Since English was the only working language used, youths from French Canada wishing to enter the Royal Military College were required to be fluent in English, a measure that amounted to exclusion that lasted well into the 20th century. A professor of French was appointed from January 1879, but it was one of the many disciplines taught. Nevertheless, the education virtues promoted by the pioneers of the Royal Military College were, and continue to be, outstanding and it is rightly considered as one of Canada's finest universities and one of the best military academies anywhere. As early as 1890, its regulations were shared with Australia for its own proposed military college (LAC, reels T4175/541; T274/1095).

The main models for "RMC" were Britain's Royal Military College and the military academy at Sandhurst and that gave it a strong infusion of British military culture. The dress uniform chosen for the gentlemen cadets was a scarlet tunic with dark blue collar and cuffs, white piping, gold buttons and lace, dark blue trousers with scarlet piping and, initially for full parade dress, the shako then worn in the British army with a white drooping horsehair plume and a gilt star with a crown at its centre and "Military College Canada" around it. The shako was replaced from about later 1879 or 1880 by the white helmet with gilt fittings. For undress, a dark blue pillbox cap piped red with a gold band and lace button on top of the crown was worn on other occasions. This uniform for gentlemen cadets was approved in December 1875 and was imported from Great Britain for many years. From 1897, a scarlet mess jacket was added.

The undress frock was scarlet with dark blue collar and cuffs edged with gold cord and gilt buttons. From 1887, it was replaced by an all-dark blue frock that, after certain minor changes, is still worn today. For sports activities, white sleeved jerseys with dark blue edging at the collar and cuffs with white shorts with two dark blue bands and dark blue knee stockings were worn by the soccer team during the 1880s and white "shaped Gymnastic Vests" were routinely ordered, for instance in October 1898 (LAC, reel T385/783). These, with trousers were white trimmed with dark blue.

The gentlemen cadets had a peculiar ranking system from 1876. At first, the usual NCOs chevrons, but made of gold lace and worn on both sleeves, were used to denote gentlemen cadet ranks:

Lance corporal: 1 chevron
Corporal: 2 chevrons
Sergeant: 3 chevrons
Sergeant-Major: 4 chevrons
Battalion Sergeant-Major: 4 chevrons and crown

In 1882 the chevrons on the sleeves were abolished. They were henceforth replaced by the distinctive system of lace bars on the collar, still in use today:

Lance corporal: 1 bar
Corporal: 2 bars
Sergeant: 3 bars
Sergeant-major: 4 bars
Battalion sergeant-major; 5 bars

Proficiency badges worn on sleeves were similar to those of other Canadian and British military personnel (please see the chapter on rank badges). Furthermore, from November 1899, gold crossed rifles badges on blue ground, without a crown, were used "as Honorary Distinctive Badges for Cadets of the Royal Military College" (LAC, reel T388/174).

Instructors from the regular British or Canadian forces kept wearing their respective unit's uniform. For instance, in July 1878, Sergeant Morgan of the Scots Guards "at present attached to the Royal Military College, Kingston" was to receive a tunic, a white jacket, trousers, boots and gloves from his unit in London (LAC, reel T247/166). The faculty, military or civilian, does not appear to have had academic robes to the college until well into the 20th century. In Queen Victoria's days, the usual dress was the patrol jacket of the professor's corps for the military members of the faculty. The civilian professors wore civilian dress although it is possible that some that were graduates from British or Canadian universities might also have the academic short black gown with their universities hoods in the colour of their field of expertise when giving classes.[92]

Gentlemen cadet Battalion sergeant-major, Royal Military College, Kingston, c. 1880-1881. This image shows the college's gentlemen cadet Battalion sergeant-major's gold chevron (points down on lower sleeve) and crown, crossed guns with another crown for a "Top 5" gentleman cadet in artillery practice and the spur badge showing he is a riding instructor. In theory, the sleeve chevrons would have been worn point up after the GO of 8 April 1881 before the college abolished them altogether in 1882. The uniform is the standard college's dress uniform except for the pillbox cap. The sitter's identity is uncertain as four individuals successively had that rank between 1880-1881. Courtesy Scott Duncan.

School Cadet Corps

School cadet organisations were often are associated with learning institution and this dates back to the French Regime. The earliest cadet corps organised in Canada was at the Quebec Seminary (now Laval University) in Quebec City in the days of New France. Its cadets were mustered for active service during the 1690 and 1759 sieges of Quebec; this unit was humorously nicknamed "Royal-Syntaxe" in 1759. From 1668, its students had

[92]The question of wearing the short gown when giving classes stems from the author's youth. In the mid-20th century; the short gown was worn by professors from England when giving classes at Queen's College, Nassau, Bahamas, which he attended and which was affiliated with the University of London (UK). The full gowns were worn at formal events. In Canada, McGill University's faculty had gowns from the 1820s and, until the 1930s, English Canadian universities customarily prescribed academic robes for its professors and students at classes and lectures. In Canadian and British military colleges, the uniform was the dress worn by military member of the faculty and all gentlemen cadets. In Canada's Royal Military College, distinctive faculty gowns for formal occasions were adopted from 1953.

(Left) Gentlemen cadet's dress (or ceremonial) tunic, Royal Military College, Kingston, 1883. This was the tunic worn by Cadet L.M. Lambe who attended the college from 1880 to 1883. Scarlet with dark blue collar and cuffs, white piping edging the front and the cuffs, gilt buttons and gold lace. The rank system for cadets shown on the collar was introduced in 1882, three vertical bars standing for sergeant, a rank attained by Gentlemen cadet Lambe in 1883. The tunic was otherwise as shown since 1876 until 1890 when the collar was slightly altered and made higher. Royal Military College of Canada Museum, Kingston, Ontario. Ph: RC. (Right) Sergeant Major gentlemen cadet's undress frock, Royal Military College, Kingston, c. 1885. Scarlet with dark blue collar and cuffs, gold braid edging the collar and cuffs, gold cord shoulder straps, gold buttons. Since 1882, the gold bars on the collar indicate the cadet's rank, five indicating the Sergeant Major. Introduced in 1876, this scarlet pattern undress frock was replaced by a dark blue frock in 1887. Royal Military College of Canada Museum, Kingston, Ontario. Ph: RC.

(Left) Gentlemen cadet sergeant-major blue frock, Royal Military College, Kingston, c. 1900. From 1887, the undress blue serge frock became the every-day dress of the gentlemen cadets. This particular example displays the various proficiency badges on the left sleeve. Royal Military College of Canada Museum, Kingston, Ontario. Ph: RC. (Above) Two gentlemen cadet sergeant-majors in their quarters, Royal Military College, Kingston, 1898. Both wear the dark blue undress uniform introduced from 1887. The lace and buttons were gold. Gold crossed canons and, seemingly, crossed rifles, are visible on the left sleeve of a gentleman cadet. The interior arrangement seem cozy. Courtesy Toronto Reference Library, pictures-r-1153.

Royal Military College gentlemen cadets in various orders of dress, 1890s. Left to right, standing: cadet in field marching order; in dress with white helmet; dress with pillbox cap; undress with dark blue serge frock; two in winter dress; sergeant in dress; two in winter dress; two in field order, one with Glengarry forage cap; winter dress. Sitting: two in sports (note the rugby ball) and possibly gymnastic dress; in serge frock. They are armed with Martini-Henry rifles and equipped with British Valise accoutrements. Royal Military College Archives, photo John Grodinsky.

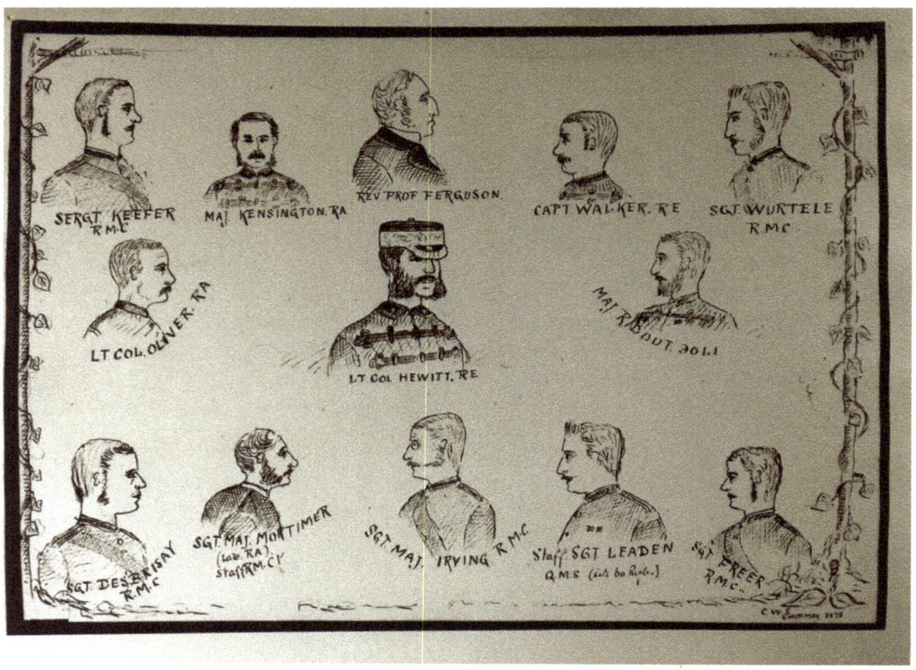

The faculty and senior gentlemen cadets at Royal Military College sketched in May 1878. Apart from Reeverend Professor Ferguson, the faculty is from the British Army: two, including Commandant Lt. Col. Hewitt are from the Royal Engineers, three from the Royal Artillery, one from the 90th Light Infantry and another from the 60th Rifles. Sketch after "C W S" Royal Military College Archives, photo John Grodinsky.

The faculty at Royal Military College, 1889. The military instructors mostly wear patrol jackets or frocks of their respective corps. Two artillery officers at rear wear mess jackets and pillbox caps. The non-military and clerical staff are in ordinary civilian costumes. Royal Military College Archives, photo John Grodinsky.

Gymnastics dress, Royal Military College, 1899. The gentlemen cadets wear white short-sleeved T-Shirt style "vests", fabric belts and trousers trimmed with dark blue, white canvas shoes. Courtesy Toronto Reference Library, Baldwin Collection, 986-7-32.

a college uniform consisting of a dark blue capot trimmed with white piping at the seams and a white waist sash. From the early 19th century, a dark blue peaked cap with white piping was worn and the "wool sash [was of] various colour". The school's uniform became, except for some distinguishing differences, the standard dress for students in all institution of higher learning in Francophone Canada until the 20th century. The seminary's cadets drilled with muskets (later rifles) furnished by the government and their officers carried swords. Amazingly, the Quebec Seminary's two company cadet corps was only officially recognised in 1880.[93]

The St. Hyacinthe Seminary, south-east of Montreal, also had one of the earliest cadet corps in the nation, which originated in the spring of 1832. It is one of the very few early cadet corps having a historical account, which gives a rare glimpse of how some of these early units operated. The St. Hyacinthe Cadets started as an informal gathering of two groups of students wielding sticks; they elected a leader with the title of "general" and, as time passed, the cadet corps became better organised. The St. Hyacinthe Seminary's student uniform was generally similar to that of the Quebec Seminary and thus dressed in dark blue with white piping, the cadets were seen drilling with wooden muskets. Martial spirit was high and, in time, their "battalion" was named the "invincibles." In 1846, a band was organised by a music student and the cadet unit now also had a colour. By then, the general had a sword as the symbol of command. The unit existed under that form until 1867, when the "Corps des réguliers" (regular's corps) was organised following the Canadian government's first call to raise cadet corps in colleges. The new organisation carried the same colours as the former "invincibles" and its four companies were named Meurons, Grenadiers, Voltigeurs and Chasseurs. They also used rifled muskets on loan from the local volunteer unit. It was only in 1879-1880 that 60 rifles ("carabines" — possibly Peabody rifles) were received from Ottawa and only then was this unit recognised as the "Compagnie d'exercice du Collège de Saint-Hyacinthe" (St. Hyacinthe College Drill Company) although it had existed since 1832. Its official recognition occurred because the Hon. Rodrigue Masson, a seminary alumni and former "general" of its "invincibles" cadets, had become minister of Militia. These rifles were used until replaced by lighter ones in 1898 and, in 1911, by 150 Ross rifles.

In 1867, the "Canadian Cadets" (which meant the French Canadian Cadets — *Cadets canadiens*) were organised at the University of Ottawa, at Laval Normal School (later part of the University) and at Napierville. The cadets at the College of Saint-Anne de la Pocatière, officially recognised in 1879, had the usual dark blue capot with white piping, but with a distinctive green sash. The student's capots, trousers and caps at the Seminary of Sainte-Thérèse, whose cadets were officially recognised in 1880, were black piped with white and had a blue sash. Another unit that predated its official 1879 formation was the Joliette Seminary Cadets that was known as a drill company in 1863 and most likely existed previously. It should be noted that not all Francophone cadet units

[93]Francis Back, "De bleu vêtu", *Cap-aux-Diamants*, 1993, pp. 16-18; Michel Bibaud, "Mon dernier voyage à Québec", *Encyclopédie Canadienne* (Montreal, 1843), I: 309.

Members of the Montreal High School Cadet Corps, 1866. The unit was formed in 1864 and had a grey uniform with blue-black facings shown as collar, cuffs, edging lace and trouser stripes. The caps were likely that colour, the belts probably black. Private collection. Ph: CL.

wore the traditional school uniform consisting of capots and sashes. Montreal's Mont Saint-Louis College's cadet corps organised in 1892 wore a French-style dark blue military uniform that featured a tunic with two rows of buttons and a kepi. Another, much less wealthy institution was Montreal's "Asile Saint-Joseph" for children, a sort of day-care centre for the children of lowly paid women workers ran by nun Julie Gaudry, a Sister of Charity of the Hôpital Général, that, in the 1880s, had a children's cadet corps. Sister Gaudry applied to the Minister of Militia for purchasing 60 children's rifles (without locks) and sabres in 1890, but this was not approved.[94]

In Anglophone Canada, cadet units seem to have appeared as fairly coherent organisations from the early 1860s. As in Francophone Canada, they were associated with learning institutions. The first officially recorded unit was organised from June 1861 at Trinity College in Port Hope, Ontario. Its uniform was reported as rifle green with red facings in April 1862 and, the boys becoming older, listed as the 9th company of the 2nd Queen's Own Rifles. Bishop's College Cadet Rifle Company in Lennoxville, Quebec, started in December 1861 and, during the 1870 Fenian Raids that occurred in the Eastern Townships, was mobilised for a short period of time, the only cadet unit to be called on active service in the Victorian era. It wore a rifle dark green uniform with scarlet collar and cuffs with a badge in front of its dark green "pillbox" style forage cap. It was armed with percussion muskets and "old style" accoutrements. In Toronto, some students of the "University & Colleges Rifle Company" were drilling in 1861 at Upper Canada College with 60 "Long Enfields" and wore a uniform of "green cloth, red facings as per circular 19 May 1860, forage cap rifle green with red stripes." It became the 11th company of the 2nd Battalion Queen's Own Rifles and continued wearing rifle style uniforms. Schools were encouraged to teach military drill to their student even if they did not form cadet units. By 1863, some 18 Ontario grammar schools reported having military drills as part of the courses given and the same was done at the Jacques Cartier Normal School as well as other schools in Montreal.[95]

---

[94]C.-P. Choquette, *Histoire du Séminaire de Saint-Hyacinthe* (Montreal, 1911), pp. 113-114; Alexis Mailloux, *Histoire du Collège de Saint-Anne-de-la-Pocatière* (Quebec, 1948), p. 58; Émile Dubois, *Le Petit Séminaire de Sainte-Thérèse 1825-1925* (Montreal, 1928), p. 88; www.armycadethistory.com; LAC, reel T274/900, note by B[enjamin] S[ulte], 22 May 1890. Hector Fabre, MP, had procured in Paris five "fusils scolaires" (school rifles) with small sabres for the Department of Militia during 1883 for consideration, but they were not ordered due to difficulties in obtaining ammunition. The rifle sent to Ottawa by Sister Gaudry was described as a "joujou" (toy) and the sabre very common,

[95] "Military Instruction in Canadian Schools", *Journal of Education, Upper Canada*, XVIII: 7 (July 1865), pp. 98-99. The Upper Canada College Cadet Cords remained affiliated with the 2nd Queen's Own Rifles for 127 years as per: https://qormuseum.org/history/cadets/

Cadet Guy Thorold of the Guelph Collegiate Institute's Highland Cadets, c. 1884. The unit was organized in 1882 and imported its Highland uniform from Glasgow (Scotland). It consisted of a black velvet doublet trimmed with white piping, royal Stuart tartan kilt and sash, white sporran with two black tails and a black Glengarry with an eagle's feather. Courtesy Don Blyth Collection, Guelph, Ontario.

The High School of Québec City Cadet Corps, 10 June 1890. this "drill company" was authorized by MGO of 4 February 1887. The uniform of this Anglophone boy's school was grey with black collar and lace. They are shown armed with old Enfield short rifles with sword-bayonets fixed. Courtesy Library and Archives Canada, PA99974.

A member of the Toronto Public School Cadet Corps. It initial six companies were organized in various schools in late 1898. Canadian military artist Derek FitzJames was a member of the corps in 1932-1934 and later made this watercolour of its uniform, which, except for white piping and cuff trefoils, agrees with its first dress described in the 8 December 1898 Daily Mail and Empire bearing in mind that those details may have simply not metionned. Anne S.K. Brown Military Collection, Brown University Library, Providence, USA. Ph: RC.

Members of Montreal's Highland Cadets with their colours, c. 1898. At left the colours presented by the Duke of Connaught and at right those presented by the ladies supporting the unit. From Chambers 1901 history of the corps. Ph: RC.

(Left) Officers of the cadet companies at the Quebec Seminary, c. 1901. They wear the seminary's uniform, which originated in the days of New France: dark blue capot trimmed with white piping. A peaked dark blue cap with a white band and piping was adopted in the 19th century. The officers were armed with swords except for the bandmaster (seated at left) who holds a baton. Album Universel, 10 May 1902. Ph: RC.

(Below) Women's Cadet Corps, St. Catharines, Ontario, c. 1890. This volunteer unit was photographed in front of the town hall. The dresses are uniform (possibly their school's costume) and they all also have a small kepi as their headdress. The drummers have what appears to be short Zouave-style jackets ornamented with braid loops. They were armed with Snider-Enfield rifles muskets and the unit had its drum corps. At left, two officers, probably of the 19th Lincoln Infantry who would have been the instructors. Courtesy Ontario Archives, Toronto, S4821.

(Above) Corporal Calder of the Vancouver Home Guard, late 1890s-early 1900s. She wears a dark-coloured militarized lady's dress. The stylish jacket is braided at the collar, cuffs and down the front. The skirt has a wide stripe at its side and she wears an artillery-style pillbox cap, features that seem to indicate that it was attached to a company of the 5th British Columbia Artillery although it could also be after July 1899 when it was converted into the 6th Rifle Battalion. Her rifle, surprisingly, is a Lee-Enfield Mk I that appeared in Canada from 1896 and not a surplus weapon. Some may have been lent to this ladies unit or even purchased privately. Courtesy of A.M. de Quesada, Jr.

62nd St. John Fusiliers' "Amazons", 1901. Patriotic feelings brought about by the Boer War were strong enough in St. John, New Brunswick, to inspire some of the daughters of the city's elite families to form the "Amazons" that were attached, perhaps in more ways than one, to the 62nd Fusiliers as shown in this image with a lone officer wearing a scarlet mess jacket and his unit's service dress cap surrounded by young Amazons. For their uniform, they chose what appears to be a khaki jacket with, probably, a red collar and shoulder straps cut along feminine lines and, seemingly, a khaki brimmed hat. They hold obsolete Snider-Enfield infantry rifles. Private collection. Ph: RC.

The Montreal High School Cadet formed in 1864 wore a "neat uniform" of grey "light Halifax tweed, with blue-black facings" and eventually obtained from the military authorities the loan of "enough muzzle-loading Enfield carbines to arm one company." In 1866, the Victoria Cadets were raised at the Ottawa Grammar School and had a grey Halifax tweed uniform with dark green facings, grey forage cap with dark green band and black leather accoutrements, while the Ottawa Cadets were organised at the Ottawa Classical and Commercial School; both corps participated in the 1 July 1867 ceremonies inaugurating Confederation on Parliament Hill. More units were organised in 1867 at Queen's University in Kingston, Belleville, Brockville and more in 1868 at Rimouski, Picton, Oshawa, Whitby, Barrie and more. These units probably mostly used their school uniforms at drills.

The 1879 Militia Regulations authorised 74 "Drill Associations in Educational Institution" to be made up of young men over 14 years of age that would qualify for issue of arms, accoutrements and other assistance by the Department of Militia. They were not to be deployed on active service. Many of these cadet companies already existed. Uniforms could be "composed of a scarlet, blue, rifle green, or grey tunic or Norfolk jacket, with blue, black, rifle green, or grey trousers, and forage cap, or shako as may be preferred. Clothing of any pattern worn by cadets or soldiers of foreign countries will not be approved." As time passed and for those who could afford it, distinct cadet corps uniforms became increasingly fashionable. By 1882-1884, the Guelph Highland Cadet Company wore a Highland uniform and was armed with surplus Peabody rifles, which were very long for boys. So were the lads of Queen's College (now university) in Kingston that asked, in January 1882, that these awkward rifles be replaced. They probably were not for a while since, in October 1885, the Mount Forest High School Cadet Company was issued Peabody rifles. The Rifle Company of Upper Canada College also had Peabody rifles with bayonets and belts in 1892 for its boys of 14 years of age an older, each of which wore "a dark blue shell jacket trimmed with white, trousers same material, and forage cap."

The posh Montreal Highland Cadet Corps formed in 1888 also had a Highland uniform (see illustrations). They also had, at least in 1895, a loan of 50 Peabody rifles until replaced by Snider-Enfield short rifles. The Montreal High School had four companies who were mentioned, in Militia General Order of 2 August 1889, in its uniform of "Grey tunic, trousers and forage cap of Halifax tweed, with black facings." Montreal's St. John the Evangelist School Cadets, also named "Lady Alexander Russel's Own" had grey uniforms "very much after the style of that of the Montreal High School Cadets, but with red facings" described in the 1890 Militia Report as a "grey shell jacket and trousers, with red facings, and stripe with a service cap to match, which, with their russet leather waist and cross belts." The Eliock School Cadet Corps, also in Montreal, wore "blue patrol jacket, trimmed with gold braid with yellow lacings, dark plaid trousers, and red and white dice-band Glengarry caps, with russet leather waist and cross-belts, the whole uniform presenting an exceedingly rich and imposing appearance." Cadet units could also be directly attached to a volunteer unit and this was the case in the 5th Royal Scots from 1890.

---

upper-canada-college-cadet-corps/

In late 1898, the Toronto Public School Board made arrangements to form a cadet corps of six companies to be organised in its schools including that the necessary accommodations for armouries at the various schools. The uniform chosen was "scarlet serge tunics, blue service caps, and blue knickerbockers" and this was worn by a contingent of 50 Toronto cadets attending a North American cadet's convention in Tampa, Florida, in February 1899. The tunic had two breast pockets and five buttons down the front, in the style of the volunteer militia's serge frock so it may also have had white piping and cuff trefoils. The organisation itself was finally officially recognised in June 1899. Cadet drill associations do not seem to have been, at least officially, as prevalent in the Maritimes, but, on 17 December 1881, the St. John School drill company was gazetted, the Halifax (NS) County Academy in January 1892 and many more thereafter including some in Manitoba and British Columbia. St. John's, Newfoundland, had some four cadet companies by the end of the 1890s (see Vol. 1, p. 248). By then, cadet units were blooming all over the country and resulted in a national organisation by the Department of Militia in the early 1900s.[96]

"Amazons" Cadets

Although Army Cadet training for girls was not authorised until 30 July 1975, there is evidence of young women's cadet corps already existing in Canada during the Victorian era. In 1884, a girl's cadet unit was organised at the Guelph Collegiate Institute in a corps known as the "Daughters of the Regiment". Another girl's unit in mentioned in Ottawa during 1891. In April of that same year, the "ladies infantry company" attached to the 16th Prince Edward Infantry Battalion was "the great attraction" at a "military entertainment" in Picton as per the *Canadian Militia Gazette* of 16 April. There was also such a unit in St. Catharine's as can be seen by a c. 1895 photo that shows all wearing what is probably their school's uniform with a peaked military style cap and armed with possibly some pattern 1853 rifles. Two officers from the local volunteer infantry unit, probably the 19th Lincoln Infantry, are also seen.

A photo of Corporal Calder, a young lady in the Vancouver Home Guard, reveals that a women's auxiliary corps was seemingly attached to its local volunteer militia unit, which was the 5th British Columbia Artillery that became the 6th Rifle Battalion after July 1899. Another unofficial feminine cadet unit that may have been raised at the time of the South African War was the "Amazon Company" informally attached to the 62nd St. John's Fusiliers in New Brunswick.[97]

Notes

Richard Arthur Preston, *Canada's RMC: A History of the Royal Military College* (University of Toronto, 1969); G.M. Betts, "Truth, Duty, Valour The Royal Military Collrge of Canada, a description of its uniforms and Queen's colour", *Tradition*, No. 53, pp. 38-40; F.G. Guggisburg, *"The Shop" The story of the Royal Military College* (London, 1892); LAC, reel T4175/132, Deputy Minister's Ledgers, December 1875; artefacts at the museum of the Royal Military College, Kingston.

[96] LAC, RG9, IIB3, MGOs concerning cadet corps from vols. 1 to 7; vol. 8, Militia General Order, 2 August 1889; Joan Finnigan, *Lisgar Collegiate Institute* (Ottawa, 1993), p. 25; Province of Ontario, *Sessional Papers* (Toronto, 1893), p. 15; Ernest J. Chambers, *The Montreal Highland Cadets* (Montreal, 1901), pp. 26, 30; *Daily Mail and Empire* (Toronto), 8 December 1898, 13 January and 14 February 1899.

97 www.armycadethistory.com; CMG, 16 April 1891.

Chapter 10

Rank and Appointment Badges

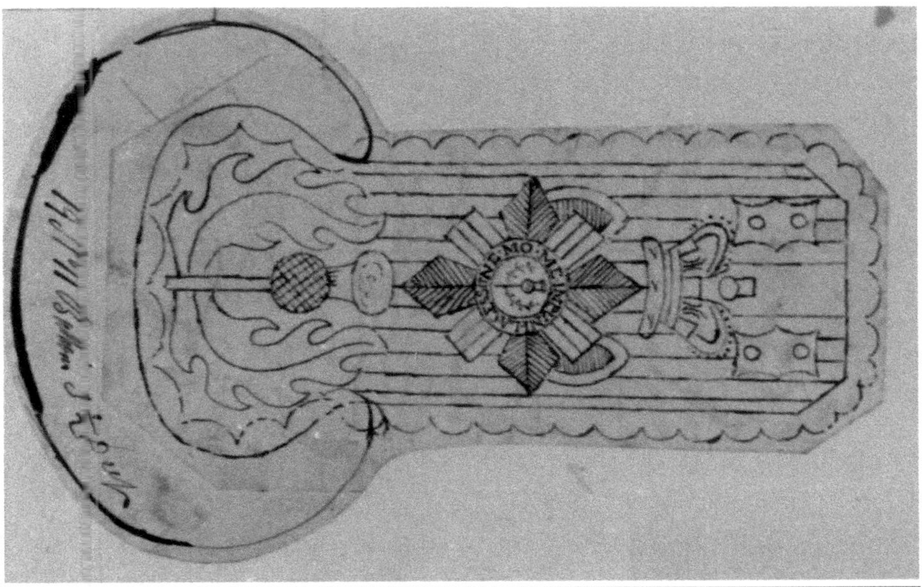

Epaulette design for a colonel of a Scot's unit, c. 1835. The strap has the crown and star, the end with the crescent has the thistle badge within; grenadiers and fusiliers had a grenade, light infantry had a bugle horn badge. From a period pattern book. A shoulder board became a dress epaulette when fringe was added, in this instance "No. 11 Bullion 3 1/2 [inch] Deep." Courtesy National Historic Sites, Parks Canada.

Badges of rank worn in Canadian units during the Victorian era were nearly always similar to those used in the British forces. There were, however, exceptions as noted below.

Officers with Commissions

General Officers

Until the later 19th century, general officers in Canada were exclusively British. From 1828, all generals had two elaborately embroidered gold epaulettes with plain fringes, the strap bearing the badges of crossed sword and baton. Rank was not denoted on the epaulettes, but instead on the coat's button placement and on the saddle cloth.

From 1855, rank badges were on the collar which had a 25.5 mm (1 inch) wide gold oak leaf pattern lace at the top and bottom of the collar. Each front end of the collar had for a:

General: silver crown and silver star.
Lieutenant General: silver crown.
Major General: silver star.
Brigadier General: silver crown and silver star, but the collar edged with narrow staff pattern lace, altered to oak leaf lace from 16 May 1860.

From 1880, the rank badges were worn on the shoulder straps and were for:

General: silver crossed sword and baton with silver crown and silver star above.
Lieutenant General: silver crossed sword and baton with silver crown only.
Major General: silver crossed sword and baton with silver silver star only.
Brigadier General: silver crossed sword and baton only.

Staff Officers

Apart from the rank badges specified in this chapter, staff officers could also have embroidery and other decorative patterns on their uniforms specific to their functions. These were the same as those worn by British army staff officers and are too numerous and complex to deal with here. Interested readers will want to consult N.P. Dawnay's study on staff distinctions (see Notes below).

Other Officers

By the royal warrant of 19 December 1768, the ranks of officers of the British forces were distinguished by fringed epaulettes and, in light troops, by crescent shaped wings or a combination of both wings and epaulettes. This rank designation system evolved in time and, especially with wings, it often also indicated the wearer's arm of service and status. For instance, officers of fusilier units — considered somewhat of an elite — also took to wearing wings.

From 1810 to 1855, the rank of field and company officers was indicated on the strap of the two thick bullion fringed epaulettes or wings as follow:

Colonel: crown and star.
Lieutenant-Colonel: crown.
Major: star.
Captain: plain epaulette with thick bullion fringes on right shoulder only.
Lieutenant: plain epaulette with thin fringes on right shoulder only.
Ensign: plain epaulette with thin fringes on right shoulder only.

Shoulder board for a lieutenant-colonel of the New Brunswick Militia, c. 1840. This item is associated with the Carleton County Militia in western New Brunswick. It would have been worn on a dark blue frock coat. The board has very dark blue cloth edged with gold lace, gold crescent and a gold crown with red coif denotes the rank of lieutenant-colonel. Carleton County Historical Society Museum, Woodstock, NB. Ph: RC.

The epaulettes of grenadier officers had a grenade on the strap and those of the light infantry a bugle horn, below the crown and/or star.

Field officers of Fusilier and Light Infantry (and some rifle) corps had wings in addition of the epaulettes. Flank company officers had wings on both shoulders with a grenade or bugle horn on the straps trimmed with bullion or fringe according to rank.

Until 1830, the epaulettes might be gold or silver depending on the unit and it would have badges of the contrasting metal (i.e. silver epaulette with gold badges). From 2 August 1830, all British regular troops had gold epaulettes and lace and United Kingdom militia units henceforth had them in silver. The stars on the strap of the officer's epaulettes were henceforth of the Order of the Bath instead of the Order of the Garter, except in corps in which a national badge had been allowed.

The change to silver for volunteers and militia in various overseas territories was not always followed upon. In Canada, nearly all cavalry and artillery volunteer units retained gold, but it seems that by the late 1850s and during the 1870s, infantry and staff officers had silver, although grudgingly, as seen in earlier chapters. When gold was allowed to some infantry units in the 1870s, the matter erupted in the press and in Parliament and was finally settled in Canada with gold being adopted from 1883.

From January 1855, the rank badges were on the collar. For field officers, the collar was laced at the top and bottom. For captains and lower ranks, it was laced at the top only. The badges were as follows:

Colonel: crown and star.
Lieutenant-Colonel: crown.
Major: star.
Captain: crown and star.

(Left) Officer's shoulder boards, Westmorland County Militia, c 1845. Having no badges, these items would have been worn by a captain or a subaltern. The boards have very dark blue cloth edged with gold lace and gilded crescents. They would have been worn on a dark blue frock coat. Webster Collection, Fort Beauséjour National Historic Site, Aulac, NB. Ph: RC.

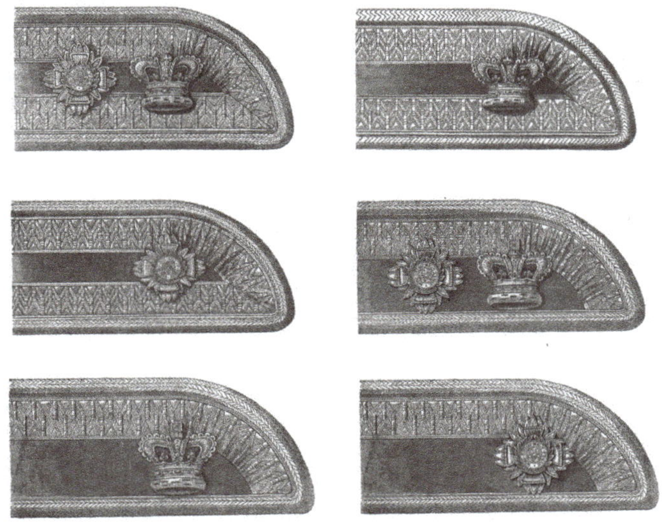

(Right) Officer's collar rank badges, 1855-1868. From left to right, top row colonel; lieutenant colonel. Middle row: major; captain. Bottom row lieutenant; 2nd lieutenant. Adapted from prints in Royal Artillery Standing Orders, 1864. Ph: RC.

New Brunswick Militia captain's rank badge, c. 1863-1868. Silver thread crown and star on the collar. Courtesy New Brunswick Museum, St. John.

Lt. Col. Joseph-G. Blanchet, 17th Lévis Battalion, c. 1865. His scarlet tunic has the gold crown on each side of its dark blue collar indicating his rank. The lace and buttons silver, the piping white and the sash crimson. Print in Marquis Le Régiment de Lévis. Ph: RC.

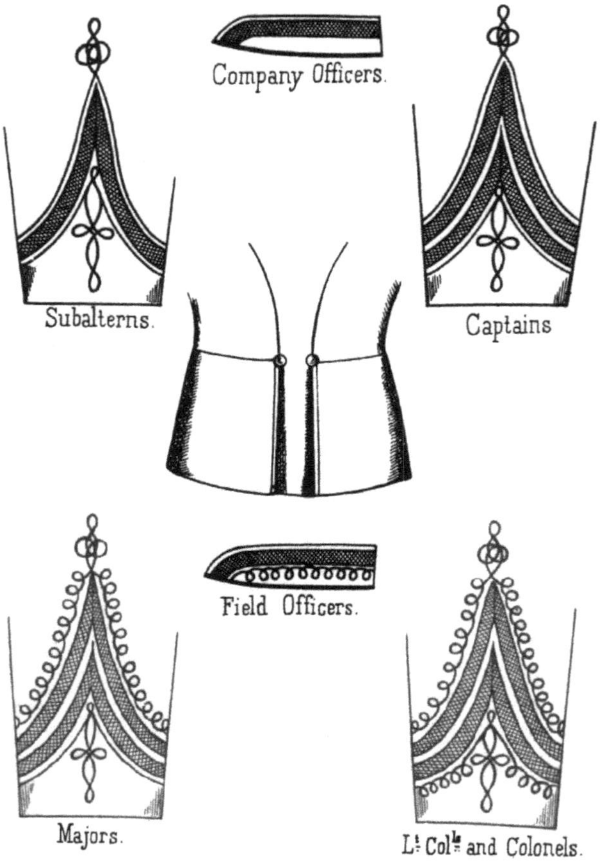

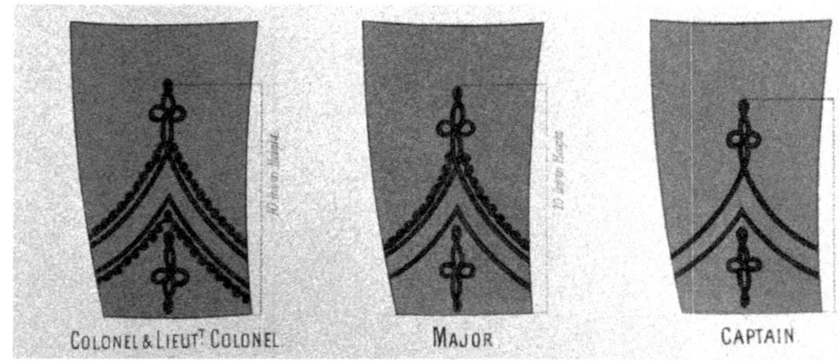

Officer's lower sleeve rank lace on the British 1872 service frock worn by some Canadian officers. Gold lace cords (shown in black) denote the ranks on the sleeves. That of lieutenant (not shown) had only one chevron. Contemporary print published by W. Jones & Co., a London military outfitter. Private collection. Ph: RC.

Infantry officer's cuff and collar rank lace and back of tunic skirt, August 1868. The cuff and skirt flaps were abolished in the infantry. Apart from crown and star rank badges being moved from the collar to the shoulder straps during the 1880s, this system has remained essentially the same for full dress up to the present time. Gazette of Fashion (London), 1 September 1868. Ph: RC.

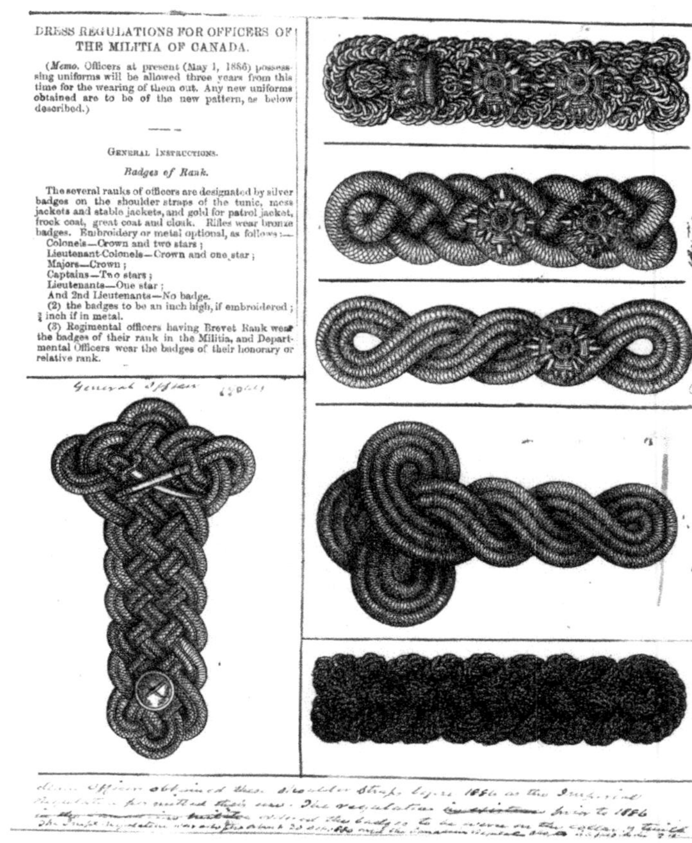

Officer's shoulder rank badges, 11 November 1885 and 1 May 1886. The shoulder knots illustrated are, at left, for general officers and, at right from top to bottom, for: hussars, artillery, infantry, engineers and rifles. Adjudant General Walker Powell also hand wrote under that "Many officers obtained these shoulder straps before 1886 as the Imperial Regulations [of 20 October 1880] permitted their use. Dark blue undress jacket and trousers, scarlet collar and wide trouser stripe, gold lace Austrian knot at cuffs, gold collar and jacket edging and shoulder cord." It had actually been permitted by Gen. Middleton from 11 November 1885. Courtesy Library and Archives Canada, C6329.

Artillery officer's undress jackets lower sleeve rank lace. From left to right: field officer; captain; lieutenant and subaltern. Engineers had generally similar lace. The dress tunic sleeve lace followed the same patterns in a more elaborate fashion. Adapted from prints in Royal Artillery Standing Orders, 1864. Ph: RC.

At left, Lt. Col. Williams, who led the charge of the Midland Battalion at Batoche on 12 May 1885; with rank badges on shoulder cords rather than at the collar. At centre, Capt. Manson, 10th Royal Grenadiers, severely wounded at Batoche; wearing patrol dress with rank badges on shoulder straps and a pre-1879 cap bearing the grenade. At right, Lt. Col. W.M. De Ray Williams commanding the 7th Fusiliers; note the grenade at the collar. Colonels Williams and De Ray Williams have the gold and crimson sash worn over the left shoulder for state occasions over their scarlet tunics. Illustrated War News, 4 July 1885. Courtesy Bruce Peel Special Collections, University of Alberta, Edmonton.

Lieutenant: crown.
Ensign: star.

The infantry and rifle tunics kept the cuff flap. The cuff had two laces for field officers and one for company officers. Cavalry, artillery and engineers, etc., company officers henceforth had cord Austrian knots at the cuff and field officers chevron cuffs.

The cord Austrian knot was favoured by members of the "Volunteer Movement" in Great Britain that was soon followed in Canada. From 1863, the Canadian infantry and rifle uniforms issued by the government had Austrian knots and this included officers.

In general, rank designation in British North America followed the British army's except for New Brunswick between 1863 and about 1868. The 24 February 1863 New Brunswick dress regulations specified a gold crown for colonels and lieutenant-colonels worn on the collar, a gold star at the top of the cuffs for majors, silver badges on the collar only for subalterns as in the British army: captain: crown and star; lieutenant: crown; ensign: star.

From August 1868, the advent of the British infantry tunic's chevron cuff brought a more precise rank system at the tunic's cuff and lower sleeve for officers. This was followed by Canadian infantry officers and confirmed in the first extensive Canadian Militia dress regulations dated 4 March 1876:

> "Field officers have figured braiding below the lace on the collar; and figured braiding on the sleeve, round, the [small tracing braid] Austrian knot extending to 11 inches from the bottom of the cuff.
> Captains have a row of braided eyes on the collar bellow the lace; and a tracing of braided eyes round the knot on the sleeve nine inches deep.
> Lieutenants have a tracing of plain braid only below the lace on the collar and round the knot of the sleeve 8 inches deep."

Infantry officer's cuff lace, c. 1868-1900s. Adapted from T. H. Holding's 1894 Uniforms…Ph: RC.

From October 1880, the British army rank badges were worn on shoulder straps and consisted of, for:

Colonel: crown and two stars.
Lieutenant-Colonel: crown and one star.
Major: crown.
Captain: two stars.
Lieutenant: one star.
2nd Lieutenant: plain.

This was not adopted right away in Canada although it appears many individual officers started using rank badges on shoulder straps like the British army. Thus, some images or tunics of Canadian officers having rank badges on shoulder straps (or knots) could be dated as early as 1881 although they were officially supposed to be at the collar. A Militia General Order of 26 November 1880 specified that Canadian officers were to adhere to the Canadian pattern previously prescribed "until changed by Order." This caused a confused debate amongst Canadian officers insofar as conforming to practices adopted elsewhere in the British Empire. It was also spurred by the fact that many officers ordered their uniforms from the United Kingdom, and these now arrived with shoulder straps or bullion knots. By May 1882, it is recorded that the officers of the 10th Royal Grenadiers obtained approval to "wear the new pattern shoulder strap on their tunic" and other units appear to have also secured such approval. By June 1885, the *Canadian Militia Gazette* reported that "although not in accordance with Canadian regulations, some battalions have obtained permission to wear badges of rank on the shoulder straps."

On 11 November 1885, it was at last announced that "the Major General Commanding has approved of the badges of rank being worn on the shoulder straps as per Imperial Regulations" adding that "Officers providing themselves with new uniform, will conform with this regulation which will also be applicable to Winter uniform at present."

The Militia Department followed suit in the dress regulations of 1 May 1886. It specified:

Colonels—Crown and two stars.
Lieutenant-Colonels—Crown and one star.
Majors—Crown.
Captains—Two stars.
Lieutenants—One star.
2nd Lieutenants—No badge.

This system lasted until May 1902 when alterations were made by the imperial authorities in London.

34th Infantry Battalion colonel's tunic cuff chevron rank lace, 1890s. Ontario Regiment Museum, Oshawa. Ph: CL.

Lt. Col. L.-Georges Desjardins, 17th Lévis Battalion, 1893. His dark blue patrol jacket has the gold crown and star on each side of its collar. Since 1886, it was to be on the shoulder straps, but later photos such as this one (much enlarged from a group image) show this might take awhile. Print in Marquis Le Régiment de Lévis. Ph: RC.

Ceremonial sash for a militia officer, c. 1868-1885. The sash was silver with two crimson stripes for militia infantry officers and worn over the left shoulder of the scarlet tunic. The white helmet in the image is of the 26th Battalion; it has gilt fittings so would date from 1885-1886 or later. Courtesy Michael Reintjes.

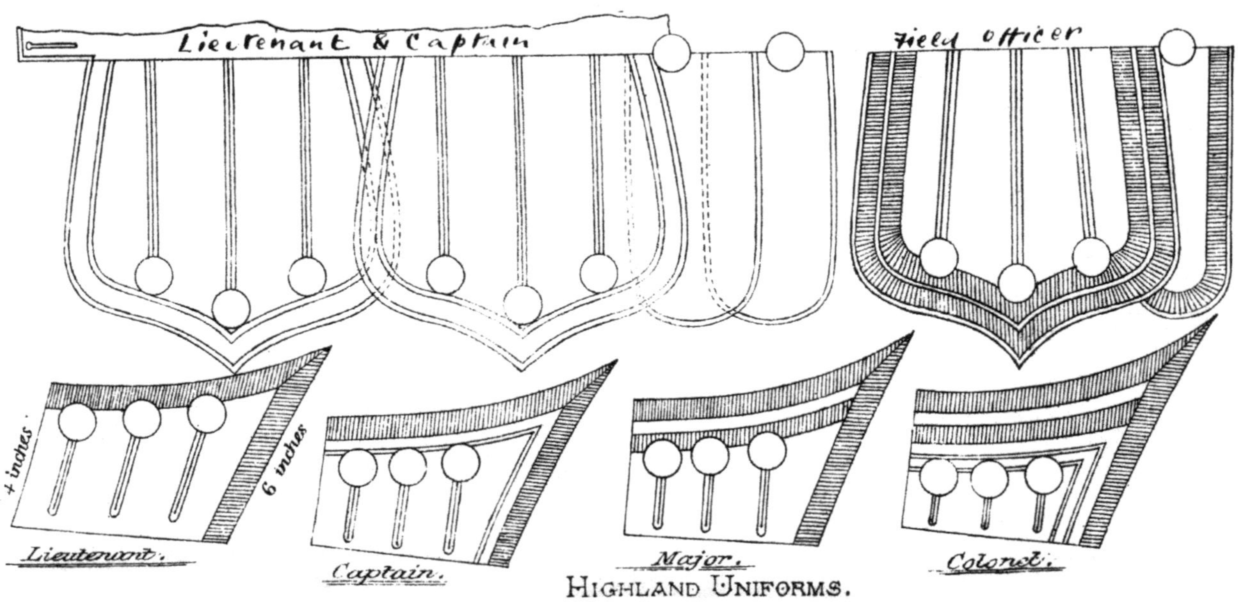

Highland units officer's badges of rank, c. 1855-1900s. Top: Inverness flaps on the Highland doublet. Bottom: cuffs of the doublet. Adapted from T. H. Holding's 1894 Uniforms…Ph: RC.

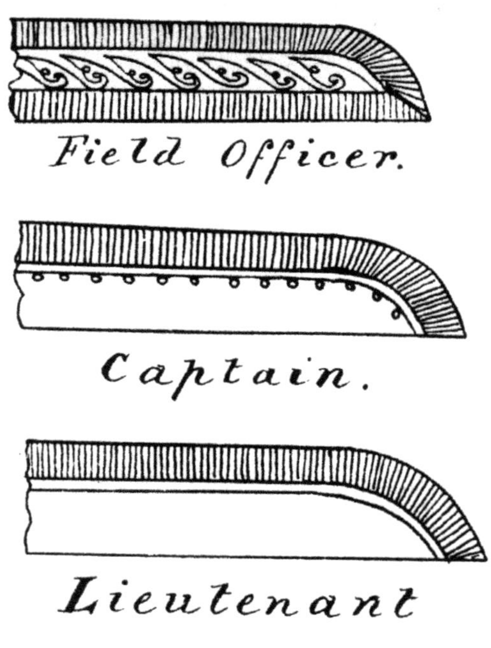

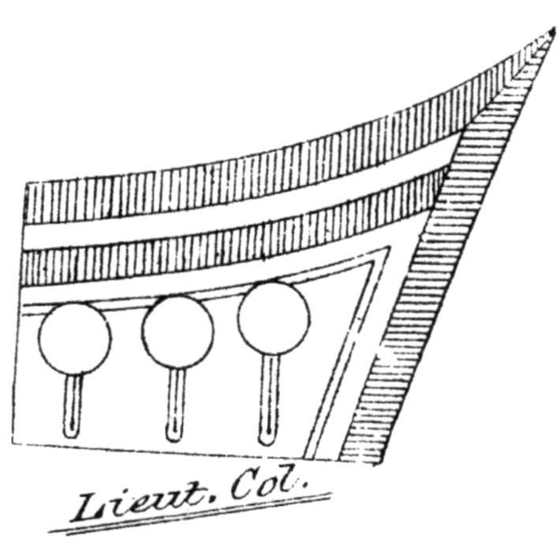

Highland units officer's rank lace on collar and cuffs, c. 1855-1900s. Adapted from T. H. Holding's 1894 Uniforms…Ph: RC.

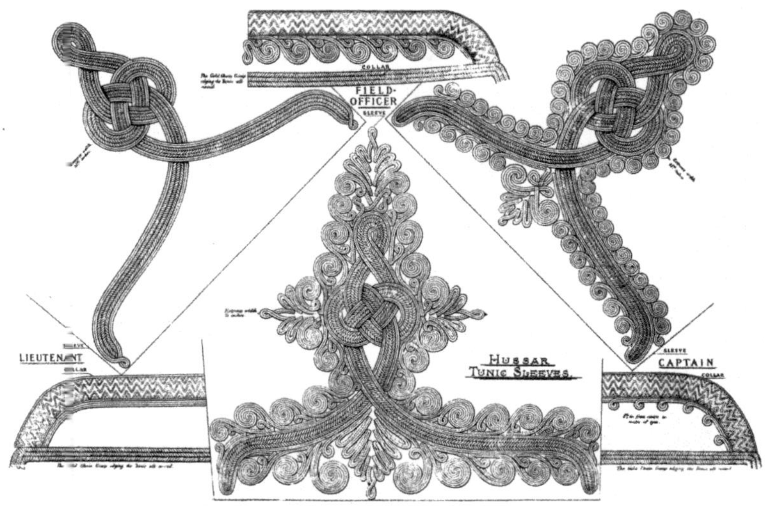

Hussar officer's rank lace at the collar and cuffs, c. 1855-1900s. Adapted from T. H. Holding's 1894 Uniforms...Ph: RC.

Officer's Sashes

From the 17th century, British officers wore crimson silk sashes over their left shoulder to denote their status as recipients of royal commissions rather than their specific rank. From December 1768 until 1 July 1856, officer's sashes were worn tied around the waist, except for Highland regiments who continued to wear them over the shoulder. From 1856, light infantry and rifle officers did not wear sashes. Thereafter, officer's sashes were worn over the right shoulder until 1902 when they were again worn around the waist. From 1867, the crimson sash was henceforth only worn in full dress.

From 1868, ceremonial sashes of gold with two crimson stripes were worn by British regular infantry officers, silver with two crimson sashes for militia officers. The silver sashes were the official ones to be worn in Canada until the colour of the lace was changed to gold in 1885-1886.

Officer's Gorget Patches

These appeared from 1889 on British army khaki service dress uniforms. Generals had them in scarlet with a loop of gold braid, staff officers scarlet with a scarlet silk worsted line and departmental officers in dark blue with loops of various colours. From 1896, when khaki started to be worn in Great Britain, many more gorget patches appeared. Medical had had blue with a black loop, veterinary a maroon loop, etc. In 1898, they were now worn, slightly larger, on the service dress dark blue serge frocks and this is when gorget patches appeared in force on the frocks of Canadian general, staff and departmental officers.

Warrant, Non-Commissioned Officers, Corporals and Privates

The most common NCO badges were adopted by the General Order of 14 July 1802: Chevrons on the right sleeve, points down, made with doubled up half inch lace, and set on backing cloth of the regimental facing colour:

Quartermaster Sergeant: 4 chevrons of lace
Sergeant-Major: 4 chevrons of lace
Sergeants: 3 chevrons of lace
Corporals: 2 chevrons of regimental lace

The lace chevrons for sergeants and above were plain white for infantry and rifles, yellow for artillery and engineer troops, plain white or yellow depending on the cavalry regiment's officer's lace colour. In practice, lance corporals also had one bar of regimental lace (WO 3/33). It should be noted that commissioned warrant officers had a double breasted coatee like those of the officers, with silver or gold lace depending on the regimental lace for officers.

Drum-major officially established in 1810, with rank of a sergeant (3 chevrons). On 6 July 1813, the rank of colour-sergeant was instituted, distinguished by a chevron with a Union flag embroidered in full colour with crossed sword (blades silver, hilts gold) at the lower pole and an embroidered Crown (in gold) above. Rifle units (who did not have colours) replaced the flag with a variation consisting of a crown over crossed swords above a bugle horn within a laurel wreath the chevron. In time, seemingly during the 1830s, the number of chevrons was increased to three and even four for a colour-sergeant-major.

12 September 1828: Steel mounted swords for officers and NCOs of Highland regiments were sealed.From 1831: Rifle units had black chevrons on both upper sleeves.

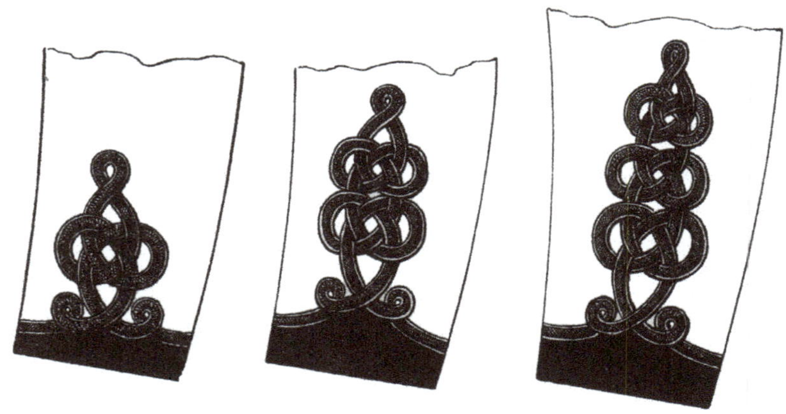

Cavalry and dragoon officer's cuff rank lace, c. 1855-1900s. From left to right: lieutenant, captain, field officers. Adapted from T. H. Holding's 1894 Uniforms…Ph: RC.

Until 20 September 1836, the lace on the infantry sergeant's coats was plain white while corporals and privates had coloured lines on their coat's lace. Thereafter, the coloured lines in the lace were abolished and the lace was henceforth totally white for privates and corporals. Furthermore, from 1836 all infantry sergeants were now to have a double breasted coat without buttonhole lace, but with white silk epaulettes, except for flank companies, light infantry and fusilier regiments which all had wings instead of epaulettes.

In 1837, the badges of rank were to be worn on both upper sleeves in fusilier, grenadier, light infantry and rifle units. Otherwise on the right upper sleeve only for line infantry.

By orders of 12 April 1845, the rank distinctions were:

Regimental Sergeant-Major: 4 chevrons and a crown
Troop Sergeant-Major: 3 chevrons and a crown
Sergeant: 3 chevrons
Corporals: 2 chevrons

Shoulder knots of a lieutenant-colonel of the Governor General's Body Guard, c. 1900. Silver with gold crown and star. Courtesy Scott Duncan.

In 1852, sergeant's swords were abolished except for staff sergeants.

From July 1858 colour sergeants of British infantry and rifle units were to wear three chevrons on the right sleeve and the Union flag with one chevron on the left sleeve. In 1859, all British army sergeant-majors had a crown over four chevrons. During the 1860s, the main senior NCOs ranks were indicated as follows:

Sergeant-Major: 4 chevrons and crown
Quartermaster-Sergeant: 4 chevrons
Band Sergeant: 4 chevrons
Sergeant Instructor in Musketry: 4 chevrons and crossed muskets
Drum-Major: 4 chevrons and drum
Colour-Sergeant (Infantry): badge with colours and one chevron below on the right arm, three chevrons on the left arm.
Colour-Sergeant (Rifles): crossed sword badge (on black cloth) and one chevron on the right arm, three

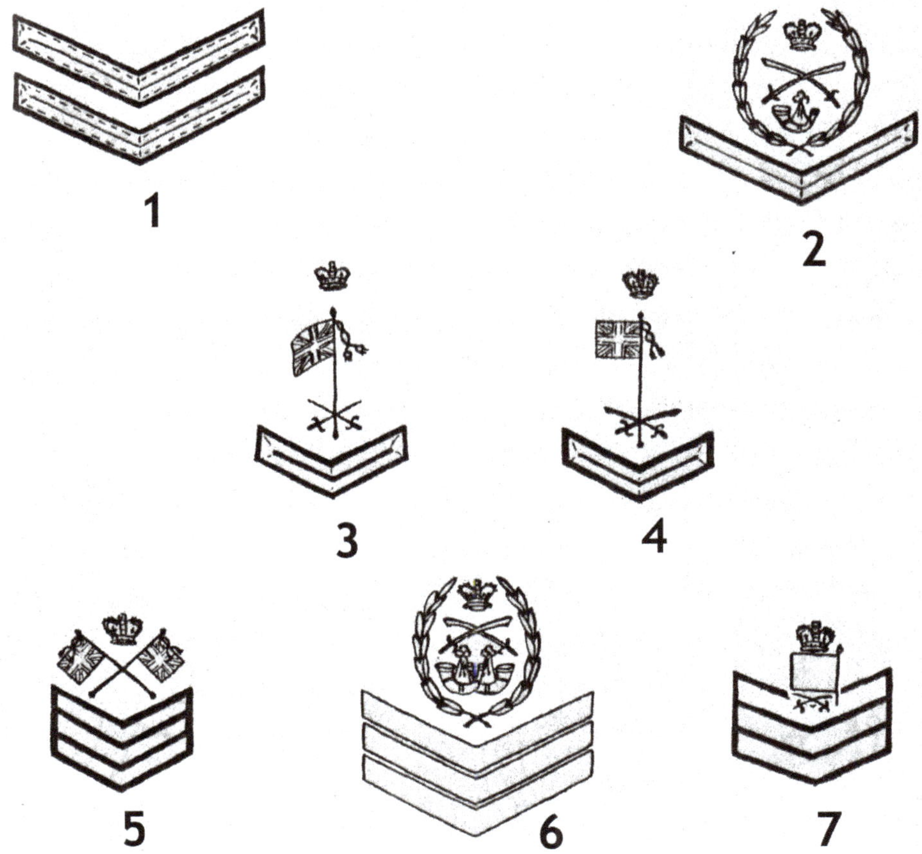

Corporal's and sergeants: 1, corporal, since 1802; 2, colour-sergeant, rifles, c. 1834; 3 and 4, colour-sergeant, infantry, 1813 and 1865; 5, colour-sergeant, infantry, 1868; 6, colour-sergeant, Rifles, 1877; 7, colour-sergeant, Foot Guards, 1815 and later. As per N.P. Dawnay, The Badges of Warrant and Non-Commissioned Officers, 1949.

Colour sergeant's badge, c. 1840. From a coat of the 8th Regiment of Foot. Fort Henry, Kingston. Ph: RC.

Colour sergeant's badges, single chevrons with colour and crown, worn on both upper sleeves from 1859 to late 1868 in the British army and probably until somewhat later in the Canadian Volunteer Militia. Detail from a drawing by Douglas Anderson. Courtesy Halifax Citadel National Historic Site, Parks Canada.

Sergeant, 2nd Queen's Own Rifles, c. 1866. He wears the three black chevrons edged scarlet on both upper sleeves. The 2nd's peculiar cuff was edged with lace for its NCOs instead of the Austrian knot at this time. Courtesy Queen's Own Rifles Museum and Archives collection and photo.

Colour sergeant-major's badge, 1868. Sketch by Percy Sumner. Anne S.K. Brown Military Collection, Brown University Library, Providence, USA. Ph: RC.

Sergeant-drummer's badge, 1860s. From a doublet of the 78th Highlanders, which served at Montreal from 1867 to 1869 and Halifax until 1871. Argenteuil Museum, Carillon, QC. Ph: RC.

Sergeant-major's gold chevrons, c. 1878. These were worn by Cadet Sergeant-Major L. Homfray Irving at the Royal Military College. Royal Military College of Canada Museum, Kingston. Ph: RC.

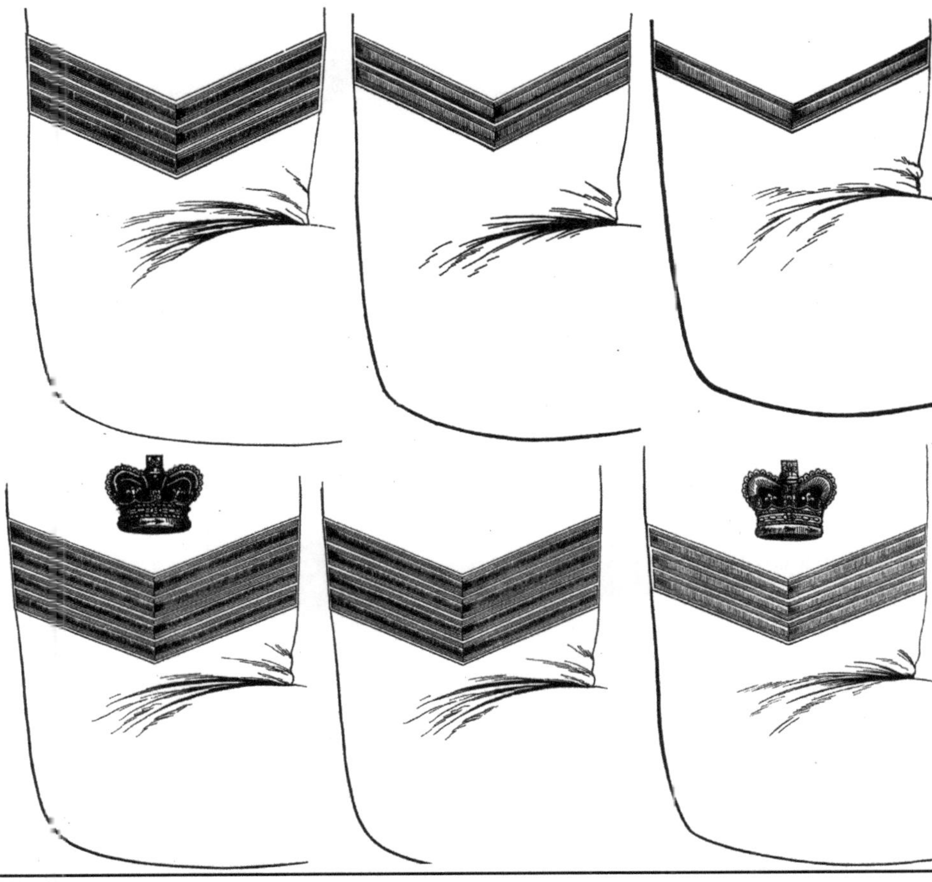

Other ranks and NCOs arm upper sleeve rank badges, 1860s-1880. From left to right, top row: lance corporal (or bombardier); corporal; sergeant. Bottom row: troop sergeant major; sergeant major; regimental sergeant major. Adapted from prints in Royal Artillery Standing Orders, 1864. Ph: RC.

chevrons on the left arm.
Bugle-Major: 4 chevrons and bugle
Bandmaster Sergeant: gold shoulder cords (no chevrons)
Orderly Room Sergeant: 3 chevrons
Paymaster Sergeant: 3 chevrons

Chevron of the above ranks were gold in the British army, but these would have been silver for Canadian volunteer infantry until the late 1880s. By then, the change to gold should have been completed.

Ranks below had cloth chevrons (including Orderly Room Sergeant and Paymaster Sergeant that only ranked as sergeants). These chevrons were white for infantry, yellow for artillerymen and cavalry and black for rifles. The cloth backing of the chevrons was usually of the facing colour. The ranks were:

Sergeant: 3 chevrons
Corporals: 2 chevrons

Dixon's 1867 Canadian *Volunteer's Active Service Manual* specified the same insignia as above for senior NCOs, sergeants and corporals and further indicated (since about 1815) one chevron for:

Bombardiers
Second-Corporals
Lance-Corporals
Acting-Bombardiers

From August 1868, the British army's colour sergeant's badge had the crown above crossed Union colours over the three chevrons worn on both sleeves. Same for rifle units, but with the crown, crossed rifles and bugle with a wreath badge. The wreath was removed from 1877.

In the British army from 1869, the four-bar chevrons (including those with crowns, etc.) were henceforth placed on the lower sleeve with points upwards. This became official much later in Canada as announced by

Rank badges worn below the elbow. 1860s-1880. From left to right: good conduct chevron; artillery brigade sergeant-major and quartermaster; artillery farrier-major (same for cavalry, but without the gun). Adapted from prints in Royal Artillery Standing Orders, 1864. Ph: RC.

General Order, Ottawa, 8 April, 1881, the badges of rank of Canadian NCOs as well as good conduct badges were to be worn as follows:

> "The chevrons of Non-Commissioned Officers of all arms of the service will, in future, be worn on the right arm only.
> All four-bar chevrons worn on tunics, or jackets, will be worn below the elbow, with the points upward, and all those of less than four bars will be worn above the elbow with the points downward.
> Good-conduct badges will be worn on the left arm below the elbow, with their points upward.
> All chevrons on cloaks or great coats will be worn below the elbow; four-bar chevrons with their points upward, and all others with their points down-ward.
> Sergeants will not wear crown badges.
> Those Non-Commissioned Officers ranking as Troop Sergeant-Majors, Battery Sergeant-Majors, Battery Quarter-Master-Sergeants, and Colour-Sergeants will wear three bar chevrons only, and crowns, or colour badges as the case may be."

Colour-sergeant's tunic, Canadian Volunteer infantry, c. 1881-1885. Note the three chevrons with crossed colours and crown introduced in Canada by MGO of 18 April 1881 (a dozen years after Great Britain). These are in silver lace up to 1885-1886, gold thereafter. Private collection. Ph:RC

From July 1882, British army Warrant Officers were to wear a crown (without chevrons) on the right lower sleeve.

There was at that time, and since the late 1850s, a multiplication of specialised ranks and recognised trades in the British army that brought about many new insignias that were worn in combination with chevrons. All these made their way to Canada and, eventually, might be seen on tunics of our volunteer militiamen. By the 1890s and early 1900s, the following badges of rank, trade and proficiency might be seen in Canada:

Master Gunner: crown and gun
Regimental Sergeant-Major (if Warrant Officer): crown
Regimental Bandmaster (if Warrant Officer): lyre and crown
Regimental Sergeant-Major (if not Warrant Officer): 4 chevrons and a crown
Regimental Bandmaster (if not Warrant Officer): 4 chevrons and lyre
Regimental Quartermaster-Sergeant: 4 chevrons
Battery Sergeant-Major and Quartermaster: 4 chevrons, cannon and crown
Colour Sergeant-Major: 4 chevrons, crossed swords, Union colour and crown
Regimental Drum-Major: 4 chevrons and a drum
Regimental Bugle-Major: 4 chevrons and crossed bugles
Quartermaster-Sergeant: 4 chevrons and eight pointed star
Sergeant Instructor of Musketry: 4 chevrons and crossed rifles
Colour-Sergeant: 3 chevrons, crossed Union colours and crown
Sergeant-Major: 3 chevrons and crown
Troop/Squadron Sergeant-Major: 3 chevrons and crown
Squadron Quartermaster-Sergeant: 3 chevrons and crown
Sergeant-Trumpeter: 3 chevrons and crossed trumpets
Sergeant-Drummer: 3 chevrons and drum
Sergeant-Musketry: 3 chevrons and crossed rifles
Sergeant-Pioneer: 3 chevrons and crossed hatchets

Infantry quartermaster-sergeant's rank badge, 1881, and Rifles' colour-sergeant badge 1887. As per N.P. Dawnay, The Badges of Warrant and Non-Commissioned Officers, 1949.

Sergeant-Signalling: 3 chevrons and crossed flags
Sergeant-Saddler: 3 chevrons and a bit
Sergeant-Band: 3 chevrons and lyre
Sergeant: 3 chevrons
Orderly Room Sergeant or Clerk: 3 chevrons
Armourer Sergeant: 3 chevrons
Hospital Sergeant: 3 chevrons
Paymaster Sergeant: 3 chevrons
Corporal: 2 chevrons
Corporal, Medical: 2 chevrons and red cross badge
Lance-Corporal and Second-Corporal: 1 chevron
Bombardier and Acting-Bombardier: 1 chevron

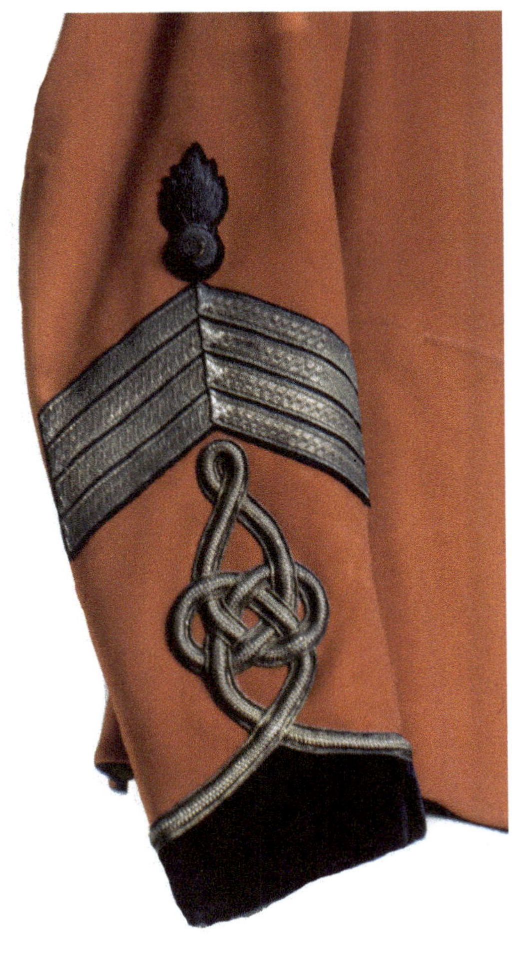

Lower sleeve of a volunteer militia engineer's sergeant-major. From 1881, sergeants of engineers had a grenade rather than a crown above the chevrons. Sheldon Kasman collection. Ph: RC.

As before, chevrons for senior NCOs were of gold (or silver in militia infantry until the late 1880s) and of white, yellow or black for lower grade sergeants and corporals depending on the arm of service. Naturally, there were exceptions and a few are given below. From 1881, engineer sergeants had a grenade instead of the crown.

On 19 May 1865, Toronto's 2nd Queen's Own Rifles were ordered to have: "For corporals, two black stripes on a red ground, for sergeants, three black stripes on a red ground; for colour sergeants, the colours worked in gold with one gold stripe beneath on the right arm, and three black stripes similar to those worn by sergeants on the left arm; for staff sergeants, three gold stripes. The quarter master sergeant will wear four gold stripes. The sergeant major will wear four gold stripes with a crown on each arm." This was altered on 8 October 1866 regarding "Gold chevrons worn by Staff Sergeants are abolished and replaced by black braid on a red background." The 31 March 1880 and 28 March 1894 regimental standing orders stipulated that henceforth: "Sergeants shall wear an Austrian knot of black square cord in place of the braid on the sleeve, and 1st Class Staff-Sergeants Tunic shall be frogged similar to a subaltern Officers, having the shoulder strap in place of the cord. Patrol Jacket: Staff Sergeants shall wear for undress a serge patrol jacket similar to the officers, but having the shoulder strap in place of the cord."[98]

For its corporals and sergeants, the 8th Rifles obtained permission in January 1882 to wear chevrons on both arms. As before, only NCOs in units distinguished by fusilier, grenadier and light infantry status were to wear chevrons on both sleeves while others were limited to the right upper sleeve. From various photos and surviving tunics, including many in this book, some of the other units also had chevrons on both arms and, in at least one case, just the left arm.

The Royal Military College's cadet's chevrons worn until 1882 were all of gold. Later on, in Highland units, when white jackets were procured for undress, the chevrons on

[98] The Queen's Own Rifles of Canada Regimental Museum and Archive at: https://qormuseum.org/history/timeline-uniforms/

those were red.

Sergeant's Sashes

On 18 February 1826, a new sash for sergeants was approved. It was of crimson wool as previously, but now had three streaks of the facings colour in the centre rather than a broad single stripe (WO 44/609). The sash was worn around the waist over the coatee except in Highland regiments whose sergeants wore their sashes over the shoulder.

From August 1845, the sergeant's sash was completely crimson and was to henceforth be two and a half inches wide. From April 1859, the sashes of staff sergeants and sergeants were to be worn over the right shoulder under the shoulder strap, its "tassels level and not to hang below the tunic" advised Dixon in 1867. In February 1889, there was a suggestion from the general officer commanding "that the silk sashes for Sergeants be exchanged for the brown leather shoulder belt and pouch" but this was not adopted (LAC, reel T4178/102).

Proficiency badges, Royal Military College of Canada, c. 1880-1884. Top and bottom: crossed sabres for use of side arms; centre: spur for equitation; left and right: crossed guns and grenade for artillery practice. The crown (here on top of the crossed guns) was awarded to the top cadet in each skill category. Royal Military College of Canada Museum, Kingston. Ph: RC.

Badges of Appointment

Bugler or Trumpeter: crossed bugles or crossed trumpets

Drummer: drum (usually on private's serge frock only, not worn by rifle units nor Governor General's Foot Guards)

Canadian officer wearing a patrol jacket with 1877 British National Rifle Association badges on his left sleeve. Such award badges appear to have been allowed on service dress jackets only. Andrew Moss Collection. Ph: CL.

Trade Badges

Farrier: hammer and pincer
Pioneer: crossed hatchets
Bandsman: lyre
Rough-Rider/Riding Instructor: spur
Medical Orderlies: red cross badge
Signaller: crossed flags
Saddler: bit
Shoeing-Smith: horseshoe
Wheeler: wheel

As seen in a few examples above, these badges were worn over the chevrons of those tradesmen that ranked as NCOs.

Good Conduct Chevrons

10 October 1836 — A chevron with its centre point uppermost to be worn above the right cuff "for a period of good conduct"; for cavalry higher up on the arm below the elbow. In 1864, the chevron was men-

Badges of target shooting competitions, 1877. These badges were authorised to be worn by the few militiamen who participated at the international marksmanship at Wimbledon and Kolapore organised by the British National Rifle Association. Detail from an image in the Andrew Moss Collection. Ph: CL.

tioned as being white except black for rifles and gold for foot guards and musicians. Artillery and engineers had a gold chevron for acting bombardiers and musicians, yellow for gunners and sappers. From 1 March 1881, a soldier could have: one good conduct chevron after two years, two chevrons after six years, three chevrons after 12 years, four for 18 years, five for 23 years and six for 28 years; these were henceforth worn on the left arm below the elbow.

Proficiency Badges for NCOs and Enlisted Men

10 March 1858 — Badges for good shooting: 1) best shot of the battalion had crossed muskets and crown worked in gold, 2) best shot of the company had gold crossed muskets only, 3) marksmen, not to exceed 100 per battalion, had crossed muskets in worsted. The badges on a backing of facing colour cloth and worn on the left sleeve just above the slashed cuff flap. There could be variations as the January 1885 Standing Orders of the 7th Fusiliers (London, ON) stipulated: 1) best shot in the regiment "will wear gold crossed guns [rifles]" and best shot in a company "will wear worsted guns [rifles]."

28 June 1865 — Badges for swordsmanship: 1) best swordsman of the regiment had crossed swords and crown worked in gold, 2) best in troop had crossed swords only, 3) second best in the troop had crossed swords in worsted. The badges on a backing of facing colour cloth and worn on the left sleeve just above the cuff.

22 May 1872, Regimental Order , B Battery, Royal Canadian Artillery — artillery marksman: crossed cannons in gold on right arm above Austrian knot; small arms marksman: crossed rifles in gold on left arm above Austrian knot

Royal Military College of Canada — rifle practice: crossed rifles; use of side arms: crossed sabres; artillery practice: crossed guns; signalling practice: crossed flags; equitation: spur. These were awarded to the top five cadets of these five military skills. The top cadet in each skill had a crown added to the badge.

Aiguillettes

These ornate shoulder cord could be worn on the uniforms of those officers that held an appointment in certain posts on the general or district staff, or as ADC's (Aide de Camp) to general officers or governors. They were worn on the right shoulder and were gold or silver according to the unit's button and lace colour.

Aiguillettes worn by enlisted men were seen particularly worn by trumpeters in regiments uniformed as cavalry or dragoon regiments, but not those as hussars. Between 1893 and 1898, the Governor General's Body Guard of Toronto obtained permission from Governor General the Earl of Aberdeen to wear aiguillettes, silver for officers and white for enlisted troopers.

Notes

N.P. Dawnay, *The Badges of Warrant and Non-Commissioned Officers Rank in the British Army* (London: Society for Army Historical Research, 1949); N.P. Dawnay, *The Distinction of Rank of Regimental Officers 1684 to 1855* (London: Society for Army Historical Research, 1960); N.P. Dawnay, "Staff Distinctions 1855-1897", JSAHR, XXXII (1954): 165-178; William Y. Carman, W.E. May and John Tanner, *Badges and Insignia of the British Armed Services* (London: A & C Black, 1974); David Linaker and Gordon Dine. *Cavalry Warrant Officer's and Non-Commissioned Officers' Arm Badges* (Military Historical Society Special Publication, Glasgow, 1997); F.E. Dixon, *The Volunteer's Active Service Manual* (Toronto, 1867), pp. 61-63; LAC, RG9, IIJ2, Vols 21, 59-62, 116, Royal Canadian Artillery, Regimental Order Books, A and B Batteries, 1871-1898; LAC, reel T4176/230, Deputy Minister's ledger, January 1882; *Clothing Regulations for the Militia* (Ottawa, 1905); *Dress Regimental Standing Orders , 2nd Queen's Own Rifles* (31st March 1880 and 1894); *Standing Orders of the 7th Battalion Fusiliers* (London, ON, 1886); W.D. Otter, *The Guide: A Manual for the Canadian Militia (Infantry)*, (Ottawa, 1905).

Lt. Col. Philippe Landry, 61st Montmagny and l'Islet Battalion, c. 1890. He led the battalion from 1885 to1901 and wears his dress uniform with a gold aiguillette indicating a staff appointment. Print in J.A. Lavoie's Le Régiment de Montmagny. Ph: RC.

Wing on a Lower Canada volunteer rifles or light infantry captain's coatee, Argenteuil County, 1820s. Gold chain on scarlet with a silver bugle badge on a gold plated disk and gold bullion fringes. This generally conforms to an 1817 order at Herbert's and a similar one of about 1820 for a 60th Rifles officer wishing to obtain a "pair gilt double curb chain wings…gilt beaded centre plat with silver bugle… bullion 1 3/8 in. long." Argenteuil County Museum, Grenville, QC. Ph: RC.

Colour sergeant's badge, 6th Hochelaga Light Infantry, c. 1862. Introduced in 1813, this senior NCOs rank badge consisted of a single chevron with crossed swords, colour and crown above. This Canadian example is typical of those used in the Victorian period. Used until 1868 in the British army. Private collection. Ph: RC.

Chapter 11

Accoutrements and equipments

Montreal Militia, 2nd Battalion officer's silvered shoulder belt-plate, c. 1820. National Historic Sites, Parks Canada. Ph: RC.

Knapsacks, haversacks, water bottles and other such items used by Canadian militiamen were those of the British army. They were not routinely issued to the volunteer or sedentary militia units, but were kept in stores and served to units called to active duty during emergencies. Storage seems to have been, besides a central depot in Ottawa, often spread in armouries across the country, especially those of volunteer battalions in cities.

The models issued to Canadians when called out on active duty were not necessarily new. Following the 1837-1838 Rebellions, Adjutant General of Militia Richard Bullock commented that it "was indeed pitiable to see the manner in which most of our Militia was equipped (I mean the Sedentary Militia) during the disturbances; old knapsacks of the late War of 1812, quite rotten, rotten accoutrements and vast numbers [of bayonet and sabre belts] without scabbards."[99]

The above comments were applicable to those many units that were called out for a few months of service. The more permanent embodied militia units that served full-time until 1850 received equipment and uniforms regularly from England. The Permanent units raised from 1871 also had good equipment. However, as seen below, this was not necessarily the case with units called up for duty in 1885 or sent to South Africa in 1899.

Foot Troops Accoutrements and Equipment

The accoutrements worn by British troops did not change a great deal from the early 1800s until the early 1850s when Enfield rifled muskets were introduced for foot troops. For most foot troops, they consisted of "cross belts" worn over the shoulders, one supporting a cartridge pouch holding 60 rounds and a bayonet belt holding a bayonet and its scabbard. The interior of the British black leather infantry pouch was altered in 1845 to have a tin magazine of three compartments holding 20 rounds each. The belts were usually of white leather for infantry units. The pouches were of black leather except for artillery which had them in white leather. It also had a belt-plate upon which the unit's badges and/or numbers might be displayed; others might be plain or with a generic badge such as a crowned "VR" cipher. The enlisted men's shoulder belt-plates were of brass, those of officers' shoulder sword belts were gilded or silvered. From September 1844, infantry coats were made with a small patent leather cap pouch that fitted behind the third lace loop (up from the waist) on the right side and only its small rounded flap was visible. This hidden little pouch became unnecessary from 1850 when the waist belt started being gradually distributed since it had the small black leather cap pouch attached to this belt.

Accoutrements for rifle units were all black and consisted of a cartridge pouch with its shoulder belt and a waist belt holding a smaller pouch (also called ball bag) and a frog for the rifle's sword-bayonet. Instead of belt-plates, riflemen simply had a brass "snake" clasp on the waist belt. Sergeants had in addition a silvered whistle

[99] LAC, RG9, IB1, Vol. 71, Adjutant General Bullock to I.M. Higginson, Military Secretary, 14 February 1844.

and chain fixed by a round plate on the shoulder belt. Officers had the same, but more ornate and also had in addition a silver badge.

The basic accoutrements worn by most Canadian volunteers were the infantry "cross belts" consisting of the cartridge pouch and the bayonet belt that could be of white or black leather. Such equipment could be issued from Ordnance stores to volunteer units that were active and had uniformed themselves. For instance, in June 1824, the Volunteer Artillery Company at Hull, Lower Canada, (now Gatineau, QC), was issued with 60 "Musquets complete India [pattern]..." with black leather slings, artillery "Swords [with] Buff Leather" belts, black cartridge pouches "Infantry Regular Pattern" with sergeant's pikes and hangers as well as artillery flannel cartridge, harness, ropes, whips, etc., for the two field guns. The Montreal Rifle Corps received 80 (undoubtedly Baker) rifles with their "sets accoutrements" in 1832; these would have been of black leather.[100]

British line infantry accoutrements, c. 1812 to c. 1853. The two soldiers at left wear accoutrements that remained basically the same until the introduction of the 1853 Pattern rifled infantry muskets.

During the 1837-1838 Rebellions, the volunteer units might have a mixture of white and black infantry belts, most likely kept to one colour in a corps or at least in a company such as in the "Queen's Own Rifles" of Prescott, Upper Canada. Col. Heriot's Eastern Township Royal Volunteers had an issue of 420 black accoutrements in January 1838. Although white accoutrements were often seen, black ones appear to have been the prevalent choice for volunteer corps, especially if made in Canada. A November 1838 government contract specified: "10,000 leather scabbards for bayonets according to pattern..., 1686 pouches, 1686 bayonet belts, 1686 musket slings...to be made of the best black leather." From 1839, the incorporated and provincial units that served until the last were disbanded in 1850 had white leather belts with black cartridge pouches similar to those of the British regular army.[101]

Nova Scotia Volunteer Artillery gilt shoulder belt-plate, c. 1828. On 28 March 1828, some 300 "Brass Breast Plates Square, and to suit the belts - with the Words 'Nova Scotia Volunteer Artillery' cast in" were ordered by the province for its militia artillery units (PANS, RG1, Vol. 440). Courtesy The Army Museum, Halifax Citadel.

1850s

On 20 May 1850, the The British army's "broad and heavy" infantry bayonet shoulder belt and belt-plate was abolished and replaced by a waist belt; its waist belt buckle was usually the round type with a crown for infantry. This universal model was worn by Canadian infantry volunteers NCOs and privates during the following decades. Officers could have more elaborate buckles with unit insignia and, in Canada, many officers opted for a "beaver" badge. The

[100] LAC, RG9, IA1, Vol. 33, Return of the Arms, Accoutrements and Ammunition issued to the Quebec Volunteer Companies from the 1st January to the 31st of December 1824; LAC, RG8, C799, Ordnance issue of 3 September 1832.

[101] LAC, RG8, C800, Required for the immediate use of the Township Royal Volunteers under the Command of Col. Heriot, 10 January 1838, issue approved on 15 January 1838; *Quebec Mercury*, 20 November 1838.

1st Halifax Militia Regiment, silvered shoulder belt-plate, c. 1835. Courtesy The Army Museum, Halifax Citadel.

2nd Halifax Militia Regiment, silvered shoulder belt-plate, c. 1835. Courtesy The Army Museum, Halifax Citadel.

1st Battalion, Sunbury County Militia, New Brunswick, gilt officer's shoulder belt-plate, c. 1835. National Historic Sites, Parks Canada. Ph: RC.

1st Battalion, Northumberland County Militia, New Brunswick, gilt officer's shoulder belt-plate, c. 1835. National Historic Sites, Parks Canada. Ph: RC.

Upper Canada Incorporated Militia officer's silvered shoulder belt-plate, c. 1840. Courtesy Scott Duncan.

Sketch of a British infantry cartridge box, 1838. during October 1838 by Henry Dearborn, an American visitor in Montréal who made notes on the accoutrements. He commented they "have a tin box divided into three compartments in which the cartridges are put in paper bundles of 20 each making 60 rounds & the name & number of the soldier is put on each bundle…", "The Dearborn Journals", Buffalo Historical Society, 1904.

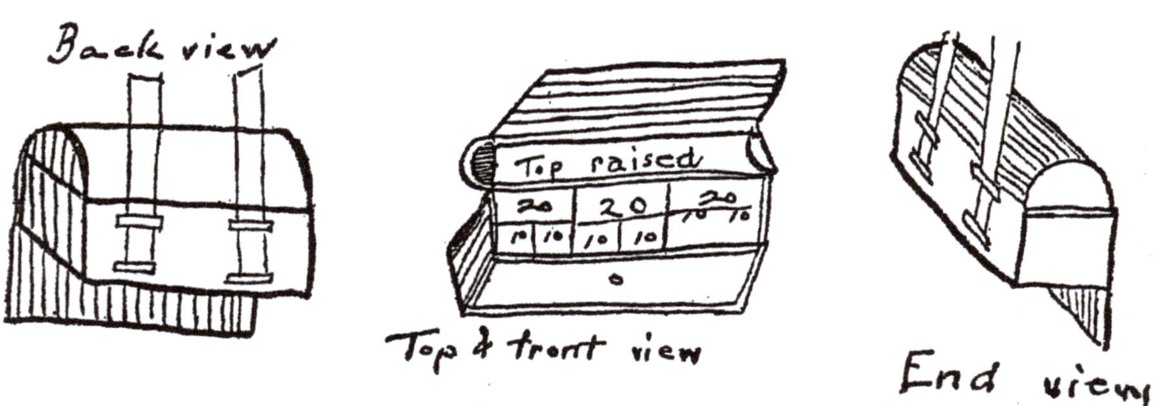

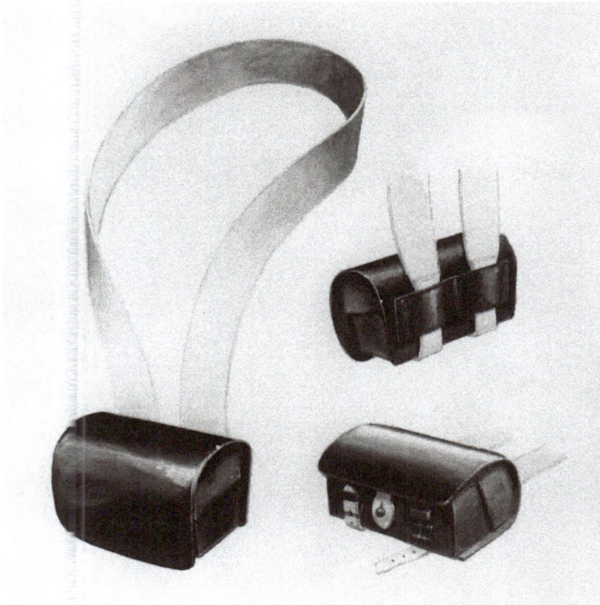

Cartridge pouch and belt, 1853 pattern. This type of pouch and its variations was used by Canadian volunteer infantrymen from the 1850s to 1890s. Drawing by Douglas Anderson. Courtesy Halifax Citadel National Historic Site, Parks Canada.

A sergeant and two privates of the 60th Rifles, c. 1832-1843. All wear the distinctive black leather accoutrements used by all rifle units armed with the Baker rifle shown. The sergeants had in addition the whistle and chain on the cartridge box shoulder belt. The same type of accoutrement continued to be used with the advent of the Brunswick rifle used from the early 1840s to the early 1850s. Watercolour by C.B. Newhouse. Anne S. K. Brown Military Collection, Brown University Library, Providence, USA. Ph: RC.

Rifle officer's pouch with the generic stringed bugle (or horn) badge in silver, 1830s. This example is from an officer's belt of the 60th Rifles but Canadian rifle officers would have worn the same type of pouch and badge. Christopher Bryan Collection. Ph: RC.

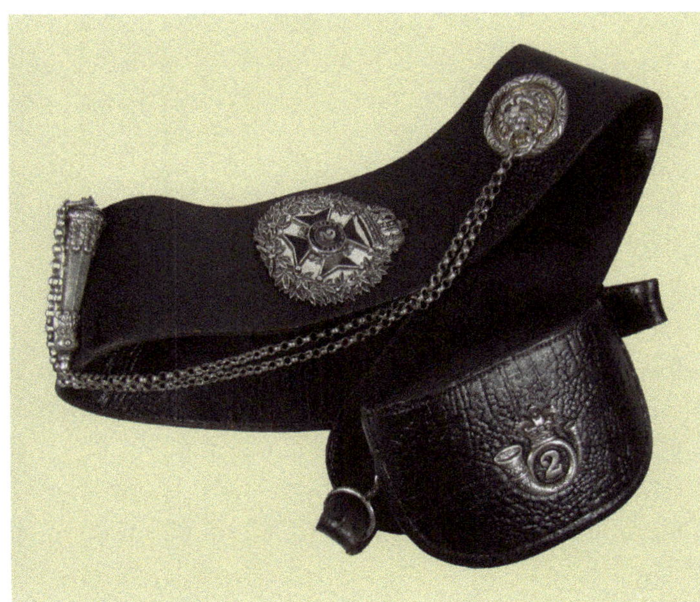

2nd Queen's Own Rifles officer's shoulder belt and cartridge pouch, last third of the 19th century. Black leather with dull silver metal whistle, chains, plate and badges. Courtesy Scott Duncan.

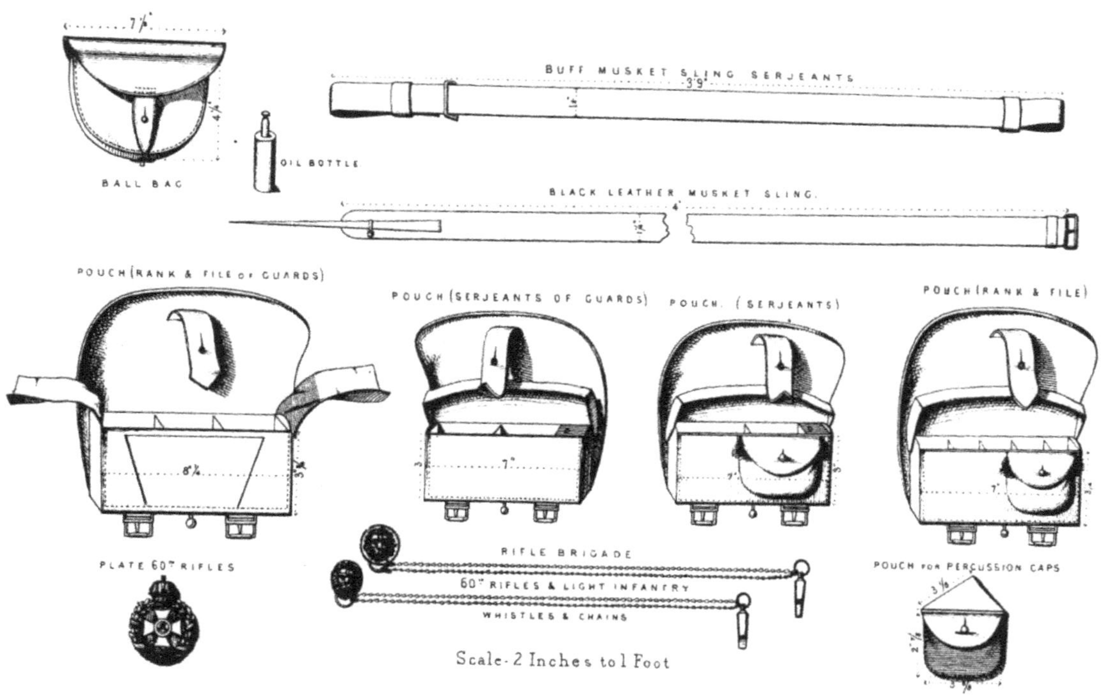

Infantry and rifles pouches, ball bags, musket slings and rifle whistles and chains, 1860s. The pouches (at right) for sergeants were smaller and with two compartments than those of the corporals and privates which had five compartments (or two depending on pattern). The Guard's pattern pouches (left) were not used by Canadian Volunteer units. Plate from Army Equipment: Infantry, 1865.

advent of the Pattern 1853 Enfield rifled musket also brought a new 40 round cartridge pouch. Rifle units henceforth had the same equipment as infantry units, but in black leather with brass snake clasps.

From late 1856, the accoutrements used by Canadian volunteer rifle and infantry units were those appropriate with the Pattern 1853 Enfield rifled musket and its bayonet that was issued from that year. This consisted of "a 40-round pouch, a pouch belt, waist belt and gun sling" with a muzzle stopper and a nipple wrench. The smaller ball pouch holding 20 rounds slung on the waist belt was also issued to many units, being gradually replaced from 1860 by the British army's new semi-circular ball pouch. Artillery units had the same equipment as the British Royal Artillery whose pouch was henceforth of black leather with a brass field gun badge on its flap.

1860s-1890s

There were initially only a few thousand accoutrements, but the numbers shot up to tens of thousands following the "Trent Affair" that occurred on the high seas on 8 November 1861. This resulted in the immediate dispatch from Britain to Canada of 20,000 Pattern 1853 Enfield rifled muskets with infantry accoutrements soon followed by 10,000 Pattern 1858 Short Enfield rifles with sword bayonets and their rifle accoutrements. The black rifle accoutrements had been all issued by 10 January 1862, only white buff belts were left in stores. The adjutant general then ruled that white belts would be blackened "on the outside and edges" for issue to rifle units in dark green uniforms.[102]

This was not always done. The Elora Rifle Company was sent to the frontier in 1866 "without greatcoat straps and had to manufacture them out of carpet binding. It was also compelled to use, upon dark green rifle uniforms, white pipe clayed belts, which could not be blackened until General Napier, who was then at Toronto, granted the necessary permission; and then the volunteers were not allowed to make the change, but old soldiers of the Canadian Rifles, with whom the Elora company was lying in barracks at Chatham, did the work and received the pay. To this no objection was made, and some of the veterans of that worthy corps fought their battles o'er again while disposing of the unexpected windfall" according to Charles Clarke's, *Sixty Years in Upper Canada*.

Back in July 1862, a further 40,000 rifled muskets were received with 45,000 sets of accoutrements. All this was intended to arm and equip the Canadian Militia. Thus, by the end of 1862, there was more than enough

[102] LAC, RG9, IC1, Vol. 114, Circular, 10 January 1862.

Infantry kit, c. 1861-1867. Canadian infantry volunteers mobilised for active duty should have been equipped like their comrades of the British regular army shown here. From 1854-1857, the ball pouch was worn on the right side of the waist belt and the small cap pouch at the centre of she shoulder belt. The greatcoat was fastened folded to the back of the knapsack and the mess tin above, the haversack worn on the left hip. The short leggings issued to British soldiers were not issued to Canadian volunteers. Print after R. Simkin in the 1890 Illustrated Military & Naval Magazine. Ph: RC.

modern weapons and equipment to equip existing volunteer units as well as those that would be raised in the future with a substantial reserve intended for the Sedentary Militia should some of it ever be mobilised. These accoutrements were similar to those used by the British regular army and consisted of an ammunition pouch with its belt, a percussion cap pouch, a waist belt and a bayonet frog.

Cartridge Pouches

As seen above, the British 1853 cartridge pouch initially held 40 rounds that were placed in its two compartments while the percussion caps had their own little pocket attached on the face of the pouch. From 29 April 1856, the pattern infantry equipment was to consist of "a large pouch, small pouch with buff belt loops, pouch belt, bayonet belt and cap pocket." Eventually, a new 50-rounds pouch with "a ball bag to hold ten rounds" was approved by the Queen on 16 June 1859 (WO 370/5). It featured a tin magazine with five compartments, each to hold a packet of 10 rounds of .57 Enfield ammunition with a small pocket to carry the percussion caps.

The old 40-round pouch in two compartments, however, did not disappear and was notably used by Canadian volunteer rifle companies armed with the Short Enfield with sword bayonet during the 1860s. Furthermore, the advent, from 1859, of the 50-round cartridge pouch with its cap pouch should have eliminated the use of the small pouch on the waist belt. However, this proved awkward to use and a new model cap pouch, this time fixed on the cartridge pouch belt, was again issued from about 1860-1863. Although there was a leather percussion cap pouch issued that was attached to the front of the shoulder belt, many images don't show it and it was officially abolished in Britain during 1867. Nevertheless, Canadian Volunteer Militia tunics continued to be made with a small cloth pocket at the front right waist to keep extra percussion caps up to the early 20th century, about two decades after it was eliminated from British tunics.

During 1867, the arrival of 30,000 Snider-Enfield rifles from England and their issue to volunteer units meant changes in the accoutrements. A Militia General Order of 23 August 1867 informed volunteers that the "50 round pouches…can be altered to contain 50 round of the new 'Central Fire Cartridges' by removing the tin partitions, with exception of that between the 20

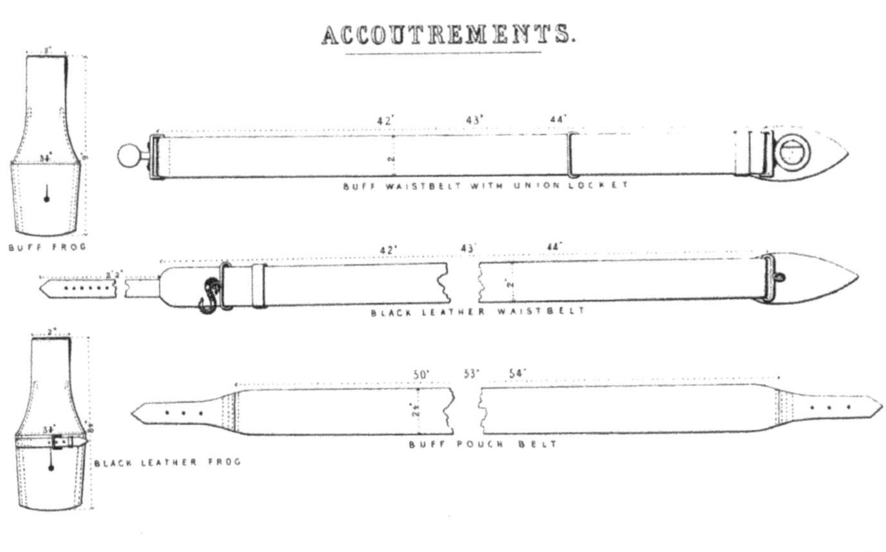

Infantry and rifle accoutrement belts and bayonet frogs, 1860s. These patterns were used by Canadian volunteer infantry and rifles until the late 1890s. Plate from Army Equipment: Infantry, 1865.

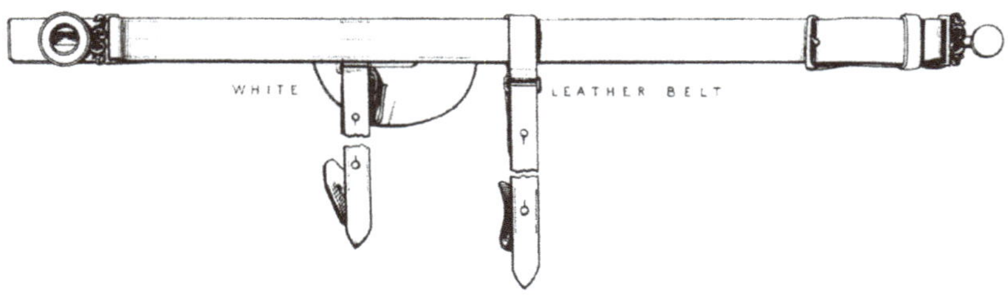

Infantry officer's white leather waist belt. Plate from Army Equipment: Infantry, 1865.

Infantry officer's belt and buckle, c. 1890. This item was worn by the colonel of the 34th Ontario Battalion. Crimson leather belt laced with gold, gilt buckle silver beaver. Ontario Regiment Museum, Oshawa. Ph: CL.

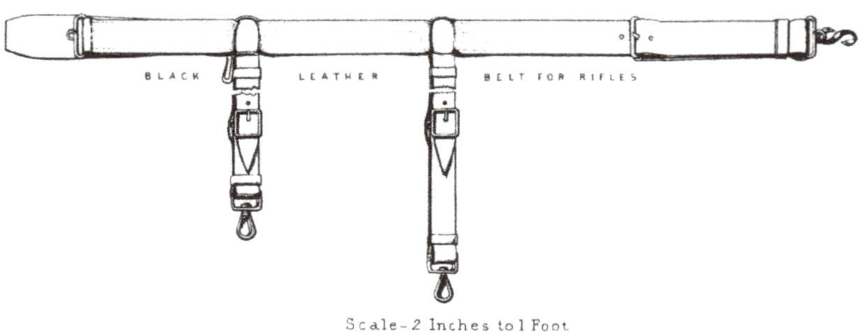

Rifle officer's black leather waist belt. Plate from Army Equipment: Infantry, 1865.

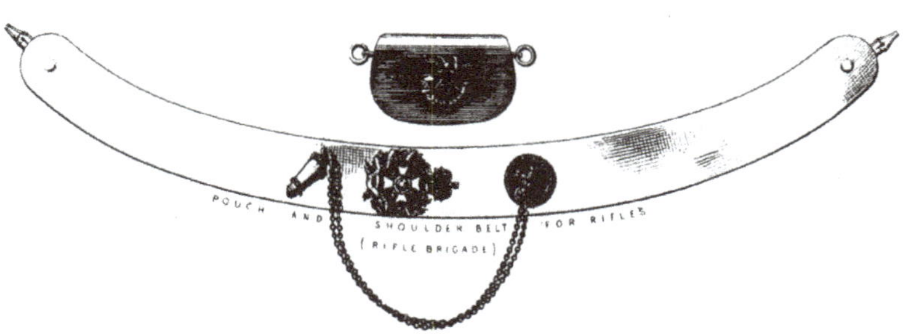

Rifle officers's pouch and shoulder belt. Apart from the Rifle Brigade's badges, this was the pattern used by Canadian rifle officers. Plate from Army Equipment: Infantry, 1865.

Royal Artillery foot battery enlisted gunner's marching order cartridge pouch, the water bottle (marked to the unit, company and rack), accoutrements, knapsack and haversack, 1860s. Canadian gunners had the same equipment. Print in the 1864 Royal Artillery Standing Orders. Ph: RC.

Cartridge pouch for Canadian Volunteer Militia Artillery, c. 1856-1870. Made of black leather, this cartridge pouch model had a peculiar brass badge consisting simply of a field gun. It had no crown or inscriptions as in the regular British artillery. This item was similar to that used by the British Royal Artillery. Officers also had this type of cartridge pouch for undress, but with gilded field gun badge, and they were used "to contain a small telescope, scale, pair of compasses and pencil" according to the 1864 Royal Artillery Standing Orders. Old Fort Henry Museum, Kingston. Ph: RC.

Rifleman James M. Austin, Company H, 49th Hastings Rifles, 1885. This company was part of the Midland Battalion that took part in the North-West Campaign. Note the non-official web ammunition belt over the shoulder. After a lithograph. Private collection. Ph: RC.

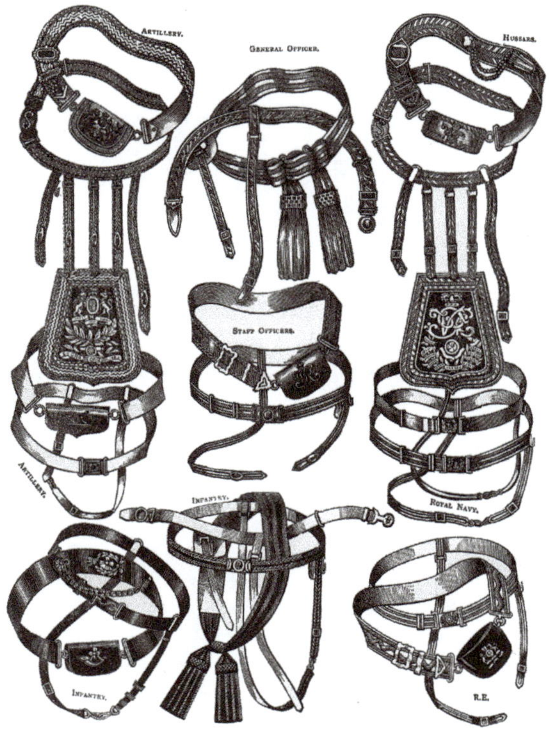

British army officer's dress and belts, pouches, sabretaches and sashes, 1880s-1890s. This remarkable plate shows the basic items for the various arms of the services. Canadian officers had identical items often ordered from British firms such as Holding's. The undress service belts are also shown for the artillery (top and centre left - the undress pouch is also shown), hussars (top right) and infantry (bottom centre). The rifles (bottom left) are mistakenly labeled "Infantry." Print after T.H. Holding's 1894 catalogue. Ph: RC.

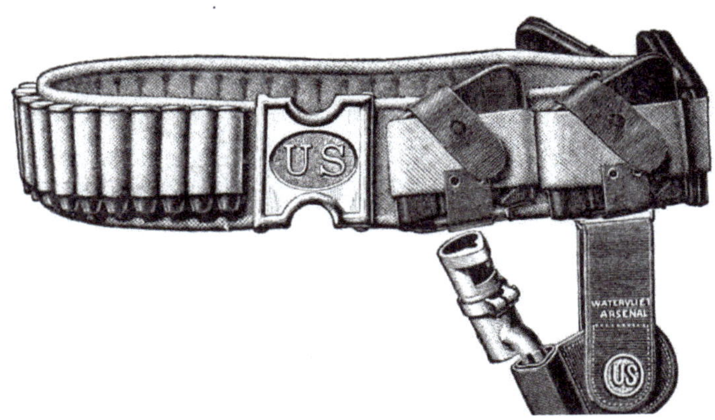

An American Mills web belt, 1880s. Some Canadian volunteers unofficially obtained web belts during the North-West Campaign. Illustrated Naval and Military Magazine, VII (1887). Ph: RC.

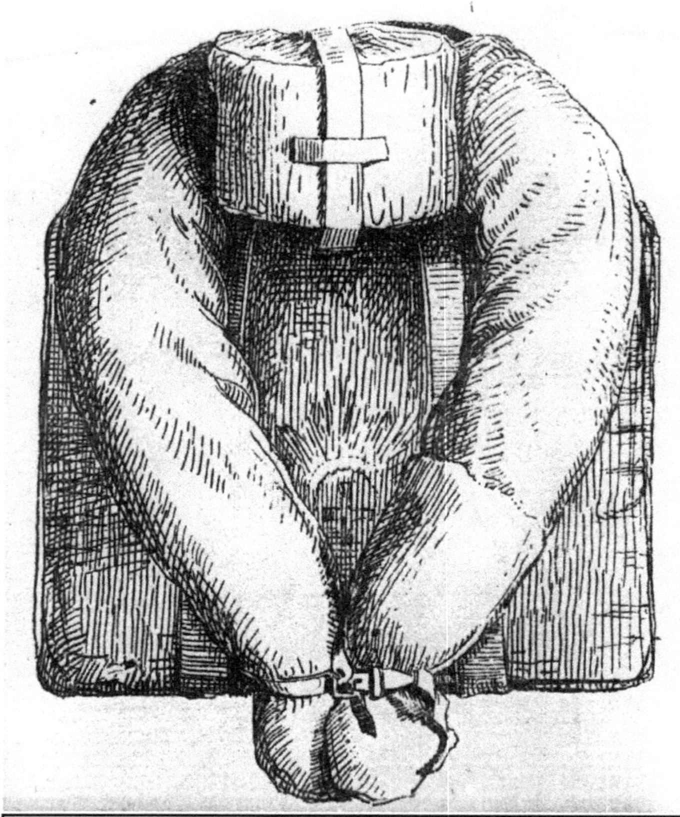

Infantry knapsack, 1824. Illustration inserted with the 28 January 1824 General Orders on knapsacks, showing how to attach the blanket or greatcoat. This was not found practical and troops soon reverted to rolling these items on top of the knapsack. Courtesy Library and Archives Canada.

and 30 rounds. If the pouches are of the old altered 40 rounds pattern, it will be necessary to remove the whole of the partitions."

The old accoutrements were retained for a number of years with these changes although new ones eventually were also received. There were now two basic patterns of Snider-Enfield cartridge pouches used in Canada: the modified 40-round pouch with no lining; the altered 50-round pouch with its two tin compartments. The belts remained unchanged. Since the Canadian Militia retained the Snider-Enfield as it basic weapon until the end of the 1890s, these accoutrements remained the same until the 1898 advent of the Oliver equipment (see below).

Thus, volunteer that took part in the 1885 North-West Campaign had the old 1860s type of accoutrements. For infantrymen, these were still of white leather and some units browned or obscured the leather so as to not provide easy targets as shown by an image of No. 2 Company of the 10th Royals. A few further improved what they had. A photo of Rifleman Austin, Company H, 49th Hastings Rifle Battalion, that was part of the Midland Battalion shows a shoulder belt with cartridge pockets made of what appears to be web or heavy linen slung over the shoulder. This type of ammunition belt appears to have been used only by some companies and was not an official general issue. Is inspiration was from a belt invented after the American Civil War by US Army officer Anson Mills. It resembled the popular "frontier belt" used by American frontiersmen, but Mills devised a manufacturing process weaving the main band and bullet loops together without resorting to stitching. Naturally, some Canadians were attracted by this handy ammunition belt during the 1885 North-West Campaign and they were not forgotten later on as an eventual replacement for the white leather belts. In 1900, Mills even gave some of the surplus belts (that had been made for US troops during the Spanish American War two years earlier) to Canadian volunteers going to South Africa.[103]

Meanwhile, since the early 1870s, cartridge boxes, bayonet scabbards and their belts became integrated with other parts of the equipment in the British army as well as in most European armies. In Canada the personal accoutrements and equipment for volunteers remained the same until the until the later 1890s and even the tiny force of regulars did not see much new before the 1880s as will be seen below.

Knapsacks

The knapsack was the equivalent of the modern "back pack" used by campers today. The model used when Queen Victoria was crowned had its genesis in 1805 as the "infamous Trotter knapsack" that lasted until 1871 in the British army and in Canada until 1898 if not later! It was a box-like affair that always remained basically the same in spite of new models bearing "improvements" that agonised generations of British servicemen and Canadian volunteers in the field. Briefly, it consisted of a square (or somewhat slightly rectangular) black painted canvas item with planks within on each side to keep it stiff, corners reinforced with black leather with two leather straps that went vertically around the knapsack. A harness of shoulder straps was attached to it. Especially note-

[103] *Illustrated Naval and Military Magazine*, VII (1887), p. 134; Douglas C. McChristian, *Uniforms, Arms, and Equipment: the U. S. Army on the Western Frontier* (Norman: University of Oklahoma Press, 2007), pp. 93-100.

worthy was its third strap that came across the chest causing medical officers to comment that this could affect breathing.

The knapsack models in use in Canada during Queen Victoria's reign were, besides those dating from the War of 1812, as follows:

Pattern 1827: 18 inches high, 17 inches wide and 3 inches deep. No greatcoat or blanket to be rolled over the top, but enclosed in a valise, but this was abolished two years later.

Pattern 1838/40: 13 inches high, 17 inches wide and 3 inches deep.

Pattern 1854: 12 1/2 inches high, 15 inches wide and 3 inches deep. Some metal fittings added. In 1856, a wooden stick was added to the upper inside edge to solidify it.

A folded greatcoat or a rolled blanket could be carried on top of the knapsack. The blanket was of grey wool and, according to a March 1899 contract, measured 60 by 84 inches (1,52 by 2,13 m as per LAC reel T386/638). A mess tin with a white linen cover, changed to black linen and strap from 1838, could also be fixed on top or at the back of the knapsack. Introduced in 1813-1814, the mess tin had a semi-circular kidney shape and remained a standard issue, with some small improvements, for Canadian soldiers until 1939.

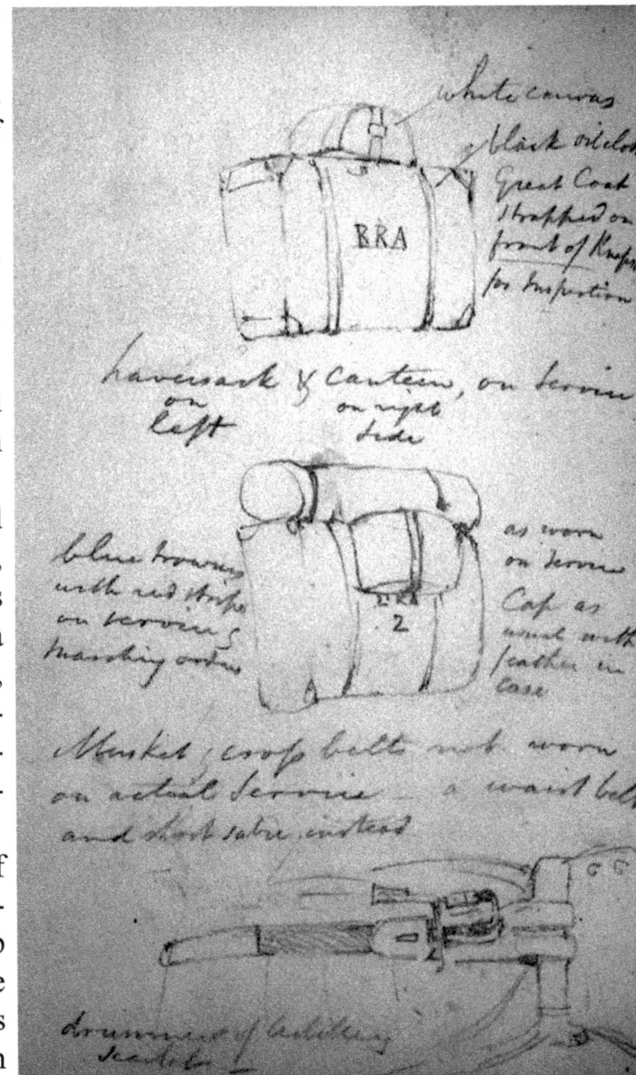

Knapsack with mess tin for Royal Artillery enlisted men, c. 1828-1845. These sketches with notes by British military artist Henry Martins show the different ways of putting the mess tin in garrison (top) and on service with the rolled greatcoat or blanket. This would also apply to infantrymen. Other notes pertain to waist belts worn by gunners when on service instead of cross belts. Anne S. K. Brown Military Collection, Brown University Library, Providence, USA. Ph: RC.

The 1854 pattern knapsack with its 1856 addition of the wooden stick was the one most used by mobilised Canadian volunteers from the 1860s. A sufficient number to equip a few companies was kept in district stores in case they were called to active service. This worked for small emergencies such as aid to civil power, but was totally inadequate when whole battalions were called up, as was the case in 1866, 1870 and 1885. From 2 January 1857, and in force according to Dixon's 1867 Canadian *Volunteer's Active Service Manual*, knapsacks were to be marked with the battalion's number "in arabic numerals 1 3/4 inches in height, painted in white, in the centre." Fusilier, light infantry and rifle battalions would have a bugle or a grenade "in white paint over the number on the back." In practice, this was probably not done unless a battalion had its own supply of knapsacks, which does not seem to have been very common. Surviving Canadian volunteer's knapsacks are usually unmarked.[104]

In June 1870, after the skirmish at Pidgeon Hill, QC, where the Fenians fled in great haste back over to Vermont leaving hundred of American rifles and accoutrements in the field, "Col. Smith conceived the idea of letting his old corps, the [3rd] Victoria [Rifles], carry Fenian knapsacks for the future as part of their accoutrements" according to the June 27 *Volunteer Review*. Col. Smith hired a carter on the ground to pick up whatever he could and he came back to Montreal with some 300 knapsacks! Thus, it seems that the 3rd Victoria Rifles secured their own supply of knapsack, which must have been used for some years. These American knapsacks were likely surplus from their Civil War. Union army knapsacks were largely somewhat like those of the British army. A few battalions are recorded as having applied and possibly issued knapsacks from militia stores at various times. We have seen reference to the 57th in April 1876, the 46th in August 1881, the Quebec Field Battery in January 1882 and a hundred knapsacks at London (ON) being unfit for service in October 1882. As for haversacks, a Militia General Order of 4 February 1883 mentioned that "Haversacks need not be worn as a general rule, except on ser-

[104] The knapsack of the 2nd Queen's Own Rifles had "QOR" painted in white as per the unit's Museum in Toronto.

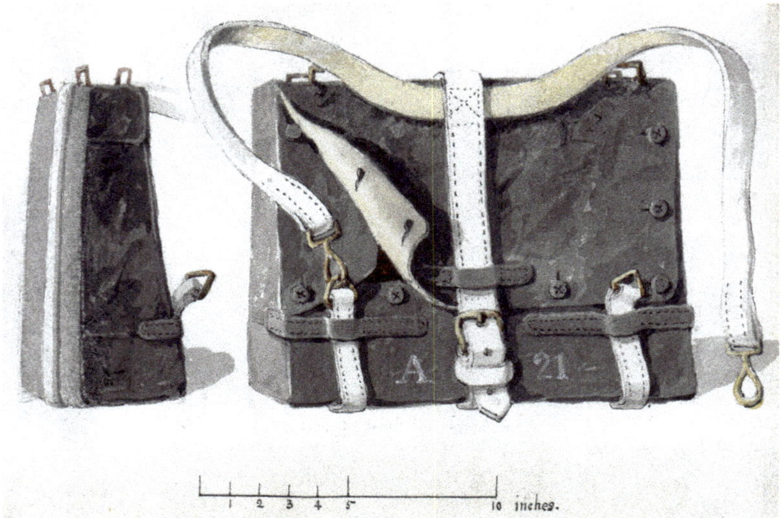

Soldier's infantry knapsack, c. 1837. Watercolour by Michael Angelo Hayes. This would be the 1827 pattern. Courtesy Yale Center for British Art, Paul Mellon Collection.

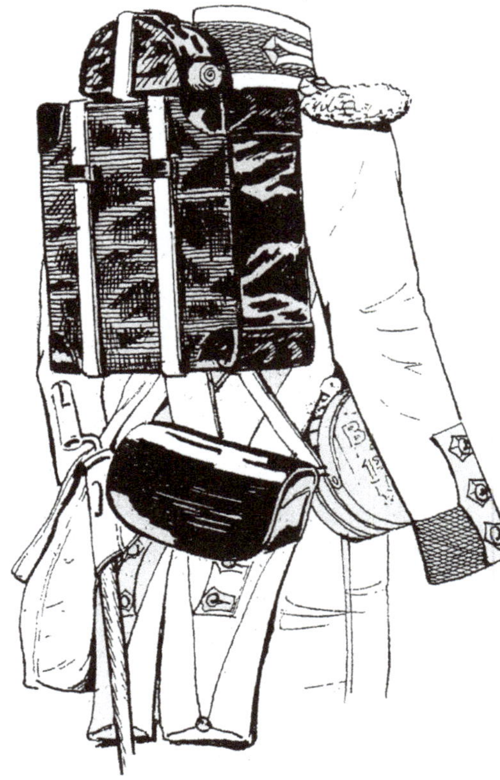

Infantry knapsack and equipment, c. 1838-1854. This was the general appearance of the British soldier's equipment. Canadian volunteers and incorporated militiamen on active duty would have been issued this type of equipment from 1838. Plate by Robert J. Marrion reproduced by kind courtesy of the artist. Ph: RC.

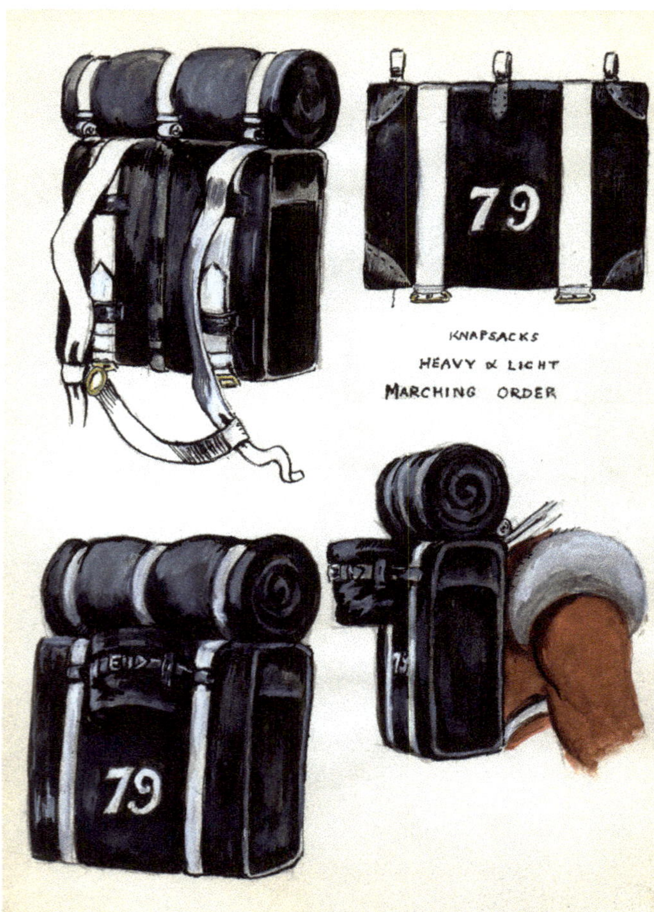

Pattern 1838 knapsack as seen in Gibraltar during 1847 by French military artist Auguste Raffet. It is marked to the 79th Cameron Highlanders Military Collection, Brown University Library, Providence, USA.

Water canteen, c. 1830s-1850s. This "barrel" type of water canteen was used since the early 1800s by the British army. It was painted light blue, stamped with "B.O." since it was issued by the Board of Ordnance. In 1855, the marking became "W.D." for War Department and the canteens were supposed to henceforth be painted black. Canadian War Museum, Ottawa. Ph: RC.

(Top right) Pattern 1854 infantry knapsack. Show with the folded blanket strapped on (No. 1) and with only the blanket with slings for light marching order. Illustration with letter of 29 March 1855 sent to Quebec. Library and Archives Canada, RG8, C1, Vol. 49.

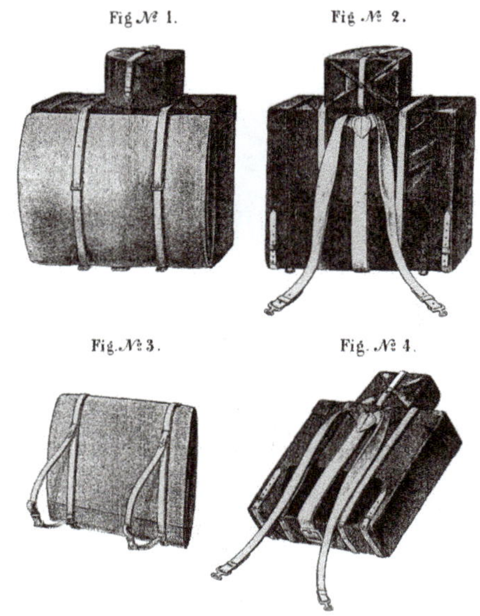

(Center right) Infantry knapsack open to show its content, 1860s. Courtesy Library and Archives Canada, C11482.

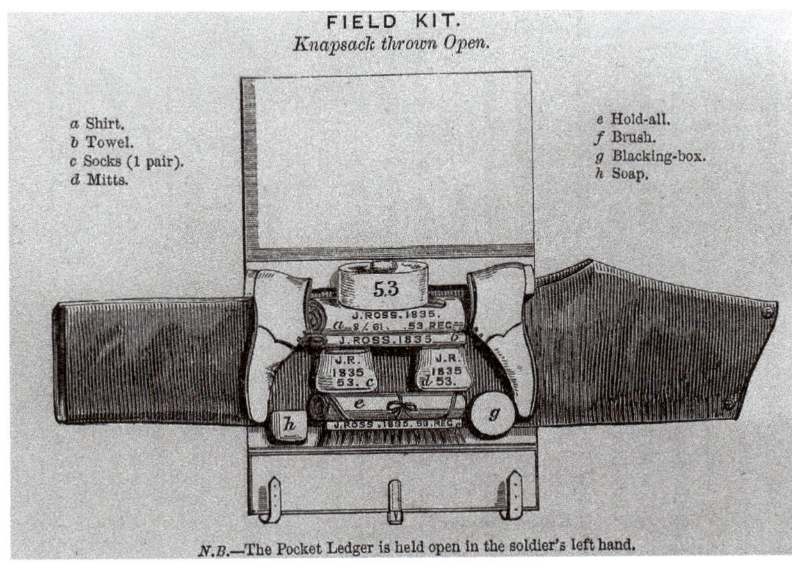

(Bottom right) British Army knapsack with straps and canteen, 1865. This was the model used in the British and Canadian forces. It had not changed a great deal since the Napoleonic Wars. This example is marked with the badge of the Royal Engineers. Print in Equipment of a Company of Engineers, London, 1866. Ph: RC.

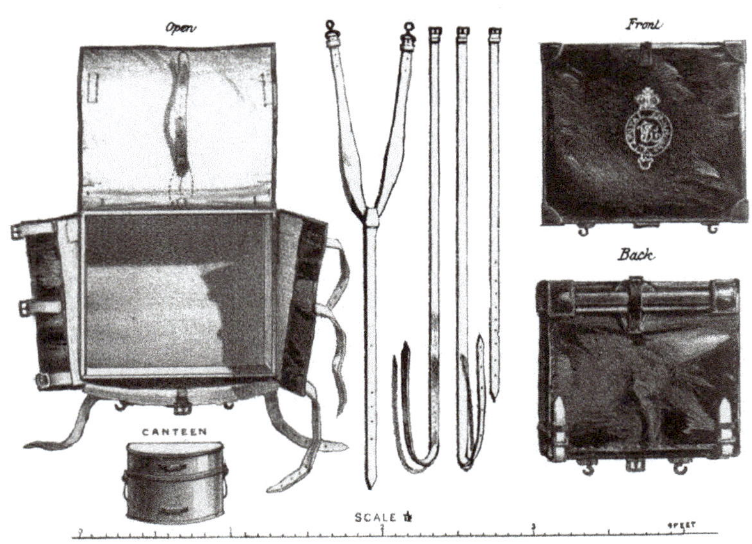

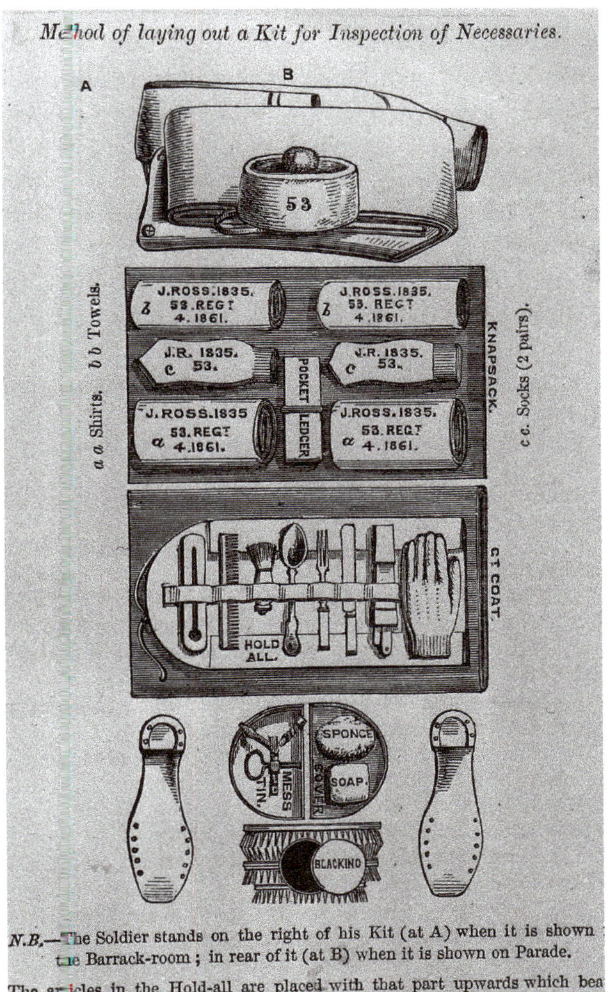

"Method of laying out a Kit for Inspection of Necessaries" in the 1860s. Courtesy Library and Archives Canada, C11483.

Infantry knapsack, 1860s. This 1856 pattern knapsack was carried by Canadian volunteer infantrymen during the 1860s and possibly later because it was not replaced until the late 1890s. Thick linen painted black, black belts, wooden bar (added in 1856) at top painted black, brass hooks, eyes and buckles. Canadian War Museum, Ottawa. Ph: RC.

vice in the field."

For the 1885 North-West Campaign, military stores were scoured for items that might available to equip the troops called up to go quell the rebellion. "Most equipment found tended to be quite old. Sergeant Major Winters recalled that the sharpshooters detachment of the Governor General's Foot Guards left Ottawa "for the front" in early April equipped with "Old style 'blacking box' knapsacks, with pipe clayed belts, and heavy pouch of Crimean days, canvas haversack, and Snider[-Enfield] .57 calibre rifle with bayonet, and heavy greatcoat [that] completed [the] outfit."

The *Canadian Militia Gazette* of 7 March 1889 reported that, at the annual sergeant's mess diner of Toronto's 2nd Queen's Own Rifles, Col. Miller gave ludicrous examples of having "the same old rotten equipment used by Wellington at Waterloo" during the 1885 North West Rebellion. On that campaign, the Queen's Own riflemen were "compelled to carry their knapsacks in their arms" until a Winnipeg saddler had repaired "these useless articles" and that the Militia Department later refused to reimburse this expense. So, the colonel took the matter to politicians "and he got the money!" Questions had been asked in the House of Commons as early as 4 May 1885 about "the worn out condition of the…knapsacks of one of the Toronto corps" but the government answered that no "information has been received by the [Militia] Department" on this matter. What was to be put in the knapsack was listed in a 28 March 1885 Brigade Order as "1 Pair marching boots. 1 Flannel shirt. 1 Under shirt. 1 Pair drawers. 1 Pair socks. Knife, fork and spoon. Soap, towel and comb. Needle and thread. Tin of dubbing."[105]

Years passed and nothing more was done. Politicians may have pressured payment for the Queen's Own Rifles' repairs, but the Militia was otherwise almost ignored in successive budgets. At best, lucky units might purchase such equipment when British or American army surplus items came up for sale.

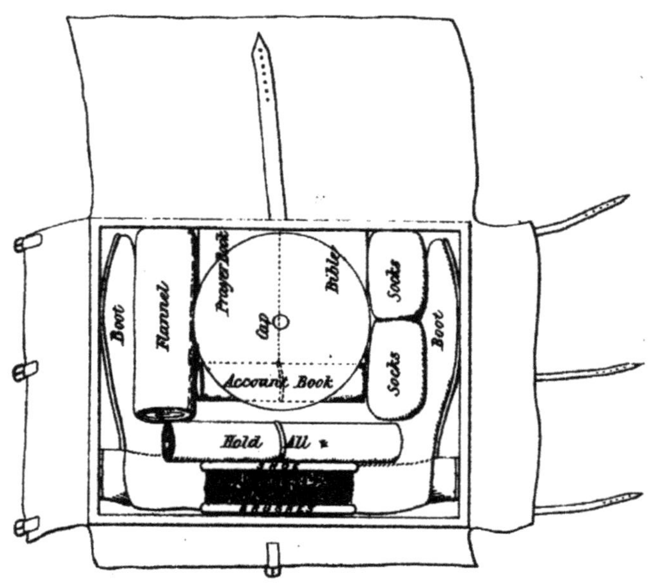

British Royal Artillery knapsack, 1864. Print in the 1864 Royal Artillery Standing Orders. Ph: RC.

Haversacks and Water Bottles

A rectangular haversacks of strong linen measuring about 18 inches wide by 14 inches high with a linen shoulder strap slung over the right shoulder was the usual type in British service until about the 1850s when a deeper and narrower model at about 9 inches wide by 10 inches high was served. Again, this was not a standard item issued to Canadian volunteers so variety amongst them was surely predominant. For 1864 Nova Scotia Volunteers, it was "to be of unbleached linen, one foot square,

[105] Thomas Edward Champion, *History of the 10th Royals and of the Royal Grenadiers* (Toronto, 1896), p. 115.

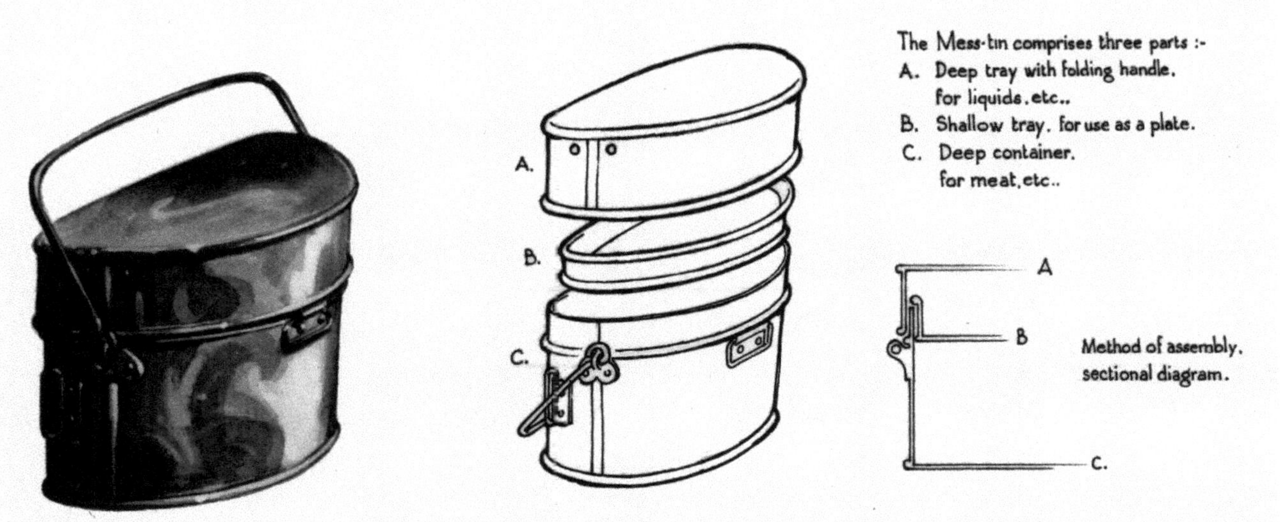

British army mess tin, 1850s-1860s. This "D" shaped or "kidney-shaped" set of eating pots and plates originated in 1814 and achieved the illustrated sophistication during the 1850s. It was made of white metal although the handle's lugs might be in brass. There were slight variations over the years, but is appears to have been the most successful military mess tin ever made; it was used by all British imperial and Commonwealth troops until the 1950s! Drawing by Douglas Anderson. Courtesy Halifax Citadel National Historic Site, Parks Canada.

with fold to button over the mouth; the strap of the same material, two inches in width — to be worn over the right shoulder strap and outside the belts; the mouth level with the hip."

The British water canteen was usually the "barrel" circular type in use since the Napoleonic Wars. It was originally painted sky blue, stamped with the Broad Arrow and the letters "BO" (Board of Ordnance) until 1855, then painted black and stamped with "WD" (War Department). From 1862, the rims around its sides were smooth, the staves no longer extending, but being flush. For Canadian volunteers of the 1850s onwards, water canteens or bottles were not included in issues and there was no truly specific model until the 1898 Oliver equipment was issued. In the middle of 1866 for instance, some 10,000 obviously American "japanned tin water bottles" were purchased by the government; these were probably the U.S. Army's model 1858, the most common type used during the American Civil War. Some were issued to units near the American border while the rest was likely kept in stores. In August 1866, the men of Hamilton's 13th Battalion were seemingly the first to receive 330 tin canteens that were then covered in leather by a local saddler. Thus, a variety of British, American, civilian or homemade canteens could be found to quench the thirst of Canadian volunteers during the second half of the 19th century.[106]

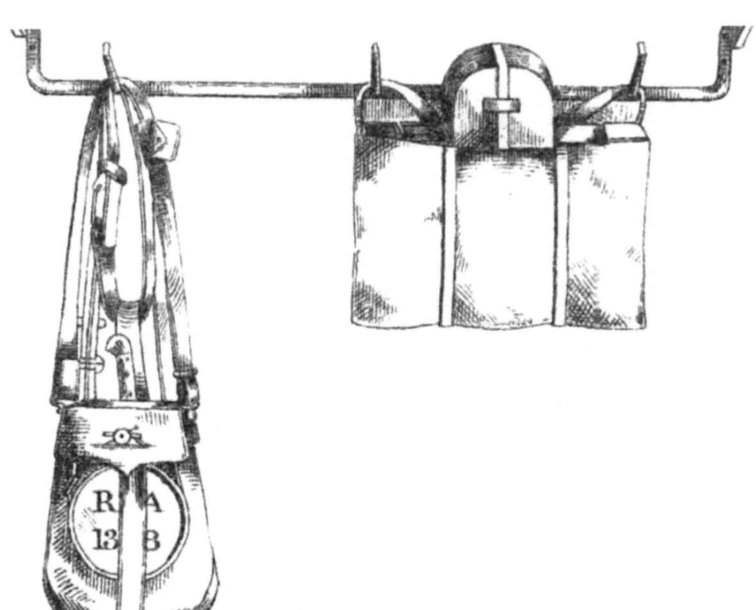

Royal Artillery field battery enlisted gunner's sword-bayonet, accoutrements, cartridge pouch and knapsack, 1860s. Print in the 1864 Royal Artillery Standing Orders. Ph: RC.

Meanwhile, unlike the Canadian Militia, the British army quickly changed weapons and, in 1873, adopted the "Valise" equipment, which included a new glass "soda water" style water canteen that was covered with leather. Another version of the valise equipment by Dr. Oliver appeared in 1878, this time with the "Italian" water bottle suspended by a buff strap. In 1882 appeared yet another version on the valise equipment, followed by the Slade-Wallace Valise equipment, in 1888, which was improved with changes in details in 1889, 1890 and 1894. We have seen reference to valise equipment being sought by the Montreal Garrison Artillery Brigade in 1882. Thus,

[106] LAC, RG9, IC1, Vol 109, W. Powell to L.T. Chesholm, 29 August1866.

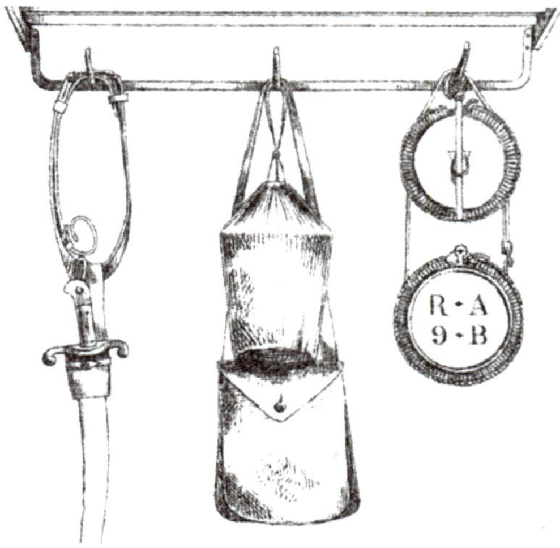

Royal Artillery field battery gunner's equipment, 1860s. From left, the waist belt holding the sword-bayonet, the linen haversack and bag, the water bottle (marked to the unit, company and rack) and canteen. Print in the 1864 Royal Artillery Standing Orders. Ph: RC.

Barrack room interior, Infantry School Corps, c. 1884. This image gives a rare view of everyday life for regular infantrymen in barracks during a winter evening. Some are cleaning or brushing various items, one is shaving and another waxes his moustache. They wear the infantry uniform so belong to the Infantry School Corps (later renamed Royal Canadian Regiment); the shirts seem of several types including a chequered one. Knapsacks and clothing are neatly placed on the forms with the winter caps on to; belts and haversacks hang from pegs behind the folded iron beds and mattresses. Courtesy National Historic Sites, Parks Canada.

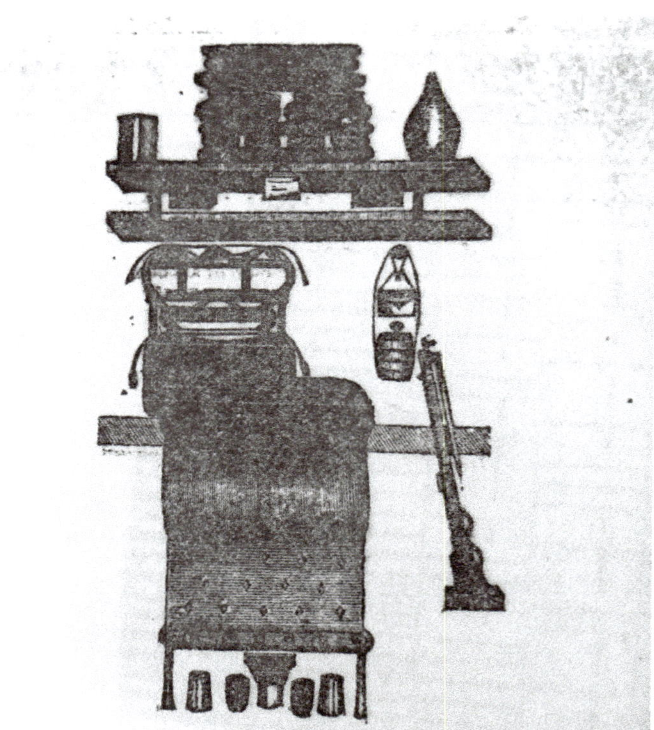

"Barrack Room Arrangement" for the Infantry School Corps from its 1884 Standing Orders. This unclear print is nevertheless probably the earliest image regarding laying out a soldiers' arms and kit for a Canadian unit. It was closely patterned after the British army's practices. Clothing and necessaries were put on top with mess tin and helmet in covers, below came ball bages, haversacks, great coat, valise, water bottle, waist belt with blankets, sheets, bed strap, palliasses and so on. The rifle with bayonet and scabbard was at the side. Bibliothèque Nationale du Québec, Montréal. Ph: RC.

Sergeant, 2nd Queen's Own Rifles, North-West Campaign, 1885. He is equipped for service in the field that includes the old British Pattern 1856 knapsack with the letters "QOR" and the mess tin covered in black on top. Since 1880, the 2nd Queen's Own Rifles' sergeant's cuffs were adorned with a black cord Austrian knot rather than the red piping and black lace common to the Canadian Pattern 1870 rifle tunic for NCOs and enlisted men. Courtesy Queen's Own Rifles Museum and Archives collection and photo.

British infantry 1873 pattern valise equipment. Print after R. Simkin in the 1890 Illustrated Military & Naval Magazine. Ph: RC.

British infantry 1878 pattern valise equipment. It was designed by Dr. Oliver and had central ventral pouch. It was issued to the Canadian Infantry School Corps with Martini-Henry rifles from 1886, but was not issued to the volunteer units. Illustrated London News, 11 May 1878. Ph: RC.

British infantry 1882 pattern valise equipment. Print after R. Simkin in the 1890 Illustrated Military & Naval Magazine. Ph: RC.

British infantry 1888 pattern Slade-Wallace equipment. Slade-Wallace equipment shown was the "British soldiers campaigning kit" that comprised, in front, two pouches that each contained eighty rounds of ammunition; at the sides a canteen and a haversack; on the back, a "valise or glazed leather bag, carried on the back, which contains the 'field kit' and twenty rounds of ammunition", the mess tin and the rolled greatcoat. Print after R. Simkin in the 1890 Illustrated Military & Naval Magazine. Ph: RC.

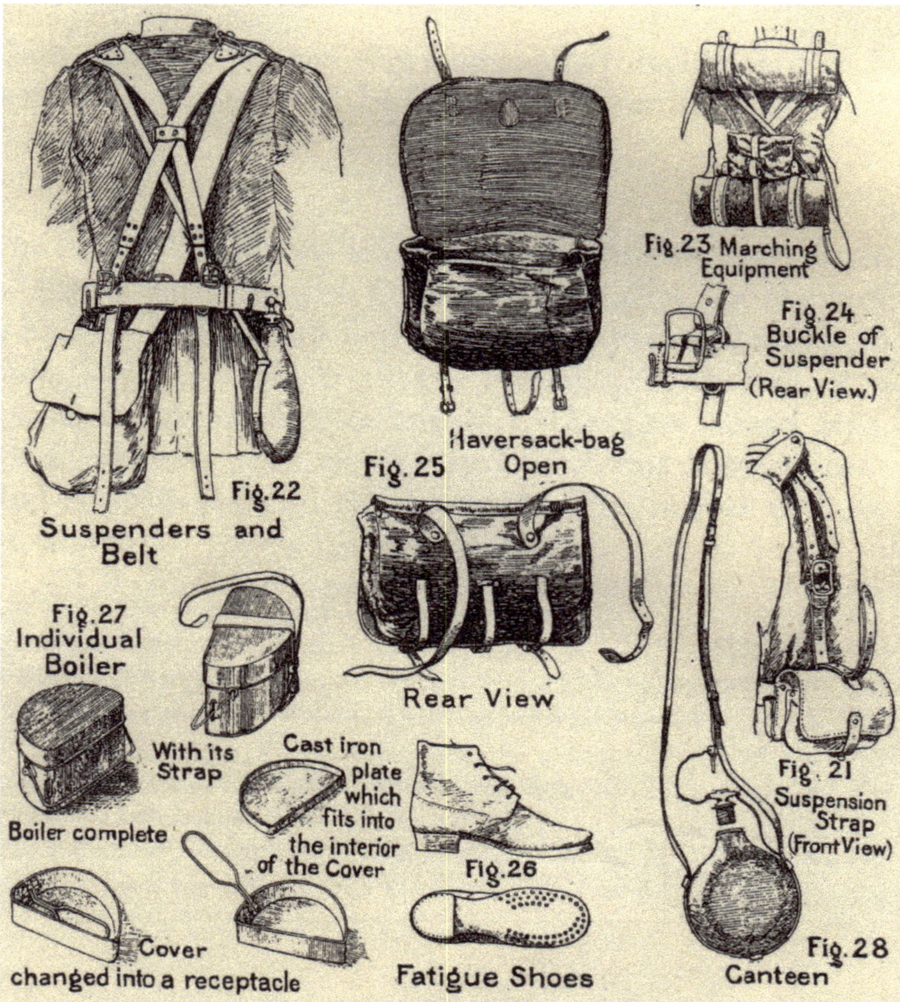

The various items making up the British army Slade-Wallace equipment, 1894-1902. This equipment was issued to Canadian infantry in South Africa. Note the rarely seen details of the "boiler" mess tin (lower left). Print after Lavisse's 1902 study. Private collection. Ph: RC.

they might be found in a few Canadian volunteer units, but, on the whole, none of these models were used widely in Canada.

It should be noted that the small Canadian Permanent or regular force was usually equipped with the more up-to-date British army accoutrements and horse furniture. The Infantry School Corps raised in 1883 was first issued with the old square knapsacks and other 1860s equipment. Three years later, it received the British Valise equipment and by the 1890s, the gentlemen cadets at the Royal Military College also had it.

Oliver Equipment

Introduced from 1878 in the British army, the Oliver valise equipment was recommended for Canadian troops from 1883. Lewis equipment was also experiment with by the Royal Canadian Artillery detachment in Quebec City during 1896, but not adopted. Meanwhile, an improved Oliver pattern set was presented in 1894 (see illustration), but it was not until 1898 that it was finally approved. Its main features were a brown leather 80-round bullet pouch carried in front, a brown leather waist belt with its bayonet frog, a back brace with various brown leather belts, a brown canvas haversack, a brown canvas valise carried on the lower back, a mess tin with is brown canvas cover, a glass pint water bottle with its holder and a white canvas kit bag.

Distribution of the 1898 Oliver equipment was moving along when, in October 1899, the 2nd (Special Service) Battalion of the Royal Canadian Regiment was organised and sailed for South Africa on 30 October. It was equipped with the Oliver equipment that featured the white canvas kit bag and the glass water bottle, this last item already cursed by the men. A coffee based mixture concocted by the troops on the ship turned the kit bag into a khaki hue of sorts while the contingent's nurses made water bottle covers with khaki material. Once ashore, the battalion plastered the stone wall of an old warehouse on the waterfront with its despised glass water bottles. Authorities, including Col. Otter who concurred with his men, turned a blind eye and "all ranks were supplied with the larger felt-covered enamelled metal covered metal water carrier in general use throughout the British Army" recalled a veteran. This was the British Mark IV pattern water bottle. However, it was only on 22 December that, to "the delight of all, British water bottles arrived for distribution" according to Private Redway.

On campaign, various adjustments were made to the Oliver equipment, the most notable of which was that its valise proved useless and was hardly ever carried in the field. The ammunition pouch was also awkward for the men were loading and firing in the prone position, the pouch in front digging into the soldier's abdomen. Over time, it appears that British Slade-Wallace equipment replaced the worn or unsuitable equipment of Canadian infantrymen. To provide more ammunition for the quicker-firing Lee-Enfield rifle, various types of khaki web bandoleers containing up to 100 extra rounds were issued to British and Canadian troops. Ammunition bandoleers were already used in the US Army, which had found them quite useful in Cuba, Porto Rico and the Philippines du-

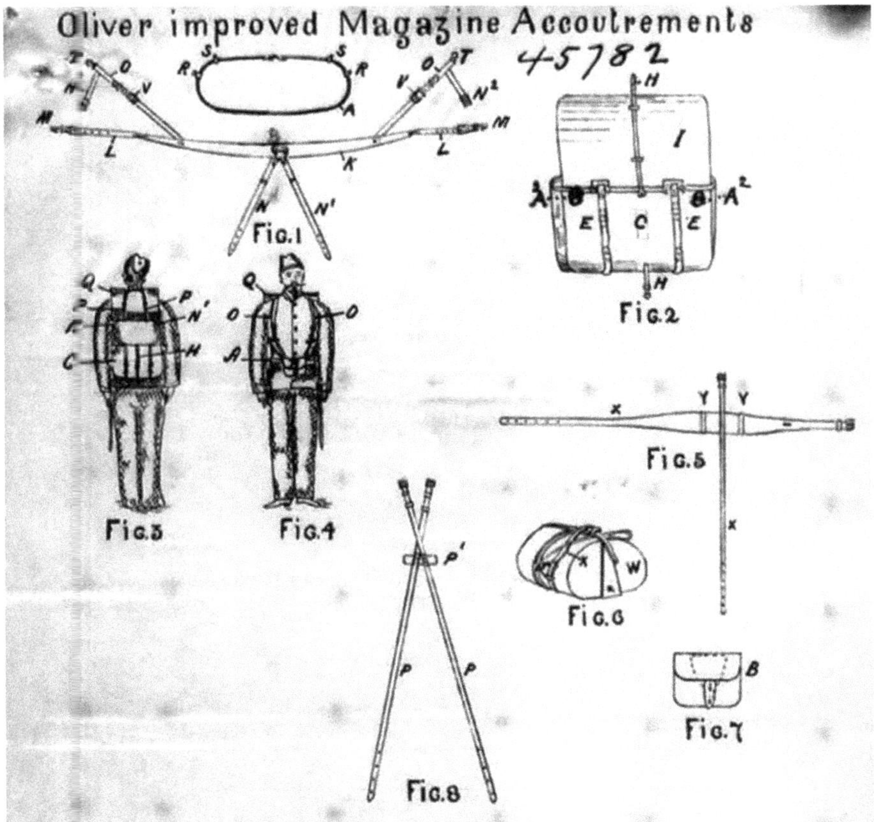

Sketch of the "Oliver Improved Magazine Equipment" patented on 12 April 1894. Fig. 1: magazine brace; Fig. 2: magazine bag; Fig. 3: back view of soldier; Fig. 4: front view of soldier. Fig. 5: canteen brace; Fig. 6: canteen; Fig. 7: reserve pouch; Fig. 8: coat brace. Patent No. 45782. Courtesy Library and Archives Canada, 2001-01023.

The brown leather Oliver equipment set up on an infantry tunic, c. 1898-1910. This was the appearance as seen from the front. The linen haversack to store rations and necessaries is slung over the shoulder. Hamilton Military Museum, Hamilton, ON. Ph: RC.

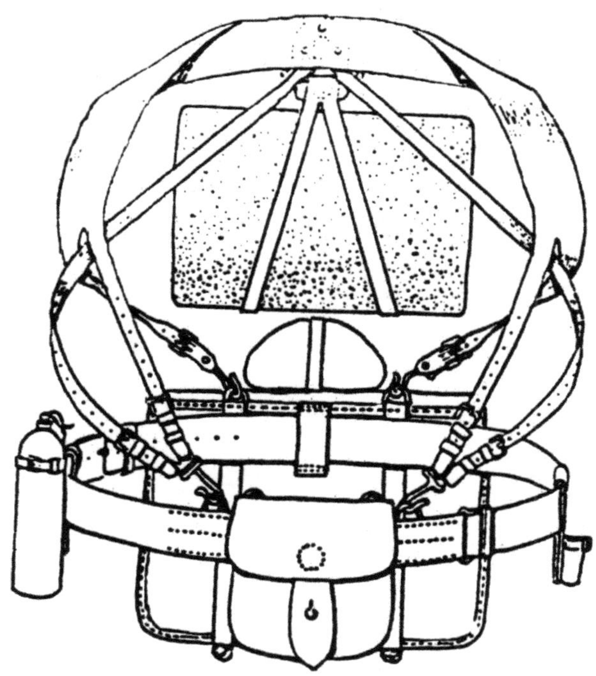

Oliver equipment assembled for marching order, c. 1899. Plate by Robert J. Marrion reproduced by kind courtesy of the artist. Ph: RC.

Oliver equipment hanging in barracks, c. 1898. The 80-round pouch is at left, somewhat hidden by the haversack belt. On the shelf are the soldier's white helmet, fur cap, folded greatcoat, blanket and extra clothes. This was most likely taken in a barrack lodging the Royal Canadian Regiment. Courtesy Scott Duncan.

Oliver magazine bag, c. 1898. Worthington Museum, Canadian Forces Base Borden. Ph: RC.

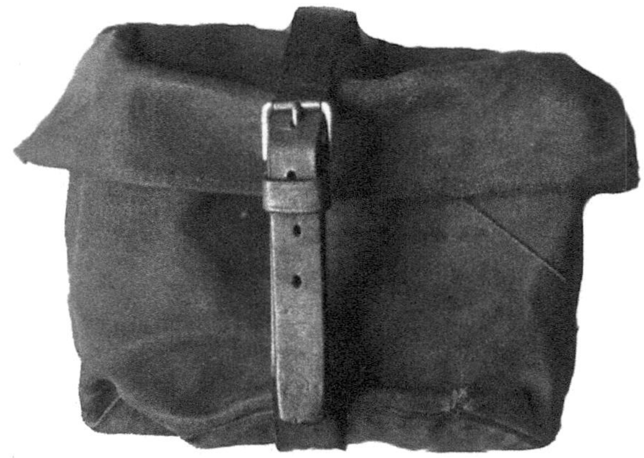

Oliver canteen pouch, c, 1898. Worthington Museum, Canadian Forces Base Borden. Ph: RC.

Private of the Royal Canadian Regiment about to leave for South Africa wearing the Oliver equipment with the khaki uniform, 1899. Front view. Courtesy The Army Museum, Halifax Citadel.

Private of the Royal Canadian Regiment about to leave for South Africa wearing the Oliver equipment with the khaki uniform, 1899. Rear view. Courtesy The Army Museum, Halifax Citadel.

Lieutenants Temple and Mason, Royal Canadian Regiment, South Africa, 1900. Both are about to go out on picket duty from Camp Belmont. They are fully kitted out with Lee-Enfield Mk. I rifles, British khaki uniforms, Slade-Wallace equipment and extra ammunition bandoleers. Each appears to have a binocular case hanging on their left side below the belt. Courtesy Library and Archives Canada, PA125698.

British Slade-Wallace equipment canteen used by Canadian troops in South Africa, c. 1900-1902. Canadian War Museum, Ottawa. Ph: RC.

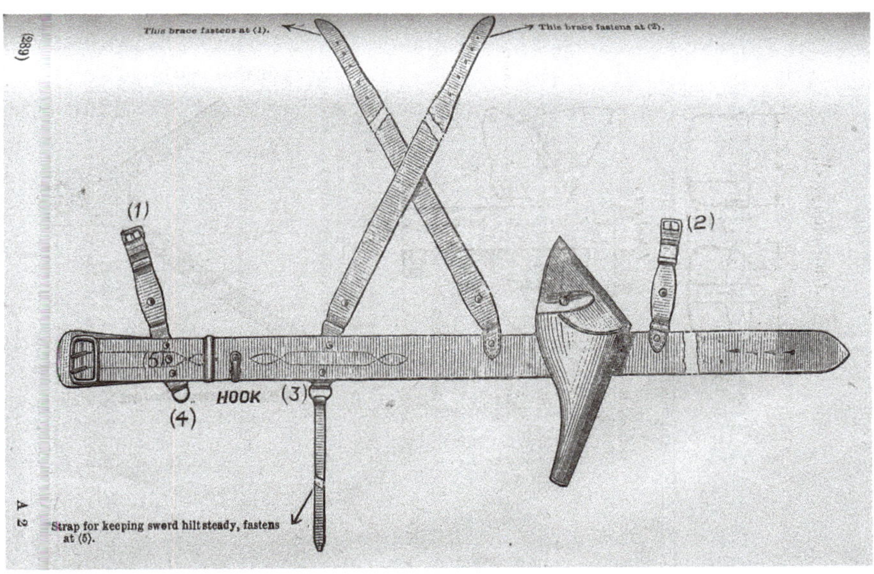

Officer's "Sam Browne" belt, 1890s. This type of belt gained much popularity with British army officers during the 1880s and 1890s since it could hold the revolver holster, ammunition pouches and had a sword frog. Officers in Canada and elsewhere in the British Empire also appreciated it on service. It was finally officially adopted in 1900. Print in List of changes of War Materiel, 28 December 1900 (London: War Office, 1901). Ph: RC.

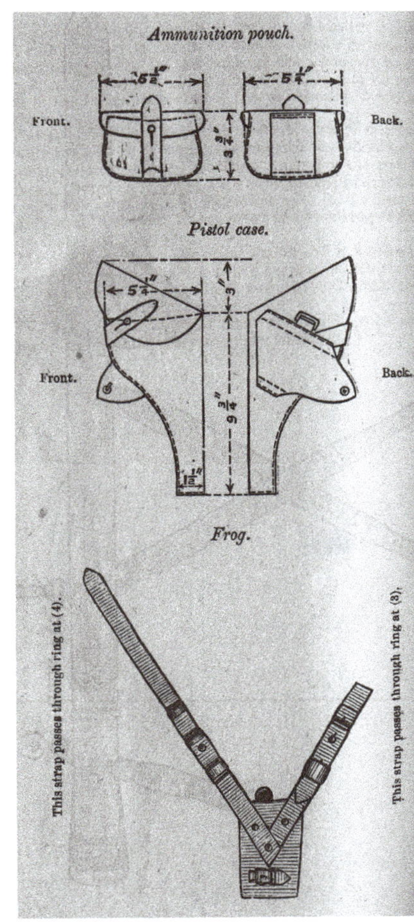

Holster and ammunition pouch of the officer's "Sam Browne" equipment. Print in List of changes of War Materiel, 28 December 1900 (London: War Office, 1901). Ph: RC.

ring 1898. Those provided to the first contingent "did not prove equal to the occasion, as the divisions soon stretched, allowing the cartridges to fall out…while the rows of brass heads in the front made a distinct and easy mark for the enemy's fire" reported Col. Macdonald in 1900. From photos, the later web bandoleer issues to infantry had no brass studs and seemed better made to hold bullets. So, the Canadian Oliver equipment had some problems when used by troops in the field, but then so did the British Slade-Wallace equipment; from 1902, both would be reviewed to introduce improvements.

Royal Montreal Cavalry officer's cartridge pouch and belt, c. 1824 to 1837. White leather sling and black cartridge box, gilt badge with intricate cipher that includes GR at centre surrounded by "Royal Montreal Cavalry". This model was replaced in 1837. Courtesy National Historic Sites, Parks Canada.

Cavalry

Canadian cavalrymen had, in principle, the same equipment as their British comrades. This consisted of light cavalry accoutrements since nearly all Canadian troops were more or less patterned after light dragoons, later hussars. The basic equipment was a waist belt that held the sabre in its scabbard. From 1855, the sword's scabbard could be attached to the Universal Pattern saddle when mounted. It is far from certain that all Canadian volunteer troopers had the sabretache, the flat leather case with a large three-pointed flap, that hung suspended by three slings from the waist belt. It was supposed to be used to carry dispatches. It was a practically useless item, but no self-respecting regular light cavalryman would be without it since it was very fashionable during the whole 19th century. The shoulder belt had a small cartridge pouch for the carbine's ammunition. Until the mid-19th century, there was a hook arrangement on this belt to hang the carbine, but this vanished after the appearance of carbine buckets attached to the saddles. The shoulder belt remained while its small ammunition box became an increasingly decorative item. The issue of Colt revolvers to Canadian volunteer cavalrymen from 1856 brought holsters fixed to waist belts to carry them until the units were re-armed with cavalry carbines from 1867.

From 1882, the British cavalrymen on campaign could be issued a brown leather bandoleer that held 50 rounds of ammunition in groups of 10 tubes that had a leather cover. The 1897 model, which was the type issued to Canadian cavalry in South Africa, had four groups each having 10 tubes, one group having six tube and another with four tubes.

Web bandoliers were also issued in 1900 to the second contingent's cavalry (1st Canadian Mounted Rifle renamed Royal Canadian Dragoons in August 1900) that were recalled by Trooper Hilder as "a heavy cotton webb bandolier holding a hundred cartridges. When a man lay down, or even threw the bandolier on his kit, many of the shells fell out and were left on the ground. The men were always refilling their bandoliers, although very little shooting was done." These web belts were judged much inferior to those issued in 1902 to the third and fourth contingents, which also received a "new waist belt of web pattern" on arrival in South Africa.

Officers

Until the Crimean War, a white leather shoulder belt with a decorated metallic belt-plate was worn by infantry officers to carry the sword, those rifle units having black waist belts for the sabre. Thereafter, the infantry officer's sword belt with its round buckle was at the waist, plain for ordinary duty and gold-laced for dress.

Rifle officers had a black waist belt with a "snake" hook buckle and a cartridge pouch (usually with a silver bugle badge) suspended from a shoulder belt that featured a silver whistle and chain as well as a unit badge. It is noted that Canadian rifle NCOs also usually had their unit's badge on the shoulder belt, unlike British NCOs who did not have the badge.

Artillery, engineers and various types of cavalry had the waist and shoulder belts, the dress versions being gold covered and pouches having embroidery or silver flaps with barges (see the outfitter Holding's 1894 plate and many illustrations in both our volumes). Officer's cartridge pouches were, like the sabretache flaps, highly embroidered although, in the cavalry, the pouch had a solid silver flap with the royal cipher in gold from the later 1850s. The pouch belt had a central line in the colour of the unit's facing. The 13th Hussars had this type of pouch and its belt, which was adopted by all Canadian officers who could afford them from the 1860s. By the end of the 19th century, the officer's cartridge pouch had revealed itself ideal to carry cigarettes...[107]

Royal Montreal Cavalry Sgt Robert Morris 1824 silouhette (Argenteuil Regional Museum Carillon QC ph RC.)

By the 1870s, officers on campaign needed more practical accoutrements and British General Samuel Browne, who had lost his left hand in combat in 1858, devised a practical brown leather belting system that he wore thereafter, notably during the 1878-1879 Afghan Campaign. This was noted by officers of a Rifle Brigade battalion who had similar belts made and, recalled Lt. Gen Wilson, then a young officer, "we wore them for the rest of the campaign" and thereafter "always wore those belts on active service." Other units followed suit and the "Sam Browne" belt eventually became universally popular with officers on active service. Although widely used, it was only officially adopted on 28 December 1900. During the 1880s, and possibly earlier, Canadian officers were also attracted using brown leather belts and, during the 1885 North-West Campaign, they had "been authorised for active service" according to the militia gazette.[108]

Royal Montreal Cavalry officer's cartridge pouch and belt, c. 1824 to 1837. Side view. Courtesy National Historic Sites, Parks Canada.

Horse Furniture

The horse furniture used by Canadian cavalry was mainly of the same types as that used by the British light cavalry. The basic type was the 1796 pattern saddle and it evolved in various details over the years. In 1856 the first Universal Pattern saddle and horse furniture was introduced, which was to apply to all the cavalry followed by the staff and infantry mounted officers in 1859 and Universal Pattern bridle in 1860.[109] "Trial and error" adjustments were made to the Universal Pattern saddle in 1872, 1878 and 1884 consisting mainly of the addition of the flat iron arch. In 1890 came the steel iron arch Universal Pattern that lasted until the next model introduced in 1902. Bridle patterns were altered in 1885 and 1890. Initially, leather pistol holsters were attached across the front arch of the saddle until the 1840s when replaced

[107] The exception to gold lace in the cavalry was the Governor General's Body Guard the had silver. Staff officers had silver lace also until the 1886 dress regulations.

[108] H.F.M. Wilson, "Sam Browne Belt", JSAHR, XIII: 4 (Winter 1934), p. 234; *List of changes of war materiel*, 28 December 1900 (London: War Office, 1901).

[109] TNA, WO 370/5, Pattern warrants of July 1856, 16 August 1856, 10 November 1856, 1 January 1859.

Royal Montreal Cavalry officer's cartridge pouch and belt, c. 1824 to 1837. Inside view. Courtesy National Historic Sites, Parks Canada.

Captain James William Brereton Rivers, 1st London (Canada West) Independent Troop of Cavalry, c. 1856. He wears a dark blue stable-jacket style with white collar and cuffs, black flat lace with a thin white central line on the breast ending in black olives, no buttons but fastened with hooks and eyes, gold lace edging and shoulder cords. He has white cartridge box and sword belts with gilt belt plates. Courtesy of a private collection and the 1st Hussar Regiment, London, Ontario.

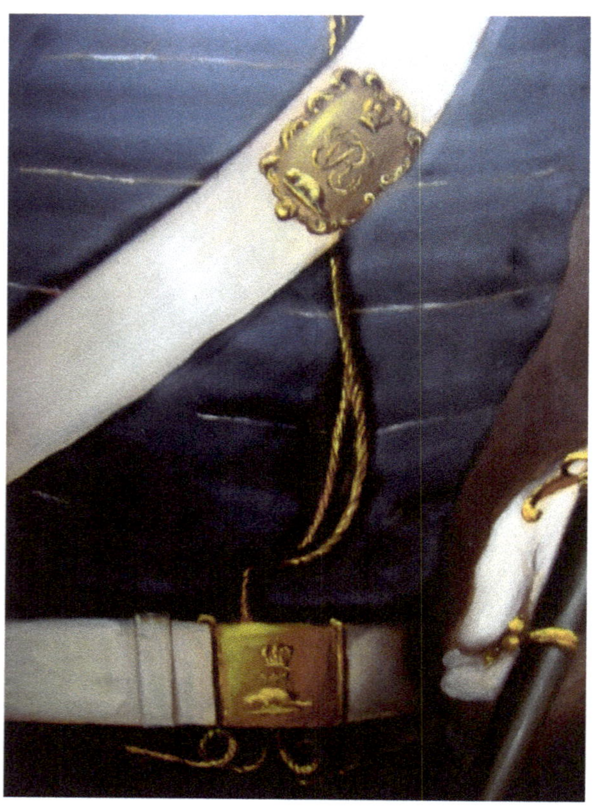

Detail of Captain Rivers' portrait showing the belt plates. Courtesy of a private collection and the 1st Hussar Regiment, London, Ontario.

Detail of Captain Rivers' portrait showing the helmet of the 1st London (Canada West) Independent Troop of Cavalry, c. 1856. It was termed a "beaver" helmet, which was obviously because of its beaver badge on its helmet and sword belt plate. Courtesy of a private collection and the 1st Hussar Regiment, London, Ontario.

Royal Montreal Cavalry cartridge box side view c1820s

Officer of the 1st or 2nd Montreal Troop of Cavalry, c. 1858. The uniform for both units was dark blue with red collar and cuffs, gold lace for officers with a shako bearing a white horse hair plume. This somewhat damaged but rare image shows that they wore the regulation dress gold shoulder belt and sword belts with the red central line and the gilt plate, chain and pickers. Private collection. Ph: Ross Wilson.

(Left) Officer's accoutrements, 8th New Brunswick Hussars, late 19th-early 20th centuries. The full dress of Canadian hussar officers, who were volunteer militiamen, was as luxurious as British regulars (when they could afford it). The accoutrements were edged with gold lace and silver chain and pickers. York-Sunbury Museum, Fredericton, New Brunswick. Ph: Courtesy Department of National Defence Library, Ottawa. (Right) Officer's accoutrements, 8th New Brunswick Hussars, late 19th-early 20th centuries. Back view. The cartridge box on the back was made of silver with gold monograms or badges. The three narrow belts at the bottom were for the sabretache. York-Sunbury Museum, Fredericton, New Brunswick. Ph: Courtesy Department of National Defence Library, Ottawa.

by large leather wallets. Leather carbine buckets became common from 1855. From 1885, sword scabbards featured two rings so they could be fastened on the left side of the saddle.

The better equipped Canadian volunteer troops and the regular provincial units on active service between 1838 and 1850 would obtain their horse furniture from Britain. The troopers of volunteer militia units had to be fairly to very wealthy and they nearly always provided their own saddles and horses at their own expense. For the less affluent volunteer troops, almost any saddlery would be pressed into service. Therefore, much variety could be expected although 500 "cavalry saddles and griddles" were imported from Britain by the Militia Department in 1862. In 1864, the Governor General's Body Guard in Toronto imported "Hungarian" saddles from England that must have been the Universal Pattern of the British cavalry. The unit's commander later wrote what happened when it was called on active duty. "During the summer of 1866, my corps formed a chain of outposts, covering a camp of observation at Thorold, from the Niagara Falls along the Niagara River to Fort Erie, and along the shore of Lake Erie as far as Point Abino, making a line of nearly forty miles. The work of constantly patrolling at night was very heavy on the men and horses, as we had not a force sufficient for the duty. I had not enough of the Hussar saddles for all the men, so some twenty were obliged to use the ordinary hunting saddles. Among the horses using the Hussar saddles, there was not one sore back, but among the others, the greater number almost immediately were laid up, and I was obliged to order that, as the men took their turn for duty, they were to use the military saddles, the men lending them to those who were without them when their turn for patrolling came. This saved further trouble." Other cavalry troops were also called out during the 1866 Fenian Raids. Many had only hunting saddles and using these for permanent duty had severe consequences for the horses' backs. If universal pattern saddles could be found, they were issued to troopers going out on patrol. By December 1866, the Adjutant General reported that the "greater part of the [22 troops of] cavalry have been supplied with new saddlery." An importation of 50 American military saddles was also made in 1866 by Captain Currie of the Queenston Mounted Infantry Company; these appear to have been the U.S. Army's McClellan saddle adopted in 1859. [110]

[110] LAC, RG7, G19, Vol. 27, arms, ammunition and saddles under orders, A. Gunn to the Secretary of the Defence Commission, Montreal, 21 August 1862; George T. Denison, *Modern Cavalry* (London, 1868), p. 94; LAC, RG9, IC1, Vol. 268, McDougall to Durie, 4 December 1866; LAC, RG9, IIA6, Vol. 8, Order in Council, 21 February 1867. The McClellan saddle was used in the American army, with various models and improvements for 84 years, in conjunction from 1936 with the Phillips saddle.

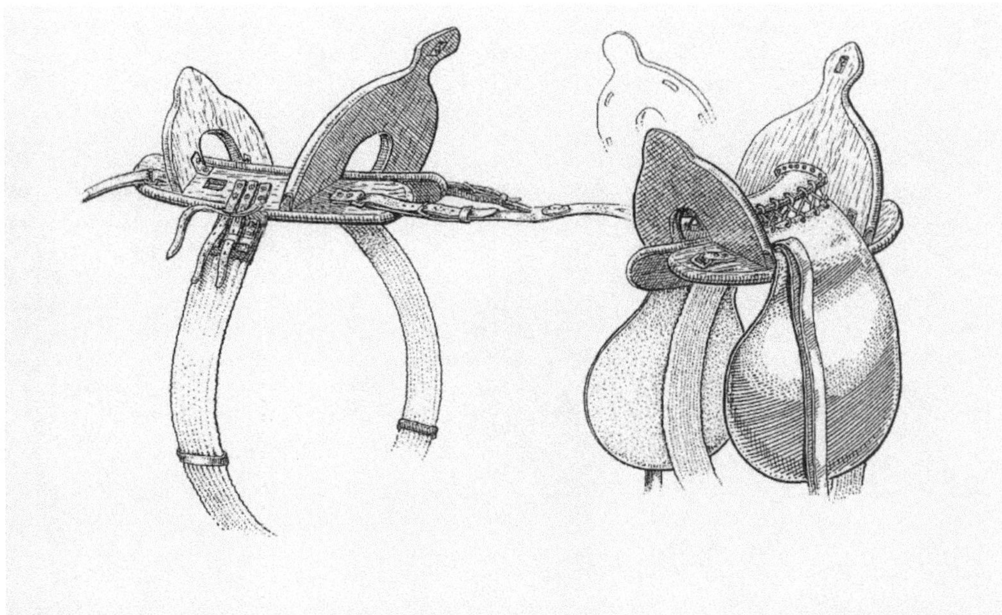

Light cavalry wood and leather saddle tree, mid-19th century. Pen and ink drawing by Ray Ahrenholz. Reproduced with the kind permission of the artist.

Over the years, Universal Pattern saddles were the usual type used in Canada. There were, however, minor modifications so that, by 1898, it had become "Steel Arch Universal Pattern Saddle, 1890 pattern, Mark III." Some improvements may have never reached Canadian volunteer cavalrymen since their saddlery, if used solely for a few weeks of militia activities, obviously lasted far longer than for regular cavalry and would continue to be used long after they had been phased out by the regular cavalry in Canada and in Great Britain. After Confederation, the Defence Department did issue saddlery to volunteer units. We have seen reference to the St. Catharine Troop of Cavalry, the Winnipeg Troop of Cavalry (March 1879), the 8th Cavalry (December 1881), the Cavalry School Corps (January 1884, which included carbine buckets) and there were certainly many other units.

During the 1885 North-West Campaign, the mounted scout units raised largely amongst cattlemen used the "cowboy" saddle if they could, also called "western", "Mexican" or "California" type. This type of saddle, which could weight some 40 pounds (18 kgs) or more, was much heavier than the 11 1/2-pound (5.2 kgs) military saddles since it was meant for very rough work. Its seat almost moulded the man's lower body and it had a high pommel in front serving as a hold for rope and was also found useful to sling a rifle by some troopers. But other types of saddles could be used too. Major Boulton recalled that, for his Boulton's Scouts, the "saddles were not alike. Being purchased at a moment's notice in Winnipeg, I had to take what I could get. I found what is known as Davis' three-guinea English saddle the best. Combined with economy, they are light, comfortable to ride in, and do not give sore backs. I used one throughout, and neither horse nor rider was a day on the sick list in a seventeen hundred mile march. Several of my men had these saddles and realised their comfort."

Light cavalry saddle with attached pistol holsters, straps and stirrups, mid-19th century. Pen and ink drawing by Ray Ahrenholz. Reproduced with the kind permission of the artist.

Some saddlery made in Quebec during 1888 for the Royal School of Cavalry was reported "quite equal to any hitherto imported from England" and was thought serviceable for up to 12 years for regular service and up to 40 yeas for volunteer troops. When 2,300 Lee-Enfield carbine were issued from 1896, the government had 500 leather carbine buckets in its stores that could be used for that rifle model; in March 1897, it purchased many more leather buckets of the latest British pattern to strap onto the saddles (MR 1888; LAC, reel T383/148-150).

The Canadian cavalry contingents sent to South Africa in 1900-1902 consisted largely of mounted rifle units and they seemingly used mainly the colonial pattern saddle except for the Royal Canadian Dragoons and the Royal Canadian Artillery of the second contingent that had the universal pattern saddle. Lt. Col. Williams commanding the 3rd Canadian Mounted Rifle found its "colonial pattern" saddlery of somewhat inferior quality with its leather "generally not so good as it should have

Light cavalry saddle with padded seat, rolled cloak, valise and satchels covered with a sheepskin, mid-19th century. Pen and ink drawing by Ray Ahrenholz. Reproduced with the kind permission of the artist.

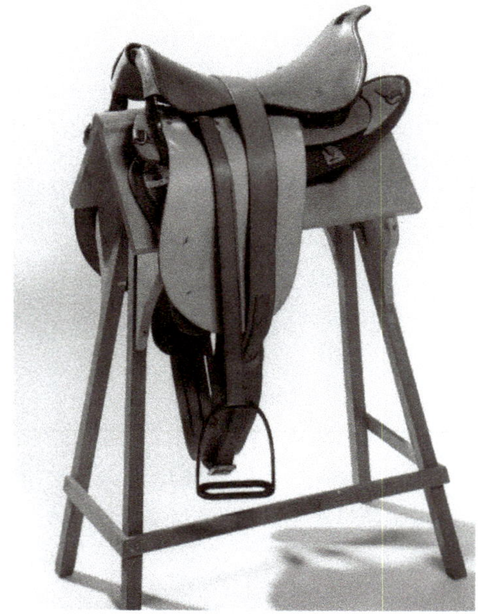

Universal Pattern saddle 1890/1898/1902. Various elements were added or changed during the 1890s on the basic 1890 model. Courtesy National Historic Sites, Parks Canada.

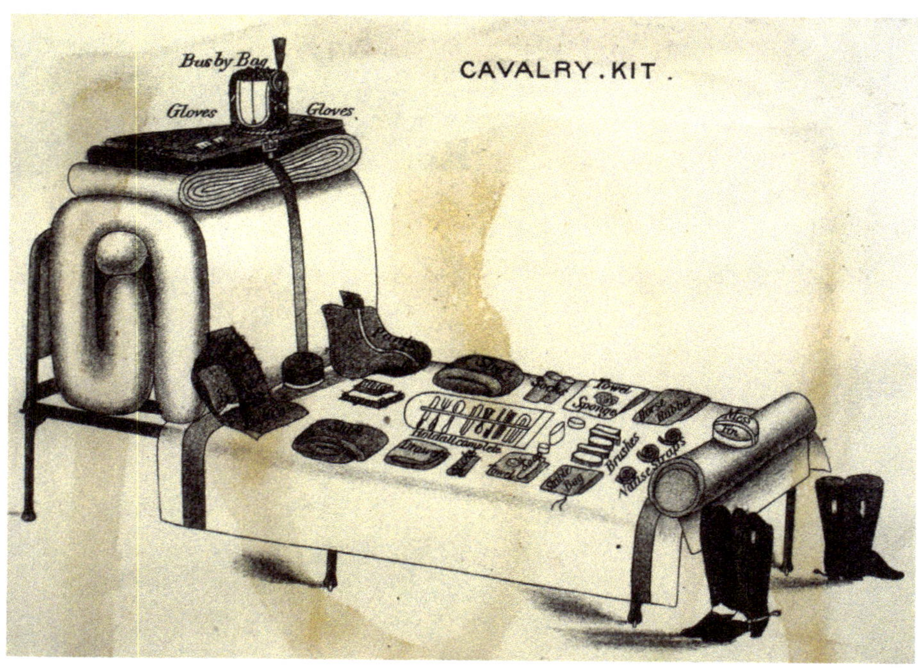

Cavalry kit, 1884. This print shows the equipment for a hussar laid out on his bunk. Troopers of the Canadian regular Cavalry School Corps would have had similar kits. Anne S. K. Brown Military Collection, Brown University Library, Providence, USA. Ph: RC.

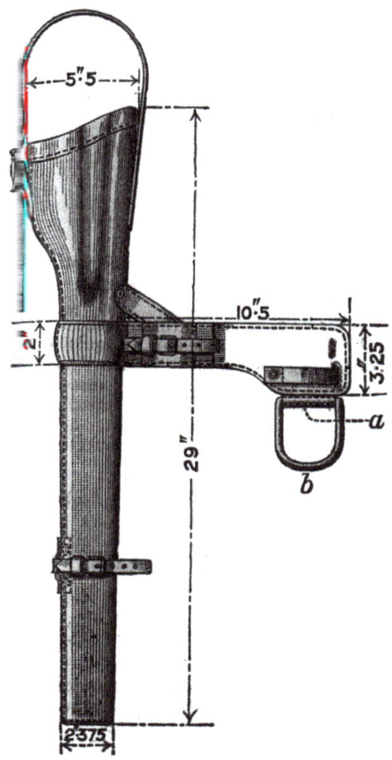

Rifle bucket for the Lee-Enfield carbine, February 1896. These brown leather buckets were usually strapped on the rear right side of the Universal pattern saddle. Print in List of changes of War Materiel, 7 February 1896 (London: War Office, 1896). Ph: RC.

Trooper of the 8th New Brunswick Hussars at summer training camp, c. 1897-1898. This image shows the carbine bucket strapped on the left side of the Universal pattern saddle. The British pattern bandoleer is slung over the shoulder. The trooper wears the pattern 1896 frock that has no pocket pleats and has dark blue shoulder straps edged with yellow. Frock issues in 1899 had pocket pleats and yellow shoulder straps. This image could also date from the early 1900s since the frock was to last at least five years. Andrew Moss collection. Ph: CL.

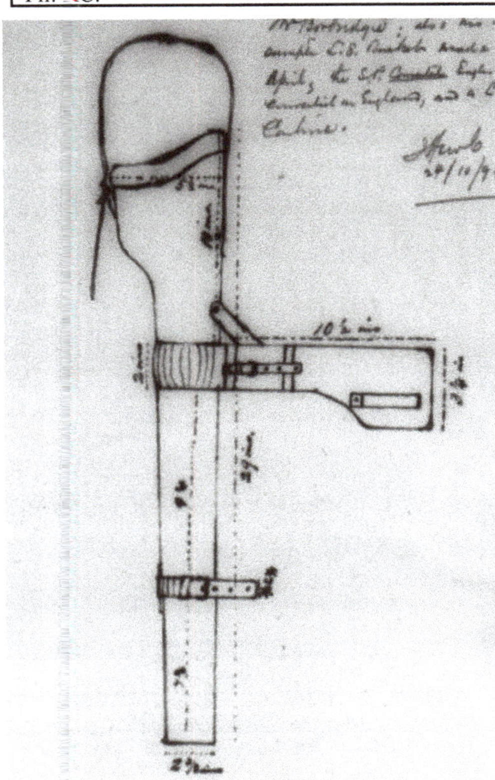

Rifle bucket for the Lee-Enfield carbine sketch with a Canadian order for 1,500 of these items, 23 October 1896. Obviously copied from the British List of changes of War Materiel of February. Courtesy Library and Archives Canada, reel T382/763.

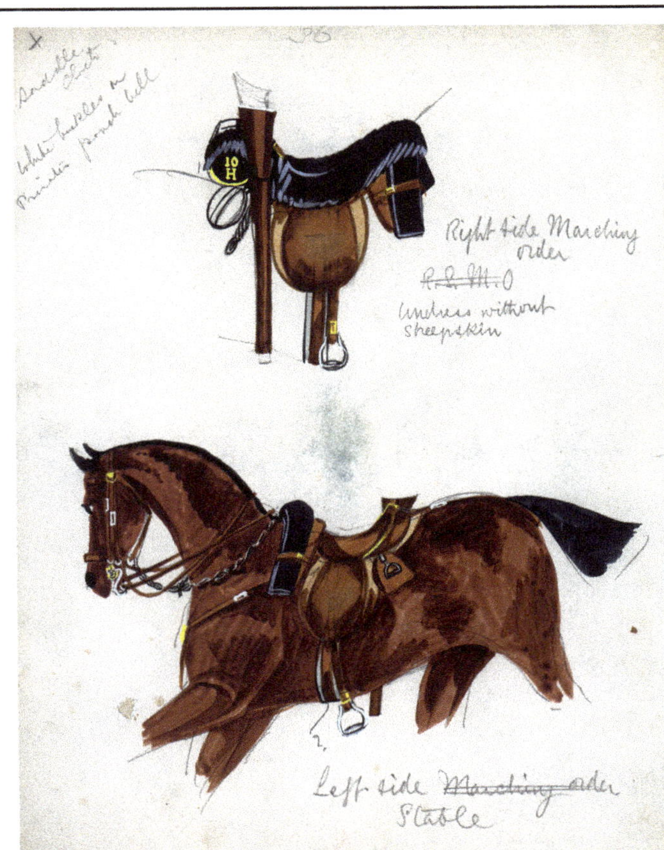

Universal pattern saddlery for hussars, 1900. At top, the sheepskin covers the seat in marching order except when troopers are wearing undress stable jackets and pillbox caps. The bottom view show the saddlery without the sheepskin and valise. Coloured sketch by MacDonald. Courtesy Anne S. K. Brown Military Collection, Brown University Library, Providence, USA.

Mounted sergeant and trooper of the 6th Hussars of Montreal, 1898. Both have the universal pattern saddle. The round valise at the saddle's rear is obscured by the carbine's butt, but it has the letter "M" (presumably for Montreal) visible. Unsigned charcoal rendering. Sergeant's Mess of the Royal Canadian Hussars, Montreal. Ph: RC.

been" while Lt. Col. Irving of the 6th felt the bridles for his colonial pattern saddles "were too light for anything like active service conditions." Trooper Hilder of the 1st Canadian Mounted Rifles (retitled Royal Canadian Dragoons from August 1900), recalled "carrying all our equipment on our saddles. The saddle's wallets contained a spare suit of underwear, one extra shirt, socks, sewing outfit, ordinary outfit and shaving equipment. Our dixie was strapped on the outside of one of the wallets. At the rear of the saddle were strapped our blankets and, in front, across the saddle wallets, were our greatcoats. We were not issued rubber ground sheets. A feed of oats was carried on our left flank, strapped to the saddle. With our rifle and a hundred rounds of ammunition, haversack and water bottle, the saddle and full equipment weighted 75 pounds…In our haversacks we carried a full day's rations. On long marches we were quite independent of our transport." On 8 January 1900, an order for 500 saddles complete of the "approved regulation United States War Department Mounted Infantry pattern" was made to Adams Brothers Company in Toronto, to be delivered at Southampton, UK, that was undoubtedly meant for a Canadian Mounted Rifles unit. This appears to have been one of the variations of the McClellan saddle. It was made of black leather rather than the brown leather used in the British service. U.S. pattern saddle blankets were also obtained at that time (LAC, reel T388/583, 624 and 627; T390/250).

Meanwhile, in 1899, nearly all Canadian volunteer cavalry units in the country received new saddles, which would have been of the Universal pattern as well as related equipment. From the late 1880s, the "colonial saddle" that originated in South Africa became popular and was manufactured "in thousands all over the British Empire" according to Tylden. As seen above, it was used by some Canadian Mounted Rifle units in South Africa from 1900. After the war, the Colonial pattern saddle was issued to the new mounted rifles units. There may have been mixture in other regiments since the troopers of the 8th New Brunswick Hussars had Universal pattern saddles while their officers had the Colonial pattern.[111]

The schabraque, a decorated parade saddle blanket, was probably not seen a great deal amongst Canadian volunteer officers due to its respectable cost. The men's would have been simpler, if they had them, of the coat colour and edged with the facing or button colour, but unlikely to have embroidery. A sheepskin could have covered the saddle for comfort. Fastened at the back of the saddle seat was a round sausage-shaped "valise" to carry various pieces of clothing and kit.

Field Artillery Saddlery and Harness

Field artillery saddlery and equipment in Canada were of the British patterns during the Victorian period. There are no known distinct Canadian patterns at that time. In March 1897 for instance, 20 "complete sets of Royal Artillery General Service Harness" were ordered for Canadian batteries (LAC, reel T383/108). The number of horses pulling the gun might vary. A light howitzer might require only two horses whereas a 9-pdr cannon would need six (please see illustrations).

Artillery drivers had the same weapons and equipment as other artillerymen. Mounted drivers further had a whip. They also a hard leather gaiter reinforced with iron strapped on to the lower right leg to protect it from being crushed by the centre pole between two horses.

[111] Douglas How, *The 8th Hussars* (Fredericton, 1964), pp. 34, 67.

Harness, saddlery and caissons for artillery, c. 1820s-1840s. Several volunteer artillery companies in Lower and Upper Canada were issued this type of equipment during the 1820s and 1830s. The piece of artillery is a howitzer. Print in Charles Dupin's A Tour through the Naval and Military establishments... (London, 1822). Ph: RC.

Markings of Accoutrements and Weapons

Board of Ordnance, War Department, Nova Scotia and New Brunswick Markings

The "Broad Arrow" with the letters B.O. denoting the Board of Ordnance was to be marked on all accoutrements, weapons and such items as blankets and soldier's greatcoats since they remained property of the government while being used by the troops, be they regulars, volunteers or drafted militiamen. On 6 June 1825, this was extended by the Board of Ordnance to "all stores belonging to this Department, whether under the denomination of Ordnance, Barrack, or Commissariat, to which such marks can be affixed, and which are not already so marked…" with barrack furniture being added the following year (WO 44/729).

The Crimean War resulted in a reorganisation of the British army uniting various part of the forces including the Board of Ordnance into one "War Department" in 1855. As a result, the letters B.O. were replaced by W.D. The Broad Arrow remained.

Saddlery and harness for artillery horse, 1892-1901. The gun are shown in service order. The artillery driver rode the horse on the left side, the right side horse having no driver although saddled and the limber with two gunners. The appearance of Canadian field batteries was generally similar except that volunteers batteries as well as the regulars of the Royal Canadian Artillery had white helmets by the 1890s. Print after R. Simkin. Private collection. Ph: RC.

Although before 1856, the provinces of central Canada did not have their own marking systems, some Atlantic provinces did. As early as 1795, Nova Scotia was marking its muskets and, one assumes, some of their accoutrements also, with the letter M and the name of the county in which it was issued, later adding NSM [Nova Scotia Militia] and sometimes using the battalion number instead of the county's name, or both the number and the county's name, by the first decades of the 19th century. This system appears to have remained until the mid-1850s. The 17 March 1825 New Brunswick Militia Act specified that "arms and accoutrements…issued from His Majesty's Stores for the use of Militia…shall be branded distinctly…with the letter M, and the name of the County…also with a capital letter to denote the Company…and number to distinguish each firelock" and accoutrement set.

The 1856 Canadian Markings

As a result of the 1855 Volunteer Militia Act, thousands of arms and their "accoutrements complete" arrived in Canada during the Spring of 1856 for issue to the volunteer units. To have some accountability, a marking system had to be devised and, on 16 May 1856, a MGO specified that each arm of the service would have the letters L.C. for Lower Canada and U.C. for Upper Canada followed by a letter of the alphabet to indicate the unit and a numeral to indicate the "rack number" of the individual weapon or equipment. Thus, say, "L.C., A, 12" would indicate the 1st Montreal Volunteer Cavalry's 12th set of arms and equipment. The letters recorded here have been compiled from two main sources: 1. The records of issue from 3 June 1856 to 13 January 1857; 2. The quarterly returns of the volunteer militia units in Canada from 1856 to 1862.

It will be seen from this compilation that the system soon started to break down and that there was increasing movement of the allowed arms and equipments, Some changed units, others acquired arms from various sources. As a result, a unit could eventually have several markings. An example is the 3rd Kingston Volunteer Rifles that, in January 1858, had 10 sets marked B, 10 marked N and 20 marked Z. There were also double letter markings that appear as early as April 1857 because in Upper Canada, authorities ran out of letters in the alphabet. Most of these markings include the letter B, probably indicating a "Class B" unit authorised in later 1856. The multiple letter markings that do not have the letter B probably denote a mixture of sources, such as arms issued to the 3rd Kingston mentioned above, rather than markings of a single weapon or equipment with multiple letters.

There are also a few inconsistencies sometimes found in the returns. For instance the letter Upper Canada letter A for a rifle unit is attributed to the Brockville Rifle Company except for a rather mysterious return of July 1857 that attributes this marking to the London Highland Rifles.

From the records examined, the system breaks down completely during 1861 and 1862. By then, many corps report "various" markings, or "numbers", or "none" or still "L.C. and numbers" as in the case of the 2nd Montreal Cavalry. This is not too surprising as, in March, 1861, DAG de Salaberry allowed the St. Vincent de Paul to "have marked by a person in your locality the sabres and belts, say from number 1 to number 30" by the most economical means.

The next step was marking according to the unit. This does not seem to have been the subject of a MGO, but rather a spontaneous action on the part of some corps commanders — at least for the Montreal Light Infantry and the Montreal Engineers. Such as they were, the May 1856 markings were surely found on certain weapons and accoutrements of some volunteers until 1867 when the whole force was rearmed with breechloading arms and their accoutrements.

Troops of Cavalry, Lower Canada

A 1st Montreal
B 1st Quebec (1856-1857)
  1st Montreal (1858-1860)
C 2nd Montreal
D St. Andrews
E Cockshire
F 2nd Quebec (1860)
  1st Laval (1861-1862)

Troops of Cavalry, Upper Canada

A 1st Frontenac

Royal Artillery harness detail, 1890s. Note that all horse were saddled. Print after the 1896 List of changes of War Materiel. Exactly the same materiel and its List of changes was used in Canada. Ph: RC.

Driver, B Battery, Royal Canadian Artillery, 1890s. This driver has the rank of bombardier or acting-bombardier as shown by the yellow chevron on his right sleeve. Watercolour by Barry Rich. Courtesy National Historic Sites, Parks Canada.

The Sydney Field Battery, Royal Canadian Artillery, c. 1890. This exceptional image shows the battery parading at full strength with gunners mounted on the right side next to the drivers on their left. A third gunner is seated in the limber. NCOs and officers are mounted individually. The officers wear the patrol jacket, their men have the tunic and all have the white helmet with the ball finial on top. Courtesy James McConnell Memorial Library, Sydney, Nova Scotia.

Enfield artillery carbine sent to Canada in 1856. Courtesy National Historic Sites, Parks Canada.

B St. Catherine's (1856-1858)
  Leed (1859-1862)
  2nd Northumberland (from April 1862)
C 1st Wentworth
D 1st Toronto
E 2nd Toronto
  1st Guelph (1862)
F Cornwall
G Napanee
H Cobourg
I 1st Essex (1858)
K St. Thomas (1858)
L 1st London
BE 2nd Wentworth
BF 4th York
BI 1st Leeds (1858)

The Broad Arrow with "WD" mark. This was stamped on weapons and accoutrements from 1855. In this instance, upside down on the butt of the artillery carbine. Courtesy National Historic Sites, Parks Canada.

Artillery units, Lower Canada

A Quebec Field
B 1st Quebec Foot
C 2nd Montreal Field
  Montreal Foot
E 2nd Quebec Foot

Artillery units, Upper Canada

A Dundas Field
B Hamilton Field
C Toronto Field
D Kingston Foot
E London Field

The letters "U C" stamped on weapons and equipment sent to Canada in 1856 stood for Upper Canada (now the province of Ontario). "L C" stood for Lower Canada (now the province of Québec). Courtesy National Historic Sites, Parks Canada.

Engineers, Lower Canada

E Montreal

Rifle Companies, Lower Canada

A 1st Montreal
B 1st Quebec
C 2nd Montreal
D 1st Sherbrooke
E 2nd Quebec
F St. Martin
G 3rd Montreal
  1st Hinchbrooke (1862)
H 4th Montreal
  2nd Sherbrooke (1862)

I 1st Granby
J 4th Quebec
K 5th Montreal
2nd Sherbrooke (1862)
L 6th Montreal
N 1st St. Vincent de Paul
O Montreal Highland
P 8th Montreal
1st Hinchbrooke (1862)
Q Sorel
R 9th Montreal
1st Hinchbrooke (1862)
S 1st Mégantic
U 5th Quebec (1860)
1st Lennoxville (1862)
L.G.R. 2nd Sherbrooke (1862)
N.A. 1st Hemmingford (1862)
MLI-A 1st Company, Montreal Light Infantry
MLI-B 2nd Company, Montreal Light Infantry
MLI-C 3rd Company, Montreal Light Infantry
MLI-D 4th Company, Montreal Light Infantry
MLI-E 5th Company, Montreal Light Infantry
MLI-C 6th Company, Montreal Light Infantry (1862)

Colt Model 1848 "Navy" Revolver. This revolver was issued in 1856 to the 1st Wentworth Troop of Cavalry as its grip is stamped (upside down!) with UC [Upper Canada], C [the troop's letter] and 28 [the rack number]. The American model Colts sent to Canada came from London, England, where they were made. Fort Henry, Kingston. Ph: RC.

Rifle Companies, Upper Canada

A Brockville
B 1st St. Catherine's
C 1st Kingston
D 2nd Kingston
E Barrie
F 1st Hamilton
G 2nd Hamilton
H 1st Cobourg
I Picton
K 1st Guelph
L 1st Toronto
2nd Toronto, 2nd Bn (1862)
M 2nd Toronto
5th Toronto, 2nd Bn (1862)
N 3rd Toronto
O 1st London
P 2nd London
Q 1st Prescott
R 1st Ottawa
S 2nd Ottawa
T Brighton
Stratford (from October 1862)
U Brampton

Butt plate marks on the 1856 artillery carbine. The letter (B) indicated the unit and the number (21) denoted the weapon's rack number. This particular carbine could have been issued to the Hamilton Field Battery in Upper Canada or the 1st Quebec Foot Artillery in Lower Canada. Courtesy National Historic Sites, Parks Canada.

V Chatham
St. Thomas (October 1857-1862)
W Galt
1st Belleville (October 1858-1862)
X Woodstock
Y Paris
Z Sarnia
1st Williamsburg (1859-1862)
AB 1st Peterboro
BB Hamilton Highland
BI 4th Toronto
BK Toronto Highland
BM Collingwood
BO 2nd Prescott
G.I. 1st Mount Forest (1862)
A.C.Q. Portsmouth (1860-1862)
B,N,Z 3rd Kingston

Imperial Markings, circa 1860-1861

In about late 1860 or early 1861, the imperial government issued regulations regarding the markings to keep track of arms issued to all volunteer units in all overseas territories assigning a letter code mark to each colony. The exact date was not included, but apart from replacing the word "Colony" for "County", it was word for word with the 19 January 1861 regulations for the British volunteers in the United Kingdom. For British North

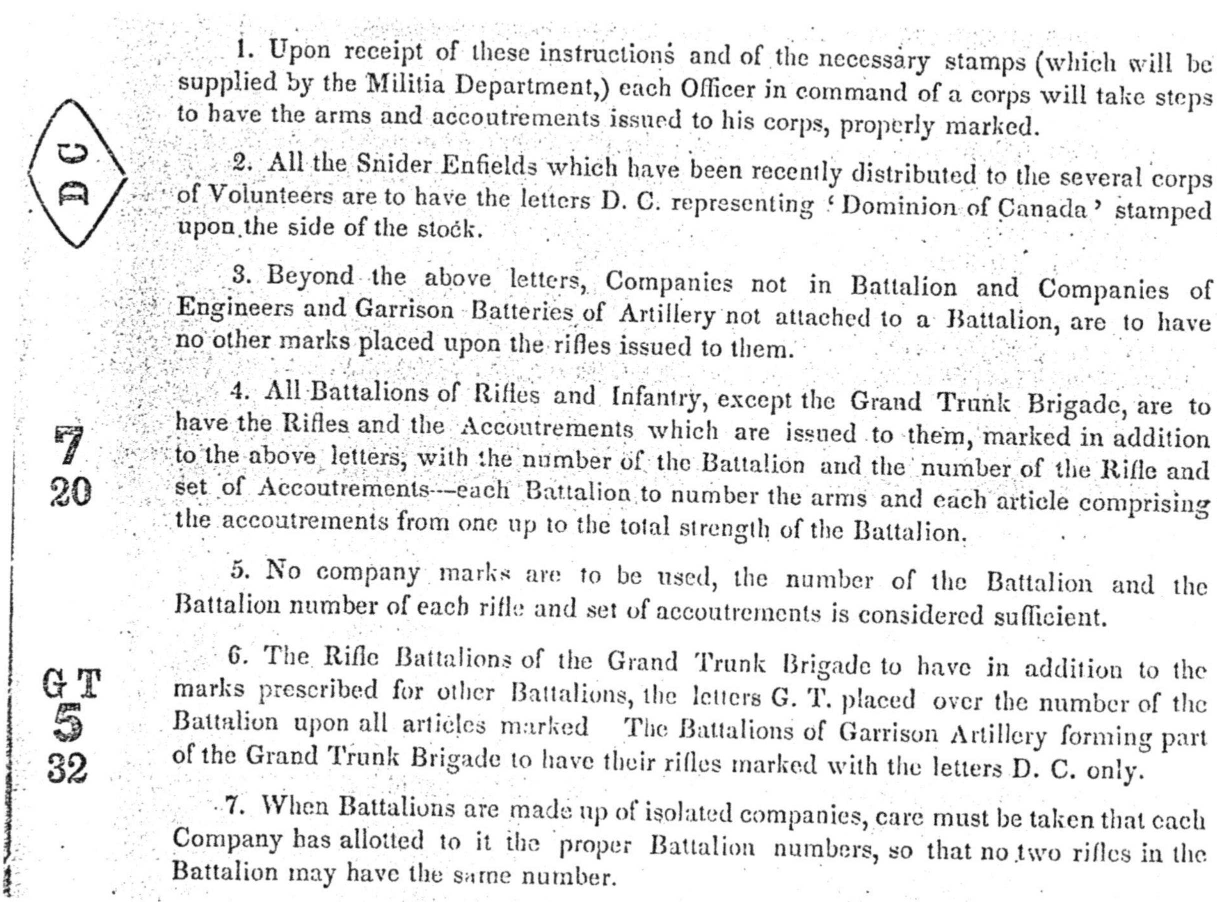

1. Upon receipt of these instructions and of the necessary stamps (which will be supplied by the Militia Department,) each Officer in command of a corps will take steps to have the arms and accoutrements issued to his corps, properly marked.

2. All the Snider Enfields which have been recently distributed to the several corps of Volunteers are to have the letters D. C. representing 'Dominion of Canada' stamped upon the side of the stock.

3. Beyond the above letters, Companies not in Battalion and Companies of Engineers and Garrison Batteries of Artillery not attached to a Battalion, are to have no other marks placed upon the rifles issued to them.

4. All Battalions of Rifles and Infantry, except the Grand Trunk Brigade, are to have the Rifles and the Accoutrements which are issued to them, marked in addition to the above letters, with the number of the Battalion and the number of the Rifle and set of Accoutrements---each Battalion to number the arms and each article comprising the accoutrements from one up to the total strength of the Battalion.

5. No company marks are to be used, the number of the Battalion and the Battalion number of each rifle and set of accoutrements is considered sufficient.

6. The Rifle Battalions of the Grand Trunk Brigade to have in addition to the marks prescribed for other Battalions, the letters G. T. placed over the number of the Battalion upon all articles marked The Battalions of Garrison Artillery forming part of the Grand Trunk Brigade to have their rifles marked with the letters D. C. only.

7. When Battalions are made up of isolated companies, care must be taken that each Company has allotted to it the proper Battalion numbers, so that no two rifles in the Battalion may have the same number.

Part of the circular dated 25 September 1867 with examples of the new marks in its margin. Ph: RC.

America, the imperial regulation comprised:

B.C	British Columbia
C^{a}	Canada
N.B.	New Brunswick
N^{d}	Newfoundland
N.S.	Nova Scotia
P.E.I.	Prince Edward Island
V^{r}.I.	Vancouver Island
V^{a}	Victoria

These letters were to be, in all territories, under the letter "V" and, except for rifle units, followed by the letter "L" for light horse, "A" for artillery, "E" for engineers and "M" for mounted rifles, and, below, the weapon's number. This was also to be stamped on the bayonet scabbard.

This does not seem to have been applied in Canada, which already had its 1856 system, and apparently little elsewhere. There is evidence that Prince Edward Island did, from a Pattern 1860 short Enfield rifled musket, one of 400 that arrived in 1861, with "V—PEI—129" engraved on the tang of its butt plate. It was not until the Summer of 1863 that all the 1,400 rifled muskets on the island were "to be marked in accordance with instructions received from the Imperial Government." This was done by Armourer Sergeant Miller of the 17th Foot and later Royal Artillery Bombardier Laird. Nova Scotia also marked some of it weapons from about 1859-1860 with N.S. and a number on the right side of the stock, but seemingly not on the accoutrements although some 6,000 sets were received from imperial stores in 1865. After Confederation in 1867, the Canadian Militia system introduced that year applied to all provinces.

1867 and thereafter

The distribution of tens of thousands of Snider Enfield rifles with their accoutrements during 1867 prompted the Militia Department to announce new instructions "relating to markings of arms and Accoutrements" in a circular dated 25 September 1867. Henceforth, all Snider Enfield rifles were to "have the letters D.C representing "Dominion of Canada" stamped upon the side of the stock. Arms and accoutrements of all battalions (except the Grand Trunk Brigade) were to bear the battalion number above the item's number. Thus, the rifles and the accoutrements sets were to have, say, 7 above (for the 7th battalion) and, say, 20 below (the numbers went from 1 to the battalion's total strenght). Grand Trunk Brigade units were to have "in addition…the letters G.T. placed over the number of the Battalion over all articles marked" except for its garrison artillery units that needed "the letters D.C. only." For all units, no company mark were to be used since the number of the battalion and the number "of each rifle and set of accoutrements is consider sufficient."

The battalion arms were to have D.C. [in a diamond] on the stock of the rifle, the battalion number in large figures and the rifle and accoutrements number. "Pouch belt, waist belt, sling and Bayonet frog, to be marked inside with large and small figures, corresponding with the number on the rifle. Pouch, Bag, Ball and Ca Pouch to be marked inside the flap." These instructions were repeated in the 1887 book of orders and regulations.

This circular was re-issued with some additions on 4 March 1870, notably that cavalry arms, accoutrements and saddlery issued would have D.C. in a diamond and "the letter C and figure or figures representing the number of the Troop; the several Troops in the Dominion to be numbered according to date of seniority of Corps." However, the troops of cavalry were instead grouped into numbered regiments. Nevertheless, the 1887 book of orders and regulations maintained the 1870 instructions.

From 1867, infantry, engineer and garrison batteries of artillery units not grouped into battalions had, in theory, no other marks than DC. Judging from markings on Snider-Enfield "two band" carbines, the artillery batteries nevertheless had markings that also may have been stamped on equipment. A number of carbines in the collection of the National Historic Sites of Parks Canada have such marks. Of some noted, one finds: HGA for Halifax Garrison Artillery, PEIGA: Prince Edward Island Garrison Artillery, QGA: Quebec Garrison Artillery and

BB probably for B Battery, Royal Canadian Artillery. Reader's wishing more details on arms markings should consult David W. Edgecombe's *Defending the Dominion: Canadian Military Rifles 1855-1955* (Ottawa: Service Publications, 2003).[112]

Notes

Jack L. Summers, *Tangled Web: Canadian Infantry Accoutrements 1855-1985* (Bloomfield: Museum Restauration Service, 1992); Pierre Turner, *Soldier's Accoutrements of the British Army 1750-1900* (Ramsbur-ry: Crowood Press, 2006); Mike Chappell, *British Infantry Equipments 1808-1908* (London: Osprey, 1980) and *British Cavalry Equipments 1800-1941* (London: Osprey, 1983); W.Y. Carman, "The Oliver Equipment", *Antique Arms and Militaria*, September 1984; Donald Featherstone, *Weapons and Equipment of the Victorian Soldier* ((Poole, UK: Blandford, 1978); "Infantry Knapsack", *United Services Journal*, No. 2. Part II, October 1831; R. Simkin, "Packs, Past and Present", *Illustrated Naval and Military Magazine*, 1890, pp. 211-218 ; G. Tylden, "Accoutrements of British Infantrymen 1660-1940", JSAHR, XLVII: 1 (Spring 1969), pp. 4-22 and *Horses and Saddlery* (London, 1965). See our *A Scarlet Coat* (Ottawa: Service Publications, 2012) for a description of War of 1812 accoutrements. For South Africa, see also: "Colonial Contingents to South Africa", SP, No. 63, Vol. 13 (1900); "Organisation, equipment, despatch and service of the Canadian contingents during the war in South Africa 1899-1902", SP, No. 35a, Vol. 13 (1903); MR, 1900; A.E. Hilder, *A Canadian Mounted Rifleman at War*, reminiscences edited by A.G. Morris (Capetown, 2000); South African War Diary of Edgar Henry Redway, Queen's Own Rifles Museum, Toronto (from its website); The Queen's Own Rifles of Canada Regimental Museum and Archive
at: https://qormuseum.org/history/timeline-uniforms/.

[112] D. Ross MacInnis, "The markings of Nova Scotia Militia muskets", *Arms Collecting*, XXXIII: 2 (May 1995): 44-48; Clive M. Law, *Canadian Military Handguns 1855-1985* (Bloomfield, 1994); René Chartrand, "The 1856 Markings of Canadian Volunteer Militia Weapons", *Arms Collecting*, XXI: 3 (May 1983): 39-46; Norman J. Crook, "Regulations for the marking and preservation of arms issued to corps of colonial volunteers", *Arms Collecting*, XVIII: 4 (November 1980): 118-124; Paul S. Lederer, "A Pattern 1860 Enfield of the Prince Edward Island Volunteers", *Arms Collecting*, XXI: 1 (Feb. 1983): 11-14; Circular, Militia Department, Ottawa, 25th September, 1867; "Marks of military ownership", *Arms Collecting*, XII: 2 (May 1974): 60-61; *Regulations and Orders for the Militia of the Dominion of Canada 1st September 1887* (Ottawa: Chamberlin, 1887), pp. 101-104. See also the fine compilation by Norman J. Crook, "Retailer's marks on the blades of Canadian officers' swords", *Arms Collecting*, XVI: 3 (August 1978): 71-83

Colour-Sergeant Curzon, 10th Royal Grenadiers of Toronto, 1885. This image appears to have been taken shortly after the battalion's return from Saskatchewan. The floral trophy at left bears the word "Batoche" in which battle the 10th took part and Colour-Sergeant Major Curzon distinguished himself for bravery. He wears the unit's late spring campaign dress: Pattern 1878 scarlet tunic, dark blue trousers, white linen peaked Havelock over the forage cap and knapsack over the shoulders. The white belts shown appear to have been browned, which was commonly done during the campaign. Courtesy Library and Archives Canada, C1992.

Chapter 12

Officer's Mess Dress

We have evoked the mess dress of cavalry, artillery and rifle units in Vol. 1. We group here that of the infantry, both regular and of the volunteer militia, the Governor General's Foot Guards peculiar mess dress, the engineers and other corps.

During the 1860s, officer's messes appeared as Canadian battalions were formed following closely British army practices, which, from 1867, included having a distinct mess dress in some battalions. The 2nd Queen's Own Rifles introduced a uniform mess jacket for its officers that year. Mess dress was meant to be a more relaxed yet still formal order of dress. It originated from the practice by army officers to dine wearing their undress "shell jacket" from about the second third of the 19th century. Indeed, the British 1864 dress regulations mentioned shell jackets and waistcoats, these last termed "mess waistcoats" by the British 1883 dress regulations. While these items were in the regulations, it was "left to an officer's discretion to have a Mess Jacket, or to economise by doing without it" due to the "fact that the War Office is rather lax as to this tasty little garment" according to tailor T.H. Holding in 1885. One could add that it also depended if commanding officers were keen on mess dress or not.[113]

Figure of an officer wearing the standing collar type mess dress, 1880s and early 1890s. Plate from Holding's 1894 Uniforms... Ph: RC.

Be it as it may, mess dress gained popularity becoming quite stylish and intricate at that time as the British 1900 dress regulations show. For the British line infantry, the mess dress jacket of all regiments now had the "roll collar" (which was a combination collar and lapel) except for the Somerset Light Infantry that had the "stand-up" collar. This was quite a change from the shell jacket fashion of the 1880s that had a standing collar and it was seemingly still the norm when Holding published his uniform patterns in 1894.

At that time, only the Foot Guards had roll collars, at least in the orders and publications. Obviously, many line infantry colonels had other ideas and their officers overwhelmingly agreed from the mid-1890s. So the War Office bowed and, by 1900, the line infantry could have a scarlet mess jacket "edged all-round with white piping" and "Stand-up or roll collar" that also meant an open waistcoat buttoning at the bottom with a white shirt and black bow tie. Other corps such as cavalry, artillery and engineers usually kept the older style with standing collar. Furthermore, by about 1901 or 1902, W.D. Vincent's *Cutter's Practical Guide* confirmed that the "general style of Mess Jackets worn by nearly all Officers is of the roll collar pattern, without buttons or braid in any part of the Jacket."[114]

In Canada, military authorities were trying to keep up with this new twist of British military fashion. In the 1870s, the regular troops consisted of two artillery batteries and the infantry provisional battalion in Winnipeg (disbanded in 1877). The kit of its officers included a mess jacket and a mess vest that would have

[113] Anthony Kellet, *The History of Messes in the Canadian Army* (Ottawa: Service Publishing, 2014), pp. 74, 80. Sergeant's messes also had mess dress in a few regiments from the late 1880s, notably the 1st, 2nd and 43rd rifle battalions

[114] T.H. Holding, *Coats: How to cut and try them on* (London, 1885), n.p.; Holding, *Uniforms, passim*; War Office, *Dress Regulations for the Officers of the Army* (London. Her Majesty's Stationary Office, 1864, 1883, 1900 [illustrated]); W.D. Vincent, *Cutter's Practical Guide to the cutting and making of all kinds of British Military Uniforms* (London: Williamson, c. 1901).

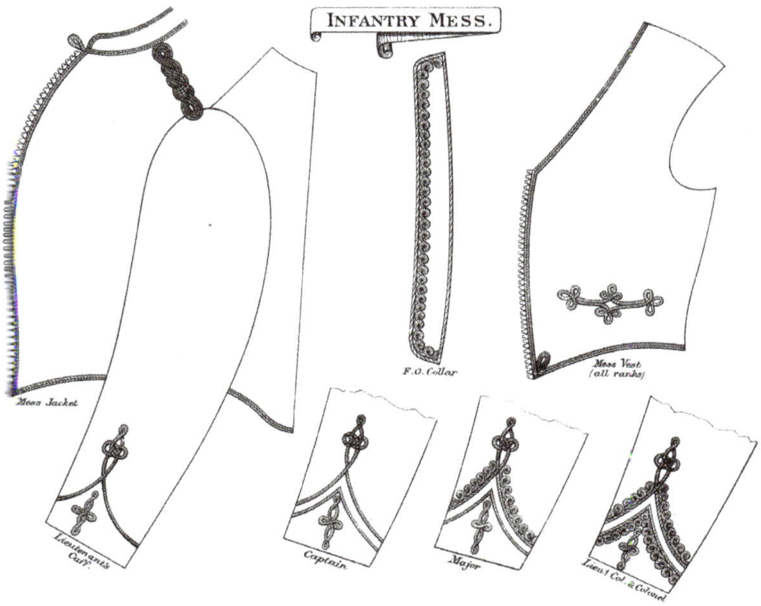

Line infantry mess jacket pattern, 1880s and early 1890s. This is the standing collar type. The rank distinctions are also shown. Plate from Holding's 1894 Uniforms… Ph: RC.

Lt. Col. Joseph John Tucker's standing collar type mess dress jacket and waistcoat, 62nd St. John Fusiliers, 1895. Scarlet jacket with dark blue collar and cuffs, gold buttons, lace, shoulder cords and collar grenade; dark blue waistcoat with gold buttons and lace. A businessman notably for Lloyd's, Tucker had quite a background as a volunteer having commanded a troopship in Crimea, become captain in the 1866 Nova Scotia Naval Brigade, rose from private to captain in the Shanghai Volunteer Corps from 1870 to 1873, major in the 62nd in 1883, lieutenant-colonel commanding in 1893 and honorary colonel in 1897. Courtesy New Brunswick Museum, St. John.

(Left) Lieutenant-Colonel's standing collar type mess dress jacket and waistcoat, 34th Ontario Battalion, late 19th and early 20th centuries. Ontario Regiment Museum, Oshawa. Ph: CL.

(Right) Lt. Col. J.J. McLennan, 59th Stormont and Glengarry Infantry Battalion, November 1898, He wears the standing collar type mess dress jacket. Courtesy Library and Archives Canada, C46601. (Far right) Infantry officer wearing mess dress, second half of the 19th century. This fascination image, difficult to date although it might be as early of the 1860s, portrays either a Canadian or a British officer, almost certainly taken in New Brunswick or Nova Scotia. His trousers bear the thin scarlet piping typical of infantry. His shell jacket, initially used as a mess jacket by officers, has its standing collar turned down and its upper front folded back to make a lapel — in effect, the ancestor of the rolled collar. Andrew Moss Collection. Ph: CL.

followed the 1876 Militia General Orders (see below) as would have the Infantry School Corps raised in late 1883 that later became the Royal Canadian Regiment. These and the 1886 regulations were meant for both the regular and the volunteer militia units

Officer of the British infantry's King's Liverpool Regiment, c. 1900-1901. He wears the infantry mess dress according to the British army's 1900 dress regulations by which each regiment had certain distinctive features. For the King's Liverpool Regiment, this consisted of white piping edging the jacket. Detail from a watercolour by E.A. Holloway. Courtesy Anne S. K. Brown Military Collection, Brown University Library, Providence, USA.

Canadian Infantry

The 4 March 1876 Militia General Orders prescribed for infantry officers as "Shell Jacket" described as: "Scarlet cloth, with collar and pointed, cuffs of blue facings; silver braid edging all round, including the top and bottom of the collar; a loop of silver braid at bottom of collar to fasten across the neck; shoulder cords as on the tunic, a row of silver studs and hooks and eyes down the front, scarlet lining" with rank badges and a "Mess Waistcoat" that was : "Blue cloth, silver braid edging round the top, down the front, and along the bottom to the side seams; the pocket-edged with braid forming crow's feet and eyes, a row of silver studs and hooks and eyes down the front." The 1886 Militia General Orders had much the same as above, but with gold braid, buttons and studs. On the whole, this was the mess dress regulation dress of the British line infantry, especially that of 1883.

It can be assumed that not all young officers in all units could bear the extra expense of this order of dress, especially in rural battalions. Perhaps a phrase in the 24 March 1892 General Orders mentioning that "the patrol jacket may be worn at mess by attached officers who are not provided with tunic or mess dress" addressed this situation. However, photos and surviving artefacts show that some officers did indeed procure mess dress.[115]

Governor General's Foot Guards

The officer's desired mess dress was to follow the pattern of the Coldstream Guards. The Canadian Militia 1886 general dress regulations were the first to specify the regiment's uniform including the mess dress. The "shell jacket" used as a mess jacket was described as of: "Scarlet cloth edged all round, including the collar with gold braid, with crow's foot in centre of back. Blue cloth collar and cuffs. The collar rolled, with six pointed stars similar to that on collar of tunic on each end, 5 inches from the shoulder seam— the cuffs pointed. Field officers have on each sleeve three rows of gold braid, Captains two rows, and Subalterns one row, terminating in Austrian knot over point of cuff." The "mess waistcoat" was of "blue cloth, gold braid edging round the top, down the front, and along bottom to side seams--edges of pockets similarly braided—crow's foot at each end, and in centre above and below, row of gilt studs and hooks and eyes down the front."

Meanwhile, in England, the Foot Guards mess jacket had become simpler and more elegant by removing the gold braid edging except at the cuffs. The Governor General's Foot Guards may have adopted this pattern in about 1890. It was of scarlet cloth with dark blue roll collar and blue cuffs. Rank was indicated by narrow gold cord at the top of the cuffs, forming a eye at the top, one row for subalterns, two rows for captains and three rows for field officers. Mess waistcoat: dark blue with a dark blue roll collar, fastened with four half inch buttons of regimental pattern in pairs. White shirt with black bow tie. The regimental badge was on the jacket's roll collar. It seems that the gold cord was extended to form a small curl on the lower side of the cuffs, but when this was introduced is unknown. On the whole, the present mess dress of the regiment offers a near-identical appearance

[115] Kellet, *The History of Messes in the Canadian Army*, p. 74.

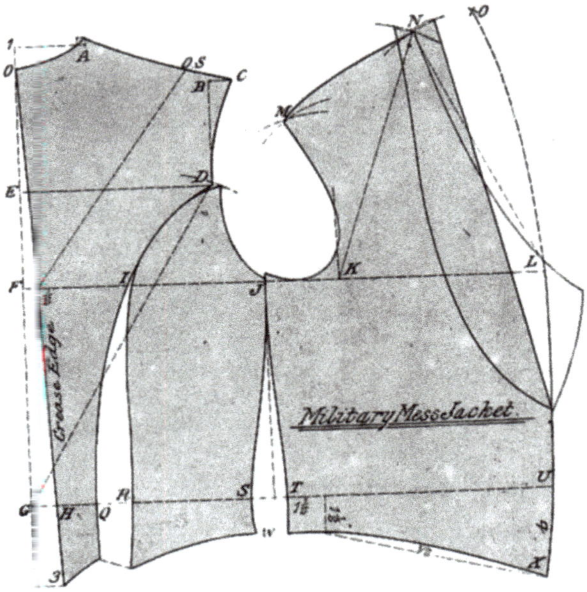

(Left) Tailor's pattern for a "Military Mess Jacket, roll collar system" for officers, c. 1900. Detail of a print from W.D. Vincent's Cutter's Practical Guide. Ph: RC.

(Right) Officer of the 2nd Coldstream Guards wearing mess dress, early 1890s. British Guards infantry regiments wore the roll collar type jacket as part of its mess dress. The Canadian Governor General's Foot Guards followed suit. Plate in T. H. Holding's 1894 Uniforms…Ph: RC.

Officer's mess dress of the Governor General's Foot Guards. As can be seen from this jacket and waistcoat dating from the later 20th century, this order of dress was almost similar to the Coldstream Guards and seemingly did not change from the late Victorian era. It is uncertain as to when the gold curls were added to the side of the cuffs. Private collection. Ph: RC.

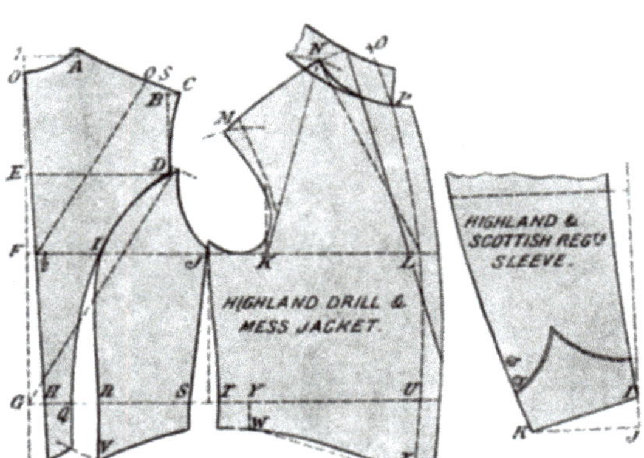

(Top) British officers in mess dress, c. 1900. The roll collar type jacket was also worn by general officers, many staff corps as well as the great majority of infantry units. Cavalry and certain other corps kept wearing the standing collar type of jacket. Print from W.D. Vincent's Cutter's Practical Guide. Ph: RC. (Left) Tailor's pattern for a "Military Mess Jacket, roll collar system" for officers, c. 1900. Detail of a print from W.D. Vincent's Cutter's Practical Guide. Ph: RC.

to what it was at the end of the 19th century.[116]

Royal Canadian Regiment

Early mess dress in the regiment would have followed what was prescribed in the 1876 and 1886 dress regulations. During the 1890s, the roll collar fashion caught up with the regiment. In July 1898, Col. Otter, after seeing some deviations in mess dress some of his by officers, reminded them that the dress in the mess was: "Officer's Mess Jacket: Scarlet, with rolling collar of blue cloth and pointed cuffs. Trimmed with blue piping around the body, collar and cuffs. Shoulder cords of double gold cord, 1/2 inch wide with miniature rank badges. Single row of four 1/4 inch Regimental buttons. Collar badges in silver to be worn on lapels. Medal ribbons to be worn on the body of the jacket and not on lapels. Officer's Mess Waist Coat: Blue cloth, quite plain, buttoning with four 1/4 inch Regimental buttons, V.R.I." Since this was a reminder, this version with the rolling collar must have been adopted earlier. We add that regimental mess waiters wore the "regimental livery (tail coat, with large brass buttons and scarlet vest and regimental trousers)".[117]

Captain Henry Reynolds Luard, Royal Engineers, 1863. He wears the mess dress that was often used as an unofficial version of undress. Scarlet jacket with dark blue velvet collar, shoulder cords and cuffs, small gold ball buttons and lace, scarlet vest with small gold buttons and lace, Oxford mixture dark grey trousers with wide scarlet stripe. Luard (1829-1870) served in British Columbia from 1858 to 1863 as adjutant and executive officer of the detachment. Courtesy British Columbia Provincial Archives, Victoria.

The Militia Infantry at large

Considering that its men were volunteers, it is surprising to find evidence that many officers did procure mess dress, which generally followed the 1876 and 1886 regulations quoted above. The prevailing early style was that of the shell jacket with its standing collar. This then gradually changed to the roll collar pattern during the 1890s, but not as much as in the British army. Col. Otter's *The Guide* for Canadian officers published since 1881 had in its 1906 edition for "Mess order" a "Mess jacket, waistcoat, and trousers, white collars and black neckties" that indicated this concerned roll collar pattern mess jackets. However, some units preferred to keep the previous standing collar style, amongst them the 13th, the 34th and the 59th battalions. Finally, the 1907 Canadian Militia Dress Regulation imposed the roll collar as the universal pattern. No doubt there may have been reluctance by volunteer militiamen to invest in a new style when the one they had was still quite attractive. Also, city units were generally more prosperous than those in the countryside who were not always very keen on mess dress.

By the British 1883 dress regulations, Highland infantry regiments were to have "scarlet shell jackets, of the same pattern as of the Infantry of the Line." The "Mess-Waistcoat" was to be "of the colour of the facing, scarlet cloth or regimental tartan" of the same pattern" as the infantry of the line. In Canada, the jacket's facing colour was dark blue for the 5th Royal Scots and the 48th Highlanders, but the 48th's mess waistcoat was noted as scarlet in 1907. Contrary to other units, Highlanders in Britain kept the shell jacket as a mess jacket and this appears to have been the case in Canada also because the Canadian 1907 dress regulations for officers prescribed it edged with white piping. However, pre-1901 evidence that Canadian Highland officers had mess dress is uncertain.[118]

[116] *The Regimental History of the Governor General's Foot Guards*, p. 240; Holding, *Uniforms…*, p. 40; Vincent, *Cutter's Practical Guide*, p. 42.

[117] Kellet, *The History of Messes in the Canadian Army*, p. 41.

[118] Ernest J. Chambers' amply illustrated regimental history of *The 5th Regiment Royal Scots of Canada Highlanders* (Montreal, 1904) does not show any of its many officers wearing mess dress.

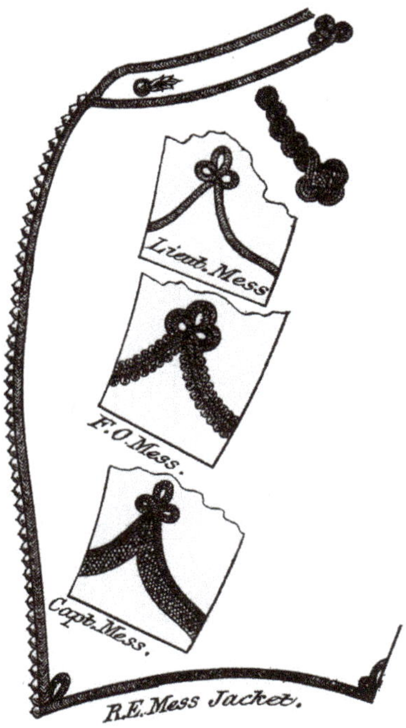

Royal Engineers or Royal Canadian Engineers officer's mess jacket with standing collar, late 19th and early 20th century. Musée du Fort Saint-Jean, Saint-Jean (Québec). Ph: RC.

Officer's mess waistcoat pattern and details of the tunic's rank badges, Royal Engineers, 1880s-1890s. Canadian officers followed these models. Plate in T. H. Holding's 1894 Uniforms…Ph: RC.

Officer's mess dress, Royal Canadian Engineers, 1960s. As can be seen, apart from the peaked cap, this order of dress has remained basically unchanged since the mid-19th century. Only the designation of ranks has since been changed by the abolition of the gold shoulder knots and cuff lace for the "pips and crowns" on scarlet shoulder straps. The elegant dark blue cape is for walking out. The white jacket with rolling collar was for summer dress and already mentioned in late 19th century regulations. In Canada, it seems to have been adopted in some units during the 20th century. Watercolour by J. H. Malion. Private collection. Ph: RC.

Engineers

Regular engineers were initially British and, even after the departure of the British garrison from central Canada during 1870-1871, a few remained at the Royal Navy bases at Halifax and Esquimalt. As seen in chapter 6, all had their versions of the dress of the British Royal Engineers. Some acquired the mess dress of the Royal Engineers.

The Canadian 1876 and 1886 dress regulations followed the British ones. Officer's mess dress consisted of a scarlet shell jacket with a standing collar, blue collar and cuffs, gold lace and small buttons. The waistcoat was scarlet with gold buttons and lace, trousers dark blue with a two inch scarlet stripe.

Medical and Veterinary Officers

Most being attached to battalions or corps wore their unit's mess dress if there was one while those attached to district staff had their British equivalent's mess dress; this was specifically ordered in April 1894 (see chapter 6). That year, the British Army Medical officers would have worn a dark blue mess jacket with standing collar edged all around with gold lace including in a point at the top of the cuffs, black velvet cuffs and collar, scarlet mess waistcoat edged with gold lace, dark blue trousers with a wide scarlet stripe for undress and gold for dress.

Apart from the above, Canadian departments and corps such as the Ordnance, Service, Pay and Veterinary corps were formed after the reign of Queen Victoria. It is uncertain if its officers had mess uniforms and if they did, it would have been of the pattern used by their British equivalents.

Chapter 13

Khaki and South Africa 1898-1901

Royal Canadian Artillery gunners, 1898. They wear an experimental uniform that was not adopted. It was made of khaki serge with, for these gunners, dark blue collar and cuffs. The equipment shown was also not adopted. This photo and a colour rendition of it were wrongly captioned "Pacific Railway Militia" in the 1898 Navy and Army Illustrated and as "British Columbia Rifles" in the Christmas supplement of Toronto's Globe newspaper. Courtesy Library and Archives Canada, C126820.

Following the outbreak of hostilities between Great Britain and the two small Boer republics of Transvaal and the Orange Free State in South Africa, Canada offered to send a contingent. The offer was accepted by Britain and the 2nd (Special Service) Battalion of the Royal Canadian Regiment arrived at Cape Town in November 1899.

This was followed by a second contingent consisting of two squadrons of the Royal Canadian Dragoons that was expanded to a war strength of 19 officers and 352 men and renamed 1st Battalion, Canadian Mounted Rifles, (but reverted to Royal Canadian Dragoons in August 1900), the 2nd Battalion, Canadian Mounted Rifles (renumbered 1st in August 1900) and three batteries serving 18 12-pounder field guns of the Royal Canadian Artillery. The 2nd Canadian Mounted Rifles (renumbered 1st), and the artillery arrived in South Africa, on 18 and 27 February 1900. The 1st Canadian Mounted Rifles (Royal Canadian Dragoons) were delayed for a month due to sickness of their ship's crew and landed at Cape Town on 26 March.

In January 1900, Lord Strathcona offered to raise in Canada at his own expense a regiment of mounted rifles. The British government accepted and the regiment arrived in South Africa on 10 April. It was officially part of the British army, but also considered a Canadian unit by nearly everyone.

On 18 and 25 February 1902, the third Canadian contingent made up of the 2nd Canadian Mounted Rifles and the 10th Canadian Field Hospital arrived at Durban, South Africa.

The fourth contingent made up of the 3rd, 4th, 5th and 6th Canadian Mounted Rifles battalions was raised from April 1902. The 3rd landed at Cape Town on 31 May, the 4th at Durban on 14 June, the 5th on 24 June and the 6th at Durban on 16 June. This was just after the war had ended by a peace agreement on 31 May.

Canadians also largely made up the Canadian Scouts which was a battalion of Mounted infantry in South Africa raised from December 1900 and the South African Constabulary during 1901, both organised by the British army, which equipped and armed them. Although officially British units, these units never quite lost their Canadian character.

Except for the fourth contingent, all the Canadian troops were engaged in many actions where they often distinguished themselves, several men winning the coveted Victoria Cross. Apart from the many Canadians that continued to serve in the South African Constabulary until 1908, nearly all others returned to Canada during July 1902.

Uniforms and Equipment

Private of the 2nd Special Service Battalion of the Royal Canadian Regiment, 1 November 1899-early 1900. He wears the first Canadian Pattern khaki uniform with Oliver equipment. Courtesy Queen's Own Rifles Museum and Archives collection and photo.

Canada was to provide its contingents with uniforms, equipment, horses and saddlery while the imperial government would provide weapons. Canadian infantry, artillery and cavalry units basically had two uniforms: the scarlet or dark blue serge frock with dark blue serge trousers, and the khaki uniform. The new mounted rifles units received a new pattern of dark blue serge frocks (please see the Cavalry chapter). The scarlet or dark blue serge frock clothed units in Canada. In South Africa, they wore khaki uniforms.

Campaigns in Africa and India had convinced the British army to abandon red coats during the 1880s and adopt khaki-coloured clothing, which helped the troops blend in with the landscape. The word "khaki" is believed to derive from the Hindi word for dust, and it was in British India that troops first wore this colour from 1846 for service and summer dress. British and Indian troops participating in the Abyssinian Campaign of 1867-1868 in Ethiopia were the first to officially use khaki uniforms as their regulation dress outside of India. The practice spread and khaki uniforms were gradually adopted as service dress in other tropical stations. The scarlet serge service jackets were slowly phased out, the last ones being worn in battle in 1885. That year, khaki became the only service dress of the Indian army and, from 1896, of the metropolitan British army.

In Canada, there was no perceived need to have khaki uniforms in a northern climate and red coats were worn during the 1885 North-West Campaign. Following the campaign, General Middleton suggested that the North-West Mounted Police should be converted to a mounted rifles corps and its scarlet uniform changed to khaki, but nothing came of it. The senior military officers nevertheless noted what was going on elsewhere in the world and, in 1898, a handsome khaki uniform prototype was made and issued to a few gunners of the Royal Canadian Artillery. This experimental khaki uniform had dark blue collar, pointed cuffs and shoulder straps, a khaki felt hat turned up on the left side with a blue-black puggaree. The gunners added the grenade at the collar. A bandoleer for extra ammunition was also proposed as part of this new appearance. An initial issue seems to have been meant for rifle corps in British Columbia. It came to public attention as the wrongly captioned "Pacific Railway Militia" in the 1898 *Navy and Army Illustrated* and as "British Columbia Rifles" in the Christmas supplement of Toronto's *Globe* newspaper (reproduced in Volume 1). Political authorities were obviously not favourable to this proposal, probably fearing increased costs, and the concept was shelved. The khaki hat found some limited favour and was approved, but only for British Columbia rifle units.

Canadian Pattern 1899 Khaki Infantry Jacket

In his 1899 report on the Canadian Militia, Maj. Gen. Hutton nevertheless recommended the adoption of a "fatigue suit of khaki, consisting of a loose jacket with stand and fall collar. Four pockets…[and] loose trousers with thigh pockets…that could also be used for campaigning of field service…" This new proposal was of course ignored by the minister of Defence, but things were about to change. When Canada offered the contingent for service in South Africa in October 1899, it was obvious that it had to be dressed with khaki uniforms for warfare on the African continent. There were no khaki uniforms in the country, yet they had to be made in a matter of weeks. The main supplier of military uniforms, the W.E. Sandford Manufacturing Company in Hamilton, Ontario, somehow managed to produce uniforms in record time patterned after what was known of the 1896 and 1897 pattern British khaki uniforms and made them with dyed canvas because there was no khaki cotton drill or wool

A group of officers, Royal Canadian Regiment, November 1899. All wear their Canadian-made uniforms. The officer's equipment is British. This image was taken on board the steamship Sardinian during its crossing from Halifax to Cape Town, South Africa. From Marquis' 1900 Canada's Sons… Ph: RC.

Officers en route to join the Royal Canadian Regiment in South Africa, 19 April 1900. Left to right: captains A.J. Boyd, A.E. Carpenter and C.F. Winter all wearing slouch hats. Courtesy Library and Archives Canada, PA23998.

Norfolk jacket, South Africa, 1900. Khaki British-made Norfolk jacket with pleated front, four brass buttons and integral cloth belt. This type of jacket is also known to have been worn by some officers of Lord Strathcona's Horse, but with breast pockets. This jacket is thought to have been worn by a member of the second contingent, probably an officer. Canadian War Museum, Ottawa. Ph: RC.

Private Alexander Sinclair, Royal Canadian Regiment, 2nd Special Service Battalion, South Africa, 1900. Photographed in light marching order wearing the British khaki uniform. He is equipped with the 1898 version of the Oliver equipment featuring the 80-round brown leather pouch worn in front that proved awkward for the men when lying on the ground on their stomach. The khaki web extra cartridge bandolier worn over the shoulder is also seen. He is armed with a Lee-Enfield Mk I .303 calibre rifle. Courtesy The 48th Highanders Regiment of Canada Museum, Toronto.

"Royal Canadian March" music poster, 1900. This advertising poster published in Toronto and Winnipeg shows members of the Royal Canadian Regiment wearing British uniforms with Slade-Wallace equipment rather than the Oliver equipment. Most Canadians thought their first contingent was dressed and equipped exactly like the British army; in this, they were nearly always confirmed by the Canadian media. Canadian War Museum, Ottawa. Ph: RC.

The 2nd Special Service Battalion of the Royal Canadian Regiment infantrymen in action at the battle of Paardeberg, 1900. In this print as in many other images, the Canadians are shown in khaki uniforms equipped with the British Slade-wallace equipment and armed with the .303 Lee-Enfield rifles. Anne S.K. Brown Military Collection, Brown University Library, Providence, USA. Ph: RC

Group of Royal Canadian Regiment officers, November 1900. Although shown in their South African khaki uniforms, the group posed at Windsor Castle where the regiment was inspected and thanked by Queen Victoria before coming back to Canada. Lt. Col. Otter stands at the centre. Several officers have purchased the new "Wolseley" pattern helmet. Canadian War Museum, Ottawa. Ph: RC

Gunner P. O'Reilly, Royal Canadian Artillery, South Africa, c. 1900. Royal Hamilton Light Infantry Museum, St. Catharines. Ph: CL.

serge fabric in Canada such as used by the British army.

The Canadian Pattern 1899 khaki infantry jacket for first contingent officers had a stand and fall collar, five buttons in front and two breast pockets with box pleats whose tops were in line with the second button. There could be minor variations such as lower front pockets and breast pocket one inch lower such as modified on the British 1897 pattern jacket. The enlisted men's khaki infantry jacket made by Sandford's and issued by the Militia Department had a standing collar, five buttons in front and two breast pockets with box pleats whose tops were in line between the second and third buttons. There were no lower pockets.

From a distance, Canadians going to South Africa generally looked liked their comrades in the other British forces. The Canadian khaki infantry uniform was as durable as that worn by the British army, but the canvas was stiff and harsh and the dye soon faded to an off-white colour after a few washings. This last feature was especially worrisome as the Boers were reputed to be excellent marksmen. Once in South Africa from November 1899, Canadians in the first contingent therefore received uniforms from British stocks as soon as they could be made available.

The 2nd (Special Service) Battalion, Royal Canadian Regiment of Infantry wore the British cork helmet covered with white fabric, already very popular amongst Canadian volunteers. It was the British regulation headdress in tropical climates. While sailing on the transport ship, the men attempted to dye them with coffee. Shortly after their arrival in Cape Town during, however, they were issued with British khaki helmet covers.

Uniform of gunner Francis E.L. Coombes, E Battery, Royal Canadian Artillery, South Africa, 1900. Khaki British army jacket khaki jacket with standing collar, brass buttons, khaki trousers and puttees, brown belt and shoes. Canadian War Museum, Ottawa. Ph. RC.

Canadian Mounted Rifles and Strathcona's Horse

The corps of Canadian mounted rifles and Strathcona's Horse were to be all "clothed, armed and equipped as [the British] Imperial Yeomanry" units at the expense of the Imperial government. The units of the second contingent were supplied in Canada with a dark blue serge uniform (for use only in Canada, please see the cavalry chapter) and "khaki frocks, mounted rifles… khaki frocks, artillery… khaki pantaloons…khaki trousers…" and "cowboy hats…with best lace leather…" that included "punching 2 holes in hats" to wear in South Africa. The "issue of khaki uniforms was made" in Halifax "together with Stetson hats" before departure giving the unit's 640 men a "smart standing" recalled trooper A.E. Hilder of the 1st Canadian Mounted Rifles (retitled Royal Canadian Dragoons in August 1900). In June, he described fellow trooper George Ballamy in Pretoria wearing a "new uniform, brown puttee leather leggings, silver spur, Sam Browne belt and a couple of revolvers on his sides, and wearing a white collar." The officers wore the same pattern jacket as before, but might also be seen in the field with enlisted men's jackets and some were also sporting a non-regulation, but popular cotton drill khaki Norfolk-style jacket. The enlisted men of the second contingent had the Canadian khaki canvass uniforms much like the first contingent, but made with better material. The jackets now had a stand and fall collar, breast pockets in line with the second button and lower front pockets with flaps. They were also issued with British-made cotton uniforms.[119]

The officers and men of the mounted rifles in all subsequent contingents were also issued with the broad-brimmed felt Stetson "cowboy" or "mounted police" hat, which probably became the most readily identifiable Canadian trademark of the war. Troopers also were issued with a greatcoat, a pair of breeches, drawers, canvas shoes, a cloth brush, a "housewife" (a small cloth folder with pockets to carry needles, extra thread, etc.) and a valise to store these items as well as blankets. As previously, reported the Deputy Minister of Militia and Defence in

[119] AGR, 1899-1900. The MR, 1900 has a list of all items embarked on board the *Sardinian* with the second contingent in March 1900; Kyle McIntyre, "Sons of good Western stock. The South African artefacts of Private Alexander W. Stewart, Strathcona's Horse", *Canadian Military History*, VI: 1 (Spring 1997), pp. 84-91; A.E. Hilder, *A Canadian Mounted Rifleman at War*, reminiscences edited by A.G. Morris (Capetown, 2000), pp. 12, 35.

Officers, 1st Canadian Mounted Rifles, South Africa, 1900. Canadian War Museum, Ottawa. Ph: RC.

Trooper of a Canadian Mounted Rifles regiment, South Africa, c. 1900-1902. The two crossed leather bandoleers show the groups of cartridge tubes with their covers. Watercolour by Ron Volsatd. Courtesy Directorate of History and Heritage, Department of National Defence, Ottawa.

Canadian officer's khaki jacket, South Africa, c. 1900-1902. This garment is typical of such items worn in South Africa. It has a stand and fall collar and pointed cuffs. The shoulder and collar badges are missing. The horn buttons could be original since some were used instead of yellow metal buttons that gleamed in the sun. Fort Henry, Kingston. Ph: CL.

Slouch hat, Canadian Mounted Rifles, c. 1902. Khaki felt with a khaki puggaree and bronzed badge. As seen by this object and in many photos, some Canadians had hats with turned up brims like the British Imperial Yeomanry. Said to have been worn by George Fabian Roberts of Victoria, BC, possibly of the 2nd Canadian Mounted Rifles. The hole at top left of the crown is said to be from a Boer's bullet. Canadian War Museum, Ottawa. Ph: RC.

"Canadian scout on the veldt", 1901. Watercolour by W.B. Wollen. Courtesy Anne S.K. Brown Military Collection, Brown University Library, Providence, USA.

Officers and guidons of Strathcona's Horse, 1900. All the officers wear khaki as they are about to leave for South Africa; the guidon bearers have their dark blue uniforms worn in Canada. From Chambers' 1906 North West Mounted Police. Ph: RC.

Officer, Strathcona's Horse, 1900. Watercolour by R.J. Marrion. Courtesy Canadian War Museum, Ottawa.

Sgt. Arthur Herbert Lindsay Richardson, VC, Strathcona's Horse, 1900. He wears the khaki jacket and Stetson hats typical of the Canadian cavalry deployed in South Africa. He earned his Victoria Cross on 5 July 1900 at Wolve's Spruit, South Africa, for saving a comrade under heavy fire from a greatly superior Boer force. Army Museum, Halifax Citadel. Ph: RC.

Lt. Col. Samuel Steele, Strathcona's Horse, 1900. Sam Steele was already celebrated as a veteran soldier and mounted policeman; he commanded Strathcona's Horse in South Africa from April 1900 to January 1901 and went on to lead B Division of the South African Constabulary until 1906. In this image, he wears the khaki uniform for officers of Strathcona's Horse. From Marquis' 1900 Canada's Sons… Ph: RC.

Trooper's jacket of Strathcona's Horse, 1900. All khaki with brass buttons. Lord Strathcona's Horse (Royal Canadians) Museum, Calgary. Ph: CL.

Hat of Strathcona's Horse, 1900. Pale brown felt with brown leather band and chinstrap. This type of wide-brimmed hat, a "Stetson" in Canada, was often called "Montana Peak" by Americans, was the great favourite with Canadian mounted troops in South Africa. Canadian War Museum, Ottawa. Ph: RC.

Troopers of the 2nd Canadian Mounted Rifles patrolling the Transvaal "Veld" in 1902. A scarce view of the men and horse's actual appearance on campaign. Some have rifles slung, others held over the saddle's front. Most troopers wear their hats with the brims horizontal, but some (further back) also have the brim turned up at the left. Courtesy Library and Archives Canada, PA170029.

Trooper, 2nd Canadian Mounted Rifles, photographed at Newcastle, Natal, South Africa, in February-March 1902. Note the two leather bandoleers worn over his shoulders that would give him 100 rounds of ammunition. The saddle appears to be of the colonial pattern. Courtesy Library and Archives Canada, PA113041.

Members of the 3rd Canadian Mounted Rifles, 1902. Detail of a photo taken at Halifax prior to the regiment's departure for South Africa. Canadian War Museum, Ottawa. Ph: RC.

November 1902, the "corps were all clothed, armed and equipped as Imperial Yeomanry, and all expense connected with their organisation was borne by the Imperial authorities."

On the whole, the quality of uniforms made in Canada was much improved for issues to the third and fourth contingents. Lt. Col. Williams of the 3rd Canadian Mounted Rifles commented in June 1902 that the "clothing and equipment issued to the men were excellent… The khaki clothing was exceedingly good and well made" adding that a British staff officer had told him that "the Canadian troops were the best clothed and equipped of any troops in the field." Lt. Col. Irving of the 6th concurred noting that the khaki frock had "plenty of pockets."[120]

Insignia and Footwear

As insignia, the members of the 2nd (Special Service) Battalion, Royal Canadian Regiment of Infantry wore a brass maple-leaf badge surmounted by the imperial crown and bearing the word "Canada" on the left side of their cork helmets. Two smaller similar badges were also worn on the collar. The Royal Canadian Dragoons, Canadian Mounted Rifles, and Royal Canadian Field Artillery wore the maple-leaf badge on their headdress. They also had brass letters on their shoulder straps to indicate the unit. The insignia for Strathcona's Horse consisted of a shield-shaped cap and collar badge and the letters SH in brass on the shoulder straps. The actual wear of insignia does not appear to have always been consistent. Conan Doyle of "Sherlock Holmes" fame noted Canadians he saw in South Africa wearing "…a maple leaf, I think, upon their shoulder straps, and the British Brigade are glad enough to have those maple leaves besides them, for they are the men of Paardeberg." Many photos show officers and men with plain collars or some with and some without badges posing on the same image.[121]

The footwear of Canadian troops in South Africa was usually similar to that of their British comrades: brown leather ankle boots and khaki woollen puttees. The boots were of lighter construction than the British, but were of better quality and durability. The bandage-like puttees were wrapped around the lower leg overlapping the top of the ankle boot to prevent dirt and stones from entering the boot. Ankle boots and puttees were worn by foot and mounted units. Officers and men in the mounted rifles and artillery also wore the formed brown leather "Stohwasser" leggings. Although issued with ankle boots and puttees, all ranks of Strathcona's Horse wore especially made by the Slater Shoes Company at Montreal consisting of high brown leather riding boots, which laced at the ankle and were called Strathcona boots. The men of the 6th Canadian Mounted Rifles were reportedly issued with "very good tan leather brogan boots.[122]

Notes

"Colonial Contingents to South Africa", SP, No. 63, Vol. 13 (1900); "Organisation, equipment, despatch and service of the Canadian contingents during the war in South Africa 1899-1902", SP, No. 35a, Vol. 13 (1903); MR, 1899 and 1900; T.G. Marquis, *Canada's Sons on kopje and veldt* (Toronto, 1900); Louis Creswicke, *South Africa and the Transvaal War* (Edinburgh, 1900), Vol. 4; Grant Tyler, *Drab Serge and Khaki Drill* (Parks Canada, 2003); Jack L. Summers and René Chartrand, *Military Uniforms in Canada 1665-1970* (Ottawa: Canadian War Museum, 1981), pp. 119-121.

[120] Reports in SP, No. 35a, Vol. 13. The 3rd, 4th, 5th and 6th Canadian Mounted Rifle regiments were, however, considered part of the British army.

[121] Conan Doyle quoted by A Canadian Officer, "The Maple Leaf in South Africa", *The Canadian Magazine*, August 1900, p. 355. For photos, see also Brendan McCoy, "Snapshotss from the South African War", *Canadian Military History*, IX: 2 (Spring 2000), pp. 72-81 and *The Canadian Magazine*, July 1902.

[122] The Strathcona boots were made by the Slater Shoe Company of Montreal, supposed to be worn on parades and walking out, but in fact often used by the troopers in the field. Report in SP, No. 35a, Vol. 13, on the 6th Canadian Mounted Rifle regiment.

Chapter 14

Naval Units

Victorian Canada also had a number of volunteer "marine" or naval units from time to time and, decades before the birth of the Royal Canadian Navy in 1910, there also was an armed and uniformed regular Marine service.

New Brunswick Sea Fencibles

The earliest such units appear to have been the "Sea Fencibles" raised in New Brunswick in the early 19th century. This type of naval volunteer unit originated in Great Britain from 1793 and became quite popular; recruited amongst sailors, it was mainly meant to serve artillery in shore batteries or on board coastal vessels. In 1831, two such corps were reported in New Brunswick at St. Andrews and at St. John "drilled in the service of battery guns" (CO 188/41). The St. John Sea Fencibles were the most important and, eventually, the only such unit existing. In 1849, it mustered well over 500 men including 28 officers, 21 NCOs and 476 privates, but with 230 absent, which was considered normal as these were sailors out to sea. According to an inspecting officer, the Sea Fencibles officers generally knew "their duty very well and the men have been drilled constantly to the Artillery exercise with field pieces which have been furnished by the Royal Artillery." Personal armament was poor since the Sea Fencibles had not been supplied "at any time...with arms and accoutrements of any kind" (WO 1/549). This unit gradually became smaller and vanished in the later 1860s.

Sailor, The Sailor's Company, or "Queens's Pets", 1837-1838. "Paid as Regulars. 2 Company of 80 Men." According to "G.S.P.", it comprised "the seamen and seafaring men who happened to be in the port of Quebec, and were enrolled under the command of Captain Rayside, a veteran naval officer." Dark brown fur cap, blue pea jacket with a white anchor badge on the sleeve, dark red and yellow striped scarf, blue trousers, brown moccasins. Note the gun mounted on a sled carriage in the background. Plate by Peregrine Pouchbelt and Rodrick Ramrod. Anne S.K. Brown Military Collection, Brown University Library, Providence, USA. Ph: RC.

According to a retrospective article in the 22 December 1869 *Volunteer Review*, "The uniform worn by the [Sea Fencibles] Officers, was the same as that worn by the officers of the Royal Navy. The N[on].C[ommissioned]. Officers were dressed as Middies [Midshipmen], and the men wore blue jackets, white pants and black glazed hats, with broad ribbon bands." This was confirmed in a 1851 report stating that "The officers are all in uniform blue dress coat, with gold epaulettes, blue trousers in winter, and white in summer, blue cap with gold band, the whole dress similar to the Naval Officers, the Naval sword, and black belt." This, however, seem to be more like the undress uniform of the Royal Navy. As for the men, they had "blue jackets and white trousers" as would be expected for sailors (WO 1/549). New Brunswick artillery officer George F. Thomson later recalled that the men's "uniform was a blue round about jacket with white duck trousers and a glazed cap. The rank of officers in those days was distinguished by epaulettes, the lieutenants wearing one and the captains two."

Queen s Marine Artillery, Kingston, 1837-1838. Blue pilot cloth coat and fur cap. Print after Barry Rich. Private collection. Pi: RC.

From the above, it would appear that the officers, or perhaps only the most senior ones, wore the Royal Navy's dress coat with long tails. This garment was initially dark blue with scarlet collar and cuffs because King William IV (1830-1837) insisted that red was the national colour and ordered the navy to discard its white facings on 10 July 1830. Thus, the officers of the New Brunswick Sea Fencibles would have taken this dress from 1833. On 30 June 1843, some six years after Victoria's accession to the throne, the colour of the collar was changed back to white, the cuff to dark blue and the cuff flap went from dark blue to white. The Sea Fencible officers must have followed suit shortly thereafter. On 21 March 1846, Royal Navy lieutenants were allowed two epaulettes instead of one on the right shoulder, this distinction being now assumed by mates. Before 1861, captains had three cuff laces and commanders two, but, after September of that year, all the above were given an extra lace when the rank of commodore 2nd class was created with a single wide lace. In March 1863, sub-lieutenants were granted a single lace of ordinary width. New Brunswick Sea Fencible officers do not seem to have ever worn the bicorn hat as they are reported with "blue [peaked] cap with gold band" that was the undress headgear of Royal Navy officers, which was worn with a dark blue frock coat. From 1856, the peaked cap had a black lace band with the crown and anchor badge encircled with gold wreath in front; the peaks of commanders and above embroidered in gold. In the case of the Sea Fencibles, the frock and peaked cap appears to have been used for nearly all occasions, possibly as far back as the 1830s.

Naval Volunteers of 1837-1838

The many units of volunteers organised in Quebec City from November 1837 included two naval companies, each of 80 men, that were nicknamed "The Queen's Pets" remembered as "comprising the seamen and seafaring men who happened to be in the port of Quebec, and were enrolled under the command of Captain Rayside, a veteran naval officer, well known as the captain of the Montreal and Quebec steamers, and afterwards as harbour-master of the port. Their uniform consisted of blue pea-jackets and trowsers, equipped with pistols, cutlasses, and a small carronade. Had they been called into action, either for land or water warfare, they would have proved a resolute, brave and useful means of defence. Their services were frequently called into requisition, hunting up concealed arms and ammunition and disaffected parties, accompanied by Robert Symes, an active and zealous magistrate. The Queen's Pets became, for a long time, quite a household word."[123]

On 15 December 1837, the Queen's Marine Artillery was formed at Kingston under Capt. Harper of the Royal Navy with other navy officers or "of the lake steam vessels, and composed of a fine and most favourable body of sailors." According to orders, the "Sergeants and Corporals will wear a white anchor on the right sleeve above the elbow. The men will have rough blue jackets and trousers, fur caps and mittens and a narrow band of white tape to be sewed on the cuffs of the left arm for the gunners and privates." An observer reported that they were indeed "clothed uniformly in blue pilot-cloth frocks, reaching to the knees, blue trousers, and a large fur cap, with belt and bayonet, pistols and light muskets, with a proportion of cutlasses and boarding pikes. The non-commissioned officers, distinguished by white anchors on the arm. They composed four companies, or stations, of fifty men each, commanded by a captain and a first and a second lieutenant; the whole under an officer of the Navy,

[123] "The Volunteers and Incidents of 1837", recollections by G.S.P., *The New Dominion Monthly*, November 1870, p. 16.

Lieutenant Harper, who had the rank of major. The very appearance of these fine fellows, who soon became adepts in the field gun and heavy ordnance exercises, and could take their share of garrison infantry duties, was a source of much uneasiness to sympathisers [of Mackenzie], who guessed they were ugly customers."As for the officers, they most likely wore the undress uniform of the Royal Navy.[124]

Royal Navy officers in undress uniforms, c. 1837. This was likely the most common uniform seen in Canada as worn by New Brunswick Sea Fencibles officers and Royal Navy officers attached to units such as Kingston's Queen's Marine Artillery. Plate after Mansion. Anne S.K. Brown Military Collection, Brown University Library, Providence, USA. Ph: RC.

Volunteer Naval Militia 1862-1867

In central Canada, the first of the volunteer naval militia groups were formed on the Great Lakes, where the Militia Act of 1862 had authorised the establishment of seven "naval companies" at Dunnville, Garden Island, Hamilton, Kingston, Oakville, Port Stanley and Toronto. Each city formed a company of 55 men, except for Toronto's which had 65 men, from among yachtsmen and Great Lakes sailors. They were to be ready to outfit and man ships quickly, and they indeed manned several small crafts transformed into gunboats on lakes Ontario and Erie. By 1865-1866, there were three "provincial gunboats" on Lake Ontario manned by the naval volunteers. From 8 to 26 March 1866, apprehended Fenian raids caused the Toronto, Hamilton, Garden Island and Port Stanley companies to be called out on active service. It was on 1 June 1866 that the Fenians crossed into Canada. The Dunnville Naval Company boarded its steamer, the W.T. Robb, and found itself in the midst of street fighting in the village of Fort Erie.

In later 1866, six "imperial gunboats" were henceforth manned with Royal Navy crews, which acted in conjunction with the volunteers. Thus much reduced the pertinence of the naval volunteers. Furthermore, clothing and equipment were never adequate, making winter drills difficult and unsatisfying. After five years, when the danger of attack from the American border had greatly diminished while several local officers who had initiated the formation of the corps abandoned their involvement, the naval volunteers experiment on the lakes came to an end.

The naval volunteers had naval uniforms inspired from those of the Royal Navy. Several illustrations show the Canadian officers, who agreed to furnish their own uniforms, wearing the navy's service dress dark blue double-breasted frock coat with dark blue trousers, dark blue cap with black visor and gold badge. In 1862, Captain W.F. McMaster of the Toronto Naval Company went a step further and is shown in the Royal Navy's full dress complete with tailcoat and bicorn hat. The NCOs were dressed as the Royal Navy's and the men had the sailor's "Clothing (similar in material and make to that worn in the Royal Navy) to the [Toronto] Company" in 1862, and reported in October as "Canadian Blue Tweed Pantaloons, Blue Melton Shirts, Blue Cloth Caps [with] Ribbon Bands" procured in March and April. They were armed with 58 long Enfield rifled muskets with accoutrements and 30 practice swords. In 1866, a boy in Toronto described its "Naval Brigade, under command of Captain W. F. McMaster; it is a fine and efficient body of jack-tars, numbering eighty men. Their uniform consists of a short blue jacket with a wide collar, blue trousers in winter, white in summer. They have a white straw hat in summer, with a black band, and the letters T. N. B. [Toronto Naval Brigade] on it. Altogether they have a very naval ap-

[124]LAC, RG9, IB1, vol. 23, Robert Jeffery Esq. to Adjutant General, Kingston, 17 March 1838; Richard Henry Bonnycastle, *Canada...,* Vol. 1, pp. 310-311.

(Left) Royal Navy captain and lieutenant, 1849. Dark blue uniform with white collar and cuff flaps, gold buttons and lace. Officers of the New Brunswick Sea Fencibles wore peaked caps with a wide gold band instead of the bicorn hat shown in this plate published by Ackerman, but the rest of their dress was the same. At right, a sailor in dark blue with a white shirt, sky blue wide collar edged with white stripes, a black cravat and a straw hat trimmed with black ribbons. Plate after Mansion. Anne S.K. Brown Military Collection, Brown University Library, Providence, USA. Ph: RC.

Officers of the Toronto Naval Company, 1862. Captain W.F. McMaster at centre is shown in the Royal Navy's full dress complete with tailcoat and bicorn hat. The officers on either side have the standard service dress for naval officers. Courtesy Library and Archives Canada, C126600.

Officer of the Hamilton Naval Company, c. 1866. The officer may be Lt. George P. Malcomson. He wears the dark blue naval pattern uniform with gold buttons, lace and cap badge. Courtesy Don Blyth Collection, Guelph, Ontario.

Officers and ratings of the Canadian Government Ship Druid, 1873. Royal Navy type uniforms are worn. The sailors wear the white summer dress with light blue collar with dark blue caps. The lace on top of the officer's cuff appears to be a curl, but, eventually, a small triangle was worn. Canadian Illustrated News, 14 June 1873. Courtesy Library and Archives Canada.

Marine Mail officer, c. 1862-1870. Courtesy National Postal Museum, Gatineau.

2nd Class Petty Officer, Royal Navy, c. 1890. Canadian Marine personnel had the same dress. Dark blue jacket with sky or medium blue collar edged with three white stripes, black scarf with white lanyard cord, red rank chevrons and badges on sleeve, dark blue trousers, dark blue cap with black tally (or ribbon) marked with the ship's name in gold. A peculiar badge seen on navy jumpers was the "watch stripe", a strip of red (on the blue jumper) or blue (on the white jumper) material worn round the top of a sleeve that showed at a glance whether a sailor belonged to the port or seaboard watch. Those on the starboard watch wore it on the right sleeve and those on the port watch wore it on the lefts sleeve, as show in this plate after W. Christian Simmons. Private collection. Ph: RC.

Southwester cap, second half of the 19th century. This foul weather headdress belonged to Canadian Marine pilot Frédéric Bouffard. Musée maritime du Québec, L'Islet (Quebec). Ph: RC.

Captain John T. Walbran, Department of Marine and Fisheries, 1890s. His uniform is similar to the Royal Navy's except for three cuff laces for the rank of captain and the gold triangle above. The cap badge is unclear, but appear to be of the Canadian Marine. Private collection. Ph: RC.

The enlisted men of the 1866-1870 Nova Scotia Naval Brigade had similar uniforms to the Royal Navy's sailors shown in this 1869 print. Anne S.K. Brown Military Collection, Brown University Library, Providence, USA. Ph: RC.

Officers, petty officers and ratings on board Canadian Government Ship Aberdeen, 1894. This rare view shows officers wearing cuff laces and one notes only Captain M.P. McElhinney (4th from left) and the third officer at right have the triangle. Petty officers have plain cuffs and the sailors in the background wearing the same uniform as Royal Navy sailors. The name on their cap tallies, however, would be CGS ABERDEEN. Private collection. Ph: RC.

pearance."[125]

The men of the Hamilton Naval Company wore, according to drummer Zealand's reminiscences, "shirts of blue flannel, with trousers wide at the bottom and tight at the knees… Caps were in keeping with the dress." They also had rifled muskets and cutlasses and were trained to handle them by Sergeant Moulton of the British 16th Foot.

The men of the Dunnville Naval Company had, since February 1862, "Blue Flannel Jacket with white trimmings and Black Pants [with] Blue Cloth Cap[s]." The other naval companies would have also secured such uniforms. When fighting the Fenians at Fort Erie on 2 June 1866, the men of the Dunnville Naval Company were reported as having "no uniform except officers" although this might have meant, for the sailors, a variety of naval clothing.[126]

Nova Scotia Naval Brigade 1866-1870

Meanwhile, in April 1866, the government of Nova Scotia launched a plan for its own naval militia, and, by March 1867, had signed up some 554 men into 17 companies, six of which were in Halifax. The Royal Navy supported the initiative with equipment and training. Following Confederation in 1867, responsibility for these units was transferred to the Militia Department of Canada, which refused to consider a separate role for naval companies and allowed them to wither away until the last four remaining were converted to artillery in December 1870.

---

[125] LAC, RG9, IC8, vol. 4, Commanding Officers Reports…1862; *The Boy's yearly book for Eighteen Hundred and Sixty-Seven* (London, 1867), p. 285. The young correspondent added that the Toronto Naval Brigade had: "They have a drill shed of their own, and two large guns, furnished by the Government to practise with. They have also boats of their own, which are very fine ones indeed. They have a fife and drum corps. They are armed with the long Enfield rifle, bayonet, and a cutlass."

[126] *Hamilton Herald*, 6 June 1903 and 20 December 1912; LAC, RG9, IC8, vol. 4, Commanding Officers Reports…1862; *Toronto Globe*, 25-26 October 1866.

In April 1866, the "regulations as to uniform, &c" for officers and masters were printed and issued at the Naval Brigade's HQ in Halifax. These called for a dress closely similar to that of the Royal Navy frock coat, trousers and peaked cap including the buttons and cap badges that were "as per Royal Navy Regulation, for all commissioned executive officers." The rank structure, however, was peculiar to the land-based militia so that the Brigade's HQ had a commandant (also called lieutenant-colonel), majors and staff surgeon, chief engineer, master of gunnery, quartermaster and adjutant. The various companies had captains and lieutenants.

Commandant: silver and scarlet crown at each side of turned down collar, four gold laces with curl on each sleeve, 1/2 inch gold braid on trousers, 1 1/2 inch wide gold lace on cap and gold oak leaves on peak.

Majors: silver star at each side of turned down collar, three gold laces with curl on each sleeve, 1/4 inch gold braid on trousers, 1 1/2 inch wide black mohair lace on cap and gold oak leaves on peak.

Surgeon: no badge at collar, two gold laces without the curl on each sleeve with scarlet velvet between the laces, 1/8 inch gold braid on trousers, 1 1/2 inch wide black mohair lace on cap and plain gold lace on peak.

Master of Gunnery: same as surgeon, but purple velvet between the cuff laces.

Quartermaster and Adjutant: same as surgeon, but white velvet between the cuff laces.

Assistant surgeon, engineer and of gunnery: same as surgeon, but one gold lace only at sleeve.

Captain: small silver crown and anchor at each side of turned down collar, two gold laces with curl on each sleeve, 1/8 inch gold braid on trousers (none in undress), 1 1/2 inch wide black mohair lace on cap and plain peak.

Lieutenant: small silver crown at each side of turned down collar, one gold laces with curl on each sleeve.

Captain J.-E. Bernier, Department of Marine and Fisheries, c. 1900-1910. Captain Bernier was photographed wearing a coat with four cuff laces (instead of three) without the triangle and three laces on the cap band with the Canadian Marine badge. Musée maritime du Québec, L'Islet (Quebec). Ph: RC.

The men had clothing similar to Royal Navy ratings. 500 sailor's uniforms "were received, 350 issued free of expense, and 150 suits remaining in store."[127]

Marine Companies in Gaspé 1869-1878

A much lesser known endeavour was the formation of "Marine Companies" in Quebec's Gaspé Peninsula. On 19 February 1869, a company was officially gazetted at Bonaventure and another one at New Carlisle. A third marine company at Carleton was approved on 11 March 1870. On the whole, the companies do not appear to have raised local enthusiasm, which was no doubt spurred by lack of training and equipment; on 27 March 1874, the Carleton company was removed from the list as was the New Carlisle unit on 2 June of the same year. The Bonaventure company went on, no doubt getting weaker, until taken off the list when its commander, Capt. Arsenault, retired on 18 July 1878. These units were issued the standard scarlet infantry uniform, not sailor's clothing.[128]

[127] Public Archives of Nova Scotia, (Halifax), RG22, vol. 25, file on Naval Brigades 1866-1867.

[128] The "Formation of a Naval Company" at Ile Verte (Québec) was mentioned in the Deputy Minister of Militia's correspondence in February 1882, but appears not to have been actually organised. LAC reel T4176/240.

Ratings of the Department of Marine and Fisheries, c. 1900-1905. All are dressed in uniforms that are similar to the Royal Navy's, in this case the white summer dress, issued from 1880. For dress occasions the jumpers had the medium blue collar with three stripes, the black scarf, the white lanyard cord as well as the dark blue watch stripes on the upper sleeve. The only difference from the Royal Navy is that they hold the Canadian-made Ross rifle and their cap tallies have "CGS CANADA". At each side is a 1 1/4 lbs gun. Courtesy Library and Archives Canada, PA123950.

Canadian Department of Marine 1869-1910

While not a strictly military corps, the Canadian government's Department of Marine and Fisheries had a few small armed steamers "employed in the service of protecting the fisheries" at sea and on the Great Lakes that were initially commanded by former Royal Navy officers. These steamers were also active against smugglers and were also the fastest way to provide mail and supplies to distant small towns on Canada's extensive coastline. From 1869, their sailors had enough military training to serve the two cannons on the Department of Marine steamer *Druid* as well as being armed with rifled muskets with bayonets, pistols, cutlasses and accoutrements. In 1893, the sailors were reported "well-drilled in the various exercises with the rifle and the cutlass, and the movements on the march." This was a precaution since the seizure of vessels could lead to resistance on the part of the culprits intercepted. There were some "stirring escapades on this service" and, in 1892, a detachment of "B" Battery, Royal Canadian Artillery, was aboard the Canadian Marine cruiser *Constance* acting as marine gunners when chasing liquor smugglers from the French islands of Saint-Pierre-et-Miquelon who were especially adverse to being arrested. The Department of Marine's military activities were passed on to the Royal Canadian Navy from 1910 and went on to become the present Canadian Coast Guard.

Initially, the officers had uniforms generally similar to those of the Royal Navy except that only the double-breasted surtout frock coat, introduced since 1847 in the Royal Navy, was worn. A 1873 print even shows officers with the round "executive curl" of the Royal Navy (since 1856) above the gold lace rank bars on the sleeves of their coats, but this was suppressed later. White trousers and cap covers could be worn in warm weather. In time, the double-breasted ship's jacket with shorter skirts, introduced in 1879 in the Royal Navy, became the standard officer's dress.

Rank designations and distinctions for officers became slightly different from those of the Royal Navy. Instead of the "executive curl" of gold lace above the coat's cuff, Canadian Department of Marine deck officers had a gold lace triangle. There were three gold laces around the coat's sleeve for captains, two for chief officers and one for second officers. From photographic evidence, the triangle appears to have been frequently omitted, especially by chief and second officers. Third officers had only the gold triangle. Engineer officers had gold lace with purple cloth piping between the bands, those below the rank of third engineer wearing the purple cloth piping only. Pursers, who served on board larger ships, wore two silver cuff laces. The officer's caps were dark blue with black peaks, white cap covers being often worn in summer. The early cap badge is unknown in detail; photos and prints also show gold stripes worn on the band. From about the 1890s, a cap badge having a gold and dark red velvet crown over a silver anchor with a wreath of gold maple leaves and the word "Canada" between the crown and the anchor was worn. When employed on Fisheries Protection duties the commanding officer was issued with a sword in token of his enforcement role. There could be varied versions of uniforms of the Canadian Marine officers as shown by photos and a cap owned by famed Arctic Explorer Captain J.-E. Bernier.

Newfoundland members of the Royal Naval Reserve, c. 1901-1902. They wear the Royal Navy's uniform in a summer version with the straw sennet hat edged with black tape, blue jacket and white trousers. Rifles and the trousers tucked into dark blue gaiters indicate they are equipped as a landing party. Detail of a photo taken in Halifax. Courtesy Dept. of National Defence, CN5002.

Between 1860 and 1887, Canadian government vessels might also have Marine Mail officers on board since these ships were often the only way to have a mail service for isolated and distant communities. These postal officers looked very much like naval officers. From 23 June 1862, they were assigned a "Navy blue" naval uniform consisting of a "frock coat, double breasted, two rows of five gilt buttons each, the "cuff to bear one stripe of gold lace 3/4 inch in width", single-breasted vest with six buttons, trousers "without stripe", Navy blue cap with "black oak leaf lace 2 inches wide. The badge to be a crown surrounded by a wreath of Maple leaves; and the letters C.R.M.S. [Canadian Royal Mail Service] under the crown, the above to be bullion." The buttons bore "a crown encircled with the letters C.R.M.S."[129]

The petty officers and ordinary sailors had uniforms that were identical with those of the Royal Navy's, both for the dark blue colder weather and white summer caps, jumpers and trousers. The collar on the dark blue and the white jumper was sky or medium blue with three white border laces, the tie black and the lanyard white. From about the end of the 19th century, the cap's black tally (or ribbon) bore the letters "CGS" for Canadian Government Ship before the vessel's name in gold letters. This was the only obvious distinction that its wearer served on a Canadian government ship.

Canadian Marine steamers initially flew the British Blue Ensign at their stern and from 1884, the British Admiralty decreed that public service vessels belonging to the governments of Dominions and other overseas territories would hoist the Blue Ensign with the area's distinctive coat of arms. The jack at the bow was the British Union Jack. Canadian Marine ships also flew the Royal Navy's long commissioning pennant, which was blue with white only at the head. This Long Blue Pennant was important as it distinguished Canadian cruisers as regular Crown government vessels.

Newfoundland Division, British Royal Naval Reserve from 1900

Although a vast proportion of its men were sailors, Newfoundland did not have an organised military naval force until 1900. The possibility of organising such a corps had been raised by the Royal Navy's First Sea

[129] LAC, RG3, Series 9, vol. 22, Circular to Marine Mail Officers, 23 June 1862; Kenneth S. Mackenzie, *Le commis de la Poste maritime canadienne 1860-1887* (Ottawa: Musée national des Postes, n.d.). These officers also served in commercial ships that had mail delivery contracts.

Lord, Admiral J. Fisher, two years earlier. In 1900 that the governor of Newfoundland, Sir Henry MacCallum, went ahead and recruited some 300 men for what became, during 1901-1902, the British Royal Navy's "Newfoundland Division" of the "Royal Naval Reserve" that had been created in Britain since 1859. From November 1900 to May 1901, 50 of the Newfoundland naval reservists went on board Royal Navy warships to participate in a training cruise. The following year, some were on board HMS *Charybdis* in Venezuela during an international crisis and the Newfoundland reservist went on to serve in many British warships later on. The officers were all British at the outset and the men of the Newfoundland Royal Naval Reserve were issued with identical uniforms as those worn by ratings in the Royal Navy.

Notes

Ernest J. Chambers, *The Canadian Marine: A History of the Department of Marine and Fisheries* (Toronto, 1904); Thomas E. Appleton, *Usque ad Mare: A History of the Canadian Coast Guard and Marine Services* (Ottawa 1968); W. Johnson, W.G.P. Rawling, R.H. Gimblett and J. MacFarlane, *The Seabound Coast: Official History of the Royal Canadian Navy 1867-1939* (Toronto, 2010); Richard H. Gimblett and Michael L. Hadley, *Citizen Sailors* (Toronto: Dundurn, 2010); W.E. May, *Dress of Naval Officers* (London, National Maritime Museum, 1966); Dudley Jarrett, *British Naval Dress* (London, 1960); Robert Wilkinson-Latham, *The Royal Navy 1790-1970* (London, 1977); W. David Parsons, *The best small-boat seamen in the navy: Newfoundland Royal Naval Reserve 1900-1922* (St. John's, NL, DRC Publishing, 2008).

Petty officers and seamen of the Royal Navy, 1900. These same patterns of uniforms were worn by petty officers and sailors of the Canadian Marine vessels and the Newfoundland reservist serving in the Royal Navy. The second-class petty officer of a landing party under arms holds a Martini-Henry short rifle and wears Slade-Wallace Valise equipment. First-class petty officers and officers were armed with revolvers. Print after Frank Dadd. Courtesy Anne S.K. Brown Military Collection, Brown University Library, Providence, USA.

Chapter 15

Auxiliary Volunteers and Opponent Forces

Although not part of the Canadian government's regular or militia forces, the were also Canadians that formed their own volunteer organisations. We have evoked in Vol, 1 the French Canadian Pontifical Zouaves and Independent guard units and round these out in this volume with the remarkable "Red Sash" Home Guards and that fought Fenians and the Voyageurs who manned boats on the Nile River in Egypt. There were Canadians who also opposed, weapons in hand, the governments of their days. The mobilisations they caused were considerable although the fighting was of minimal scale compared to civil and revolutionary strife in other countries. In retrospect, the judgement of History has largely vindicated the Patriots, Métis and indigenous warriors as men who fought to protect their rights from abuse. The hanging of Métis leader Louis Riel created profound uneasiness in French Canada so that the enthusiasm of English Canadians to fight in South Africa to protect the rights of the British Empire was largely ignored in francophone Canada. However, unlike in many other nations, the majority Canadian society's "two solitudes" has always spurned systematic and unchecked violent action as a means to an end because it would result in the atrocities of civil war.

The "Red Sashes" Home Guard in position at the Québec-Vermont border shortly after repulsing the Fenian raid at Eccles Hill on 25 May 1870. In spite of being very blurry, this photo nevertheless shows the red sash across the shoulders of many of these local volunteer in civilian clothes. Brome County Historical Society Museum, Knowlton, Quebec. Ph: RC.

Auxiliary Volunteers:

The "Red Sash" Home Guards 1866-1870

Following the Fenian Raids of 1866, some farmers of the Missisquoi area whose land were near the Vermont border were so enraged at the Canadian Militia and the British army's failure to protect their farms, that

(Above left) Members of the "Red Sashes" Home Guard around a Fenian guns captured at Eccles Hill in May 1870. Most have their red sashes across the shoulder. Brome County Historical Society Museum, Knowlton, Quebec. Ph: RC. (Above right) Lt. Col. Frederick Denison, Nile Voyageurs, 1884. This studio photo shows him wearing a Norfolk jacket, most likely grey, that was probably made in Canada with Bedford cord breeches and a white helmet with its white pugaree. He is armed with a revolver and a sword that are attached to the Sam Browne belt. This general type of dress was typical of the British officers during the Nile Expedition. Canadian War Museum, Ottawa. Ph: RC. (Bottom) "The Iroquois pitches his tent in the land of the Pharaohs" in October 1884 at Wadi Halfa, Egypt. Print after F. Villiers in The Graphic (London), 22 November 1884. Ph: RC. (Left) Officer and boatmen, Canadian Voyageurs, Egypt, 1884-1885. Plate by Eric Manders. By courtesy of the Company of Military Historians and the artist.

is to say, stop the Fenians at the border, that about 40 of them formed their own private militia called the "Home Guard" led by Asa Westover and Andrew Ten Eyck of Dunham. They bought 40 Ballard hunting rifles with 30-inch barrels and .44 calibre in Massachusetts for armament. They had no uniforms, but settled on a wide red sash worn across the shoulder to distinguish themselves from the Fenians. In 1868, rumours of a Fenian incursion rose their number to 65 and they organised border patrols and an efficient information system.

Meanwhile, another upset local citizen was Brown Chamberlin, editor of the *Montreal Gazette* and member of Parliament for Missisquoi County. He was also a volunteer militia officer. Influential and determined to defend his native area, he became commander of the 60th Missisquoi Infantry Battalion organised in 1869. On 25 May 1870, a Fenian force crossed the border. The Home Guard and the 60th were waiting and easily drove it back across the border at Eccles Hill. Two Fenians were killed and more were wounded. The Canadians were unscathed. The members of the Home Guard wore their civilian clothes with the red sashes during the engagement. Not only was the "invasion" repulsed, but the volunteer militia was vindicated as a viable defence force. This was the last incursion attempt by the Fenians along the border of southern Quebec. Thus, the "Red Sash" Home Guard eventually faded away.

Boats manned by Canadian Voyageurs set on the Nile from Wadi Halfa towards the first rapids of the southern Nile. The Voyageurs generally steered the boats and can be seen mostly wearing primed hats. There were the perils for the Canadian Voyageurs and, indeed, six perished in the cataracts. Print after F. Villiers in The Graphic (London), 22 November 1884. Ph: RC.

Canadian Voyageurs in Egypt, 1884-1885

Upper Canadian Patriots at the battle of Montgomery's Tavern, north of Toronto, 4 December 1837. They had no uniforms, but some used armbands as shown in this early 20th century print after C.W. Jefferys in Stewart Wallace's 1915 The Family Compact. Ph: RC.

In the early 1880s, Egypt was in turmoil and Great Britain intervened militarily during 1882 to secure the Suez Canal, the strategically vital link between Western Europe, India and the Far East. Once in Egypt, the British had to cope with the rebelling southern province of the Sudan under a mystic leader known as "The Mahdi" whose many followers besieged its capital, Khartoum, where a famous British Victorian personality, Maj. Gen. C.G. (nicknamed "Chinese") Gordon, commanded the Egyptian garrison. In 1884, an Anglo-Egyptian relief expedition was organised under Lord Garnett Wolseley. He had led the 1870 Red River expedition to Manitoba and, now facing going up the Nile River with his army, recalled the good logistic services of the rugged Canadian wilderness boatmen called *Voyageurs*.

From August 1884, some 400 boatmen were

Patriote leader Dr. Jean-Olivier Chénier tells the men who had no arms to take the muskets "of the men who will be killed" during the battle of St. Eustache on 14 December 1837. Note the European-style dress worn by Dr. Chénier with the rural men wearing the traditional Canadian costume, usually a grey homespun capot with a blue "tuque" (cap). Chénier, shown armed with a hunting musket, was killed in the battle and, regrettably, his body suffered indignities, a witness recalling that it was "so badly mutilated that he was almost cut into four pieces, his heart [cut out and] on the outside. It was a sickening spectacle..." Print after Tire-Boignet in Jules Verne's Jean sans nom. Ph: RC.

hired and, led by Brevet Lt. Col. Frederick C. Denison, six officers, a sergeant and 16 foremen, went to Egypt and joined Wolseley's army to man some of its 800 32-foot long whaleboats. In early January 1885, the army and its whaleboats were progressing steadily southward on the Nile in spite of its rapids and cataracts. However, on 26 January, Khartoum's garrison was overwhelmed by the Madhi's numerous warriors and Gordon perished with many of his men. Only three days later, the advance party of Wolseley's army was in sight of Khartoum. It was too late and, eventually, the relief force went back to Egypt. The Canadian Voyageurs came back to Canada less six men drowned in cataracts and ten dying from other causes. None were killed in action, but some were present at the battle of Kirbekan that added a clasp to the campaign medal they all received.

Not being a fighting unit, the boatmen were not armed, but they had uniform clothing procured in Canada before leaving for Egypt. This consisted heavy grey tweed Norfolk jackets and trousers, blue shirt, grey wool undershirt and drawers, socks, soft grey hat and a pair of beef hide moccasins. The foremen had lighter suits of better quality. Once in Egypt, they were issued white helmets, but, according to *The Graphic* of 22 November 1884, they looked "very slovenly in their suit of woollen tweed, half moccasins, and regulation white helmet, but strong, sturdy looking fellows they are, and when in their blue flannel shirts with sleeves tucked up, they look good enough for any boating requirement." The officers were armed with revolvers and wore clothing typical of the Egyptian campaign: a khaki or grey military tunic or jacket, brownish Bedford breeches, dark blue puttees and white helmet. However, as Denison noted, there could be a great deal of variety in the British army's officer's dress in Egypt, some wearing jerseys, flannel shirts, felt hats, boots without socks and even one sporting Indian sandals.[130]

Opponents:

"Patriotes" and Patriots 1837-1838

In Lower Canada, the "Patriotes" were usually dressed in the traditional costume worn by French Canadian farmers, artisans and tradesmen that consisted of a wool "tuque" cap, usually blue or ded, a hooded "capot" coat tied at the waist with coloured sash, a waistcoat, trousers and leather moccasin-like boots or shoes. Much of this was of homespun cloth, usually grey, but sometimes other colours such as brown, because of the "Patriote" policy of boycotting English manufactured textiles. The more affluent and educated gentry, merchants and professionals dressed in the usual styles worn by gentlemen in Europe although some might wear homespun garments like their men. Wolfred Nelson was seen wearing a capot and a tuque. Dr. O'Callaghan's capot featured two maple leaves on the shoulders acting as epaulettes. There were no "Patriote" uniforms as such, the distinctive Cana-

[130]Wayne Colwell and Eric I. Manders, "Canadian Voyageurs, Nile Expedition 1884-1885" MC&H, 26: 1 (Spring 1975), pp. 32-34 (MUIA 399); *Records of the Nile Voyageurs 1884-1885*, C.P. Stacey, ed. (Toronto, 1959); Micheal Barthorp, *The British Army on Campaign 1816-1902 (4): 1882-1902* (London: Osprey MAA 201, 1988).

"Wooden Cannon Used by the Canadian Rebels loaded with Buck Shot [and] Small pieces of lead & Iron in 1837 & 1838" is written on this remarkable object. Canadian War Museum, Ottawa. Ph: RC.

A group of Lower Canada "Patriotes" sketched by Mrs. E. Ellice at Beauharnois in 1838. They all wear civilian clothing typical of rural Lower Canada and are badly armed. Courtesy Library and Archives Canada, C13392.

A Patriot Hunter leader bearing the flag of the Onondaga Hunters, late 1838. He wears a typical men's civilian costume, most likely a dark blue coat, and holds the flag of his unit, the Onondaga Hunters of upstate New York, which was blue with yellow letters, star, eagle, the words "Canada Liberated" and the name of the unit. A crimson sash and a sword denote his officer's status. Reconstruction at Fort Wellington National Historic Site, Parks Canada, Prescott, ON. Ph: RC.

A wing that once adorned the coat of patriot leader Nils von Schoultz who led the abortive raid on Prescott in November 1838. It is dark blue with silver embroidery and very likely of American origin. It is in the style found on American militia uniforms. Courtesy Fort Wellington National Historic Site, Parks Canada, Prescott, ON.

Detail of the battle of Ridgeway, 2 June 1866. At left, Fenian troops wearing green double-breasted coats with sky blue trousers and light blue (but actually dark blue) US Army type peaked caps. The green flag has a gold harp with the letters I R A. Print published at Buffalo, NY, in 1869. Courtesy Library of Congress, Washington.

dian costume they wore providing sufficient distinction. Their arms were poor. F.X. Prieur recalled some men in his St. Timothée parish had muskets, "most of which dated from the time of the French, others were armed with iron forks and of sickles transformed into sabres." At Vaudreuil, west of Montreal, some indeed tried to make sabres and bayonets from scythe blades and others even attempted to construct wooden cannons.

The 1838 "Patriotes" covertly organised themselves as "Frères Chasseurs" groups with secret signs of recognition and a rank structure for officers. In Lower Canada, they were just as badly armed, or worse, as during the previous year and had no uniforms. A unique illustration of Patriotes in Beauharnois during 1838 shows them wearing traditional garments and bearing various makeshift pikes and old muskets. Charles Hindenlang, a former officer in the French army, was one of the leaders captured in 1838. Before his execution, he willed his cap, his green overcoat, a waistcoat and the boots he wore at the engagements of Napierville and Odelltown to a mute Montreal prison guard. His companion Pierre Decoigne was said to wear a French army uniform at Odelltown.[131]

The 1837 Upper Canadian Patriots were badly armed with a few muskets and pitchforks so they marched towards Toronto to seize its arms magazine. They wore their usual civilian costumes that were of European inspiration. Some might have the usual gentlemen's clothing while others, perhaps most, had what Samuel Thompon recalled as "their rustic clothing" at the battle of Montgomery's Farm where they were repulsed by loyal militiamen. It would appear that some wore armbands in late 1837, something also done by those loyal militiamen who wore civilian garments.

In the United States, the "Brother Hunters" were organised during 1838. They were made up of expatriate Upper Canadian patriots joined by many adventurers of American or other origins. They strived to take over Canada and declare it a republic. Organisers in the Detroit area had uniforms. One self-appointed recruiting officer was seen with "a numerous staff were seen parading the streets with an entirely new, costly, and dashy undress uniform" although he turned out to be something of a fraud. However, at least one hunter captured on the schooner *Ann* near Fort Malden "was armed, and wore the same uniform" qualified as "the rebel uniform, with the stars, the insignia, as he termed it, of my command, blazoning on my breast." It was not further described, but its style was probably American-inspired. In 12 November 1838, a group of mostly American Brother Hunters landed at Windmill Point near Prescott under the leadership of Nils von Schoultz, a former Polish officer, but they were surrounded by the Canadian and British forces and, after a five day siege, surrendered. Von Schoultz wore American style wings on his coat.[132]

[131] Francis Back, "L'étoffe de la liberté", *Bulletin d'histoire politique*, X: No. 2 (hiver 202): 58-71; F.X. Prieur, *Notes d'un condamné politique* (Montreal, 1884), p. 11; Russel Bouchard, "Les armes de la rebellion de 1837-1838 au Bas-Canada", *Journal des Armes*, I, No. 1 (1979): 1-15; Robert Sellar, *The History of the County of Huntingdon* (Huntingdon, 1888), p. 562; Gilles Laporte, *Patriotes et Loyaux* (Quebec, 2004), pp. 220, 255, 272, 297.

[132] Samuel Thompson, *Reminiscences of a Canadian Pioneer for the last fifty years* (Toronto, 1884), p. 126; Edward Alexander Theller, *Canada in 1837-38: showing, by historical facts, the causes of the late attempted revolution and of its failure...* (Philadelphia 1841) Vol. 1, pp. 206, 224.

Fenian "Flag of Green patriotic song" music sheet, 1866. The flag is shown as being all green with a golden harp in its canton and the song is dedicated to "the Fenian Brotherhood Throughout the World." Canadians will be surprised to see that the music was "arranged by Calixa Lavallée" who is remembered as the composer of the music for "O Canada" written in 1880, which became the national anthem. Born in 1842 near Montreal, Lavallée was something of a musical prodigy and, from 1857 lived in the United States, was wounded at the 1862 battle of Antietam serving as a musician in the 4th Rhode Island Volunteer Regiment and thereafter pursued an international career in music. The Fenian Brotherhood obviously commissioned a renowned musician as did Lieutenant Governor of Quebec Théodore Robitaille when he asked him to write the music for Adolphe-Basile Routhier's poem "O Canada". As for Lavallée, who does not seem to have been greatly interested in politics, he passed away in Boston during 1891. Courtesy Anne S.K. Brown Military Collection, Brown University Library, Providenc, USA.

Fenian officer, c. 1866. He is seemingly wearing the 1866 dark green coat and black trousers with small shamrocks embroidered at the collar and cuffs as well as narrow edging lace going and a wide lace trouser stripe. He appears to be holding a dark coloured hat or cap, but this item is quite indistinct. Courtesy Don Blyth Collection, Guelph, Ontario.

(Right) Fenian infantryman, 1870. Dark green jacket with yellow lace edging and brass buttons, sky blue trousers, dark blue cap with white piping and dark green band edged with red. The jacket and cap are after original items captured at Eccles Hill on 25 May 1870 and presented to Prince Arthur, one of Queen Victoria's sons, who was then serving in Canada. Plate by Ron Volstad. Courtesy History and Heritage Directorate, Deparment of National Defence, Ottawa.

(Left) United States Army forage cap worn by a Fenian raider, 1870. Dark blue with brass badge, black peak and chin strap. Canadian War Museum, Ottawa. Ph: RC.

Fenians 1866-1870

Brass "I.R.A." (Irish Republican Army) button from the 1870 jacket. Courtesy National Historic Sites, Parks Canada.

The Fenians raids of 1866 and 1870 that originated in the United States were the somewhat violent expression of expatriate Irish-Americans towards anything British allied with the fanciful objective of conquering Canada and exchanging it for Ireland's independence. The Fenian's Irish Republican Army was certainly the best-armed and best-uniformed para-military force raised in the United States to invade Canada. Green being the national colour of Ireland, it was the obvious hue for the uniforms of the Irish Republican Army that sought to invade Canada.

Seemingly in 1866, the Fenian army adopted a uniform that apparently of a completely dark green double-breasted frock coat, the officers being prescribed this dark green frock coat, black trousers, black hat and a green sash. The shoulder straps had rank badges: two silver sunbursts for a major general, one for a brigadier general, a silver phoenix for a colonel, a silver shamrock for a lieutenant colonel and a gold shamrock for a major. There probably were further details in this regulation, but the original copy has been lost.[133]

Most officers probably never obtained such uniforms and wore much simpler garments. Indeed, from observer's accounts, the Fenians involved in the 1 and 2 June 1866 raids were dressed in various ways. In Fort Erie, six year old James Stanton was "was eating breakfast, when he heard a knock at the door. His mother went to open it and there on the threshold stood a stern-looking soldier in a green uniform…" For their part, Buffalo, NY, reporters had seen that "the Fenians were young men, most of them 20 to 27. Many wore black hats. Others favoured the stovepipe hat, then so popular among well-dressed gentlemen." One reporter had visited the Fenian Camp in Fort Erie "and found two green flags flying and rumours flying as well. The invaders presented a motley appearance, some wore artillery jackets and others wore sober black broadcloth dress coats. There was a 'decided preference for green'." A Canadian captured by the Fenians later recalled that their "uniform consisted of a green shirt, with brass buttons, dark trousers, a black, soft felt hat, with wide brim. Over their shirts they wore dark civilian coats which served the purpose of overcoats, and which had been used to conceal their uniform before crossing the river, It was owing to this fact that the general impression prevailed that the Fenians were not in uniform. Most of the superior officers wore the dress proper to the rank they had held in the American army during the Civil War." The captain of his escort, who had been in the Confederate army "wore his old regulation kepi, a long black frock coat, with a belt outside, in which was stuck a sword without a scabbard." That the Fenian officers largely were their old Civil War uniform is corroborated by N. Brewster, a Canadian volunteer guarding Fenian prisoners. He was "surprised to see that the only ones among them wearing uniforms was that of the U.S. Army. A Captain among them wore the full fatigue rank of his rank…" Before the battle of Ridgeway, some 40 Fenians

United States Army surplus cartridge box with belt and "US" brass buckle captured from a Fenian invader at Eccles Hill, Quebec, on 25 May 1870. Brome County Historical Society Museum, Knowlton, Quebec. Ph: RC.

[133] A copy of this regulation was sent by spies to Canada and was on hand in Ottawa by 15 January 1867. It was later consulted by historian William D'Arcy and briefly described in his *The Fenian Movement in the United States 1858-1886* (New York, 1947): 229-230, the reference being given as Edwards to Godley, January 15, 1867, G 21, Vol. I, P.A.C (Public Archives of Canada now renamed Library and Archives Canada). Unfortunately, the archives appear to have lost this volume as all attempts by archivists to retrieve it failed.

Louis Riel with Métis and First Nations warriors, 1885. This print was composed from composite images of Riel (mounted at centre), Dumont (right) facing Chief White Cap, at left Chief Big Bear sitting next to Métis Beardy holding a rifle. Illustrated War News, 2 May 1885. Courtesy Bruce Peel Special Collections, University of Alberta, Edmonton.

Gabriel Dumont, Métis leader, c. 1885. Most historians note that if Dumont had been the supreme military leader of the Métis, the campaign would have been much more difficult and bloody for the Canadian volunteers sent to suppress the rebellion. Photographed in what appears to be a hunting dress consisting a wide brimmed hat and a deerskin jacket and holding a rifle. Courtesy Library and Archives Canada, PA178147.

(Above) Metis shirt and pants, mid-19th century. These now very rare garments made of moosehide decorated with glass beads and thread were typical of the Canadian Métis dress. Canadian Museum of History, Gatineau. Ph: RC. (Right) Métis coat, c. 1885. Coat made of animal hide and fur, elaborately decorated with multicoloured glass beads and fastened with brass buttons. Said to have belonged to Louis Riel. Canadian Museum of History, Gatineau. Ph: RC.

A Métis rifleman, 1885. Many Métis might also wear European style dress as seen by prints and photos of the period. Some were also well armed with modern rifles. Reconstruction by Barry Rich. Courtesy National Historic Sites, Parks Canada.

from Buffalo joined General O'Neil's force wearing blue U.S. army uniforms.[134]

The court proceeding held in Toronto during October 1866, which could be read in *The Globe*, brought out some testimony relative the Fenians' arms and appearance when they crossed the Niagara River from the USA to Canada on 1 June. A captured Fenian gave interesting testimony as to how and when the attackers were armed so as to avoid American authorities: "When we left the [US] land we had no arms; as we crossed the [Niagara] river arms were given out. We got them in American water; the ammunition was already distributed to us on the American side; saw no swords issued nor revolvers; saw some men have revolvers, but said they were their own; bayonets were issued also; there was ammunition in boxes also found for us on the Canadian side." Canadian witnesses generally described the Fenians as dressed in civilian clothes. One stated that "some of the officers wore a uniform something like the United States uniform. I saw a good many officers with swords on." Another stated: "We found 40 rifles in or near the creek in good order, marked U. S. on the bayonets, stock and barrel. On the locks were 'Bridesburg, N. Y.' We found provisions and clothing and loose cartridges on the camp-ground. Some of the Fenians were in United States uniform. Some wore green jackets and red shirts, but no general or peculiar uniform. They had 7 or 8 green flags; on one of which was a crown surmounted by a harp; on others there were harps. They had also drums beating."

Although the 1866 raids failed, the Fenians planned to invade Canada again. The American authorities let them plan their aggression on Canada while they were in the United States. Thousands of interested volunteers signed up, money flowed in to their organisations and they contracted for weapons and uniforms. In 1867, they claimed to have 20 regiments gathering over 11,000 men. They now also had Canadian spies in their midst who reported their observations directly to the prime minister's office in Ottawa. A report that Fenians were acquiring Spencer breech-loading rifled carbines and muskets prompted the Canadian government to immediately obtain Spencer and Peabody breech-loading rifles. Canadians also learned that the Fenians were using Emory Upton's system of infantry tactics, the standard US Army manual. A close watch was also kept on the uniforms and accoutrements ordered by the Fenian organisations. In January 1867, one spy reported having seen "37 or 38 boxes…packed with clothing" that were uniforms "as follows: green jackets with yellow braid facings, blue pants with green stripe, grey overcoat, caps same as U.S. soldiers" and every man getting a breech-loading rifle. This was corroborated by a report of March 1867 stating that "nearly 60,000 uniforms were made during the last year" consisting of "green pants, green jacket and cap with brass buttons and yellow facings." Such numbers attracted interest in the press. On 21 March 1867, the *New York Daily Tribune* announced that Fenian President Roberts had contracted for "20,000 fatigue caps of blue cloth United States pattern, with narrow green bands.[135]

In May 1870, the Fenians attacked again, this time from Vermont into Quebec and were easily repulsed by Canadian volunteers. Remarkable surviving garment were a Fenian soldier's jacket and a cap captured at Eccles Hill or Trout River, that was presented to Prince Arthur, one of Queen Victoria's sons, who was then serving

134 *Fort Erie Times Review*, 23 July 1936 and *Buffalo Courier Express*, March 1975 with thanks to Brian Reid; W.H. Ellis, "The Adventures of a Prisoner of War", *Canadian Magazine*, July 1899, p. 202; E.A. Cruikshank, "The Fenian Raid of 1866", *Welland County Historical Society Papers and Records*, vol. II (1925), pp. 21, 76. For more data, please see the plate and text by Christoph Mueller, "The Fenian Irish Army of Liberation", MC&H, Fall 2015, MUIA Plate No. 936.

135 Peter M. Toner, "The military organisation of 'Canadian' Fenians 1866-1870", *The Irish Sword*, X (Summer 1971), pp. 26-37; LAC, MG26, A, vol. 59, 7 January, 16 March and 20 December 1867 reports. See also Wayne Colwell (text) and Francis Back (plate), "Irish Republican Army (Fenian Brotherhood), Vermont/Quebec Border, May 1870", MC&H, 1988. MUIA Plate No. 619.

Métis and Indians firing at the riverboat Northcoote on the Saskatchewan River, 1885. Contemporary print. Private collection. Ph: RC.

Orange Free State Artillery, c. 1899. This detail from a larger group photo shows its generally dark blue uniform imported from Germany to have been almost similar to the German Imperial Artillery except for badges. This consisted of dark blue dress coats with black collar and cuffs, brass buttons, orange piping, black trousers and black leather helmets with brass fittings. *Army and Navy Illustrated*, 16 December 1899. Ph: RC.

Orange Free State Artillery in the field, c. 1899-1900. The gunners wear a pale khaki service dress with a slouch hat as they serve a Krupp 75 mm field piece. This unit had three guns engaged at the Boer victory of Stormberg fought on 10 December 1899. In Vallentin's 1903 *Der Burenkriege.*

Boer cavalrymen of the Orange Free State overcome British troops at Stormberg, on 10 December 1899. It was one of the Boer victories that led to Canadian troops being sent to support the British in South Africa. This watercolour by German artist M. Plinzner from eyewitness accounts shows the Boer troops dressed in basically civilian dress, the coats mostly dark blue, brown or grey with brimmed hats and boots, much as they would wear on their farms. In Vallentin's 1903 *Der Burenkriege*.

Boers in Battle at Mafeking, 1900. The Boer's siege of Mafeking took place from 13 October 1899 until 17 May 1900 when its British garrison led by Col. Baden-Powell was relieved. The Boer riflemen shown in this image by Dutch photographer Jan van Hoepen have no uniforms, but are well armed with plenty of ammunition. Courtesy Rijksmuseum (Amsterdam), RP-F-F01113-U.

in Canada. These items went to England's Royal Artillery Institute until repatriated to Canada. The cap, now vanished, was noted by William Y. Carman so that its appearance is known. The jacket is dark green edged with yellow lace with a lace buttonhole at each side of its collar, the brass buttons bearing the letters "I.R.A." and the its cut is inspired from the US cavalry jacket of the time (see illustration).[136]

Métis Combatants 1885

The Métis population in Manitoba and Saskatchewan is mainly recalled as being involved in the 1870 and 1885 "North-West Rebellions" against the Canadian government. The 1870 event was resolved without any fighting by government troops sent to Manitoba where a regular garrison was maintained until 1877. The 1885 rebellion was much more serious, occurred in present-day Saskatchewan and involved certain First Nations as well as Métis communities. On the whole it must be pointed out that the great majority of the estimated 35,000 Métis and indigenous population stayed clear of the fighting between the more militant members of their communities and the government's troops.

There were no uniforms, nor any desire to have such garments by the Métis and those indigenous warriors who shared their struggle. The Metis combatants campaigned in their own civilian clothes, as they would have when out hunting. From photos and artwork of the period, this usually consisted of a wide-brimmed hat, a check shirt, perhaps a waistcoat, a cloth frock coat or a deerskin jacket, trousers of corduroy or other sturdy material and boots or moccasins. Some of the jackets could be decorated with fringes and indeginous-style bead embroidery.

The Métis used their own personal weapons in combat. These consisted of hunting knives and various types of rifles. Amongst 75 rifles surrendered at Batoche were Winchesters, Snider-Enfields and even a US M.

136 *The Volunteer Review*, 15 May 1871. The jacket "taken from a Fenian made prisoner during the raid made into Canada" is now in the collection of National Historic Sites, Parks Canada.

President Steyn of the Orange Free State and Christiaan De Wet with his "Commando", July 1900. Initiator of "Commando" raids, De Wet is considered the most formidable leader of the Boer guerrillas. He successfully evaded capture on numerous occasions. This watercolour by German artist J Schlattmann from eyewitness accounts shows the Boers dressed in basically civilian dress and well armed. They are escorting President Steyn to safety from pursuing British troops. In Vallentin's 1903 Der Burenkriege.

South African Boers from a "Commando" group, c. 1901. These irregular volunteers were extremely effective light troops. They were self-reliant on the prairies, excellent horsemen and fearsomely good target rifle shots. In their hands, the German Mauser rifle, their usual standard weapon, proved to be superior to the British Lee-Enfield Mk I rifle. They had no uniforms. Photo by a German photographer reproduced in Vallentin's 1903 Der Burenkriege.

1873 Springfield carbine. Colour Sergeant C.F. Winters of the Governor General's Foot Guard recalled that, at Cut Knife Hill on 2 May 1885, the Métis and First Nations' rifle fire "was quite good, but they were greatly hampered by lack of ammunition." He also mentioned that the Métis all had "the latest type (1884) Winchester repeating rifle, had an average of only 30 cartridges apiece", while the great bulk of the indigenous warriors "had only smooth bore muskets and shotguns" with hardly any ammunition.[137]

South African Boers 1900-1902

The Boer republics of Transvaal and the Orange Free State had tiny regular armed forces mainly consisting of artillery corps that could not resist for long against the British forces. The Transvaal artillery wore Austrian-inspired dark blue uniforms with light blue facings. The Orange Free State artillery had imported dark blue uniforms from Germany that were nearly identical to the German Imperial Army's artillery. Both units were armed mainly with German Krupp and French Creuzot guns.

The real military strength of these republics resided in their hardy and self-reliant rural volunteers who were, on the whole, fine horsemen and fearsome marksmen. They were usually mounted and organised into "Commando" units of varying strength — up to a thousand men — that might gather by the thousands or operate as a smaller group of men on the "Veld", the South African prairies. Each Commando was made up of as many as six "Wyk" (ward) under a captain. Boer armies gathering thousands of men were seen during the first year of the war, but they could not prevail against a numerous and well-equipped professional army. Besieged towns with many British residents such as Mafeking were relieved and, by the middle of 1900, all the main centres in the Boers republics were secured by the British forces. Thereafter, Boer resistance continued as the Commandos resorted to using guerrilla tactics in the vast expanses of the countryside.

Boer farmers who made up the Commandos wore their own clothing and used their own horses and saddles. While they did not have uniforms, a June 1902 report by Lt. Ryan of the 2nd Canadian Mounted Rifles pointed out that members of a Boer force might be aged from 13 to 75 years old; he was impressed "strongly not only by their fine physique, but also by their generally smart appearance. It was most noticeable that about one third of the whole were dressed in homemade suits of leather, capable of almost endless wear." Apart from various types of firearms used for hunting, their main military weapon was the excellent German Mauser rifle procured before the war that, in the hands of men used to target shooting since childhood, made them outstanding fighters. They also used captured Lee-Enfield rifles and Lt. Ryan noted these presumably captured rifles armed "about one-half" of the 635 Boers he saw surrendering.[138]

---

[137] *Telegrams of the North-West Campaign 1885*, Desmond Morton and R.H. Roy, eds., The Champlain Society, (Toronto, 1973); C.F. Winters, "The North-West Rebellion, 1885: A Memoir by [Governor General's Foot Guard] Colour Sergeant (later General) C.F. Winters," Daniel S.C. Mackay, ed., *Saskatchewan History*, XXXV: 1 (Winter 1982), p. 10.

[138] "Organisation, equipment, despatch and service of the Canadian contingents during the war in South Africa 1899-1902", Parliament of Canada, 2-3 Edward VII, *Sessional Paper* No. 35a, Vol. 13 (1903). The best study on Boer uniforms and dress is by: Erwin A. Schmidl, "Uniforms of the Boer Forces 1899-1902", *Military Illustrated, Past & Present*, No. 23 (February-March 1990) and No. 26 (July 1990).

Conclusion

The late and regretted by historian George F.G. Stanley once observed a paradox of Canada: it has an extensive military history made an unmilitary people. He was and remains absolutely correct. Canadian armed forces in peacetime have always been and continue to be quite modest in relation to the country's growing population. The other part of the paradox is that, from the 17th century, Canadians have been involved in a myriad of wars and battles requiring, in urgent times, considerable mobilisation and have served in all parts of the world. Although redoubtable on any battlefield, Canadians generally get out of uniform as fast as they can when the war ends and go back to their civilian lives. Militarism is not a national ethos.[139]

As seen in these volumes, the Volunteer Militia was the most numerous and visible portion of Canada's defence organisation. Its varied units were everywhere in the country during Queen Victoria's reign; it was a period during which the old compulsory militia system broke down and the "volunteer movement" appeared, notably strongly in Great Britain, the Unites States and, of course, Canada. It was fashionable to volunteer in the militia. A strong recruiting aspect was to dress the volunteers in attractive uniforms and Canadians had the advantage to rely on British military fashions — amongst the world's best — for guidance. During the Victorian era, the smartly uniformed volunteer militiamen were often seen parading in cities at civic events and the free concerts given by regimental bands were highly popular with the general public.

The Victorian era passed and so did dazzling military dress from 1914, but not totally. Colourful Victorian style full dress uniforms are still worn. In summer, tourists in Quebec City can see a scarlet-coated guard at its citadel while a large contingent in full dress changes the guard daily at Ottawa's Parliament wearing very colourful uniforms that have not changed much since the 1870s (see chapter 3). All this military pageantry, evoked in these two volumes, is a vivid testimony to the great and proud national heritage earned in battlefields all over the world by unmilitary warriors. Their deeds — wearing red, blue, green or khaki — have forged the very existence of Canada as a nation.

---

[139] George F.G. Stanley, *Canada's Soldiers The Military History of an Unmilitary People* (Toronto: Macmillan, 1960), pp. 417-418.

Appendices

Bellow are a some of the more important dress regulations issued from 1838 to 1892. In general, they concern the dress of officers more than that of NCOs and enlisted men; those of 1876, 1886 and 1892 concern only the officers.

Appendix 1

Upper Canada Militia Dress Regulations 1838

Upper Canada Militia General Order, Toronto, 23 April 1838

"For the General Staff.

The uniform prescribed for the staff of the Regular Service, excepting that the Epaulette, Embroidery, lace and mountings of sword belt are to be silver.

For the Cavalry.

A description or pattern of the uniform already adopted or ordered to be prepared for the several Troops to be submitted for the approbation of His Excellency the Lieutenant-Governor within six months after the date of this order, or after the issuing of any future order authorising the embodying of any new Troop.

For the Artillery.

The uniform prescribed for the Royal Regiment of Artillery, excepting that the Epaulettes and lace are to be of silver, and that the chako is to be of blue cloth of the regulation pattern, with a mohair band two inches wide, having on the front a Crown, embroidered with Silver and the Royal Cypher beneath. For undress the chaco is to be worn with an oilskin cover, buttoned at the sides.

For the Infantry.

Full dress. The same as prescribed for the Militia of Great Britain, with the exception of the chaco which is to be the same as the Artillery already described.

Undress. A blue frock coat single breasted with nine buttons down the front and a Prussian collar. The shoulder straps of silver cord one eight of an inch in diameter doubled.

Memorandum.

Officers of Militia having held Staff situations in the Incorporated are to wear the uniform prescribed for similar appointments in the Regular Service, excepting that the Epaulettes, Embroidery, lace and mountings of all kind to be silver."

MGO 168 Toronto 24 July 1838

...with a view to the convenience & accomodation of the officers of Militia is pleased to promulgate that the undress Uniform of Officers, that is a blue frock coat single breasted with nine buttons down thefront & a Prussian Collar the Shoulder straps of silver Cord One Eight

of an inch in diameter doubled with sword and belt will be sufficient for all purposes of parade or ordinary duty.

MGO 203 Toronto 24 October 1838

"...formation of four battalion of Incorporated Militia...

4th Government will furnish each NCO, drummer (or bugler) and private without charge the following articles: one jacket, one pair of half boot, one pair of cloth trousers, one cap, one stock, one greatcoat, one knapsack. The four latter articles to be returned at the expiration of the period of service.

(...)formation of a Militia Troop of cavalry at Toronto consisting of (1 Cap, ---, 50 pvts) [same items for cavalrymen] …except that viz: each man to receive a cloak and valisse, in lieu of a greatcoat and haversack.”

MGO 216 Toronto 3 November 1838

(...) “Memo: the Officers of the Militia throughout the province will consider Scale straps on the undress as consisting a portion of their uniform.” (LAC, RG9, IB3, Vol. 6)

Appendix 2

Notes on 1837-1855 uniforms by Walker Powell, Adjutant General of Canadian Militia in the later 19th century.

"1837/39 The uniform authorised to be worn by officers of the organised militia infantry was similar to that of the royal regiments of the infantry of the regular army viz: the full dress uniform for captains was a scarlet cloth coat with blue facings, a gold epaulette on the right shoulder, white linen trousers in summer, dark grey cloth at other times, a crimson silk sash around the waist. Leather scabbard sword gilt mountings, and a shako with ball tuft (white & red).

Undress. Blue cloth frock coat, dark grey trousers, up to 1815 blue cloth cap of Wellington shape, crimson sash over the left shoulder & under the right arm. White metal buttons. Sword belt around the waist over the frock coat.

Many volunteer corps provided its own uniform which differed in cut & marks in different portions of the country, but as many as could do so adopted the pattern worn by the regular troops of their arm. That worn by infantry corps of volunteer militia at Quebec 1837/39 were fanciful in design in winter but intended for comfort and indicate the nationality of the wearer. Types of those wearing coats made from blankets (in colours) of facings and ornaments in 1838-9 can be seen in the Militia Museum in Ottawa. One of the regiments at Montreal were uniform thereon (…) picture in the book [,] the totality wore the same uniform a[s] Royal Artillery with whom they trained and did duty – viz: blue tunic and trousers (dark in colour) scarlet or red facings and stripes (…) side seams of trousers. When a battalion of sedentary militia was called up for temporary duty, the NC Officers and men did not as a rule wear uniform but were distinguished by an arm badge or other device, mostly by a piece of white cotton or red flannel tied around the right arm.

When the volunteer corps were embodied for service with the regular troops of Great Britain in Canada they were auxiliaries and received pay[,] subsistence and shelter from the military chest. If the Provinces required the services for local purposes they were to be paid for services so required. The arms for the infantry were the usual pattern flintlock muskets issued on loan by Great Britain. Some of the corps of infantry enrolled in Upper Canada for continuous service as long as required during the rebellion 1837/38 were uniformed in blue pilot cloth coats and trousers, buffalo skin caps with black leather peaks. These corps were stationed at different places in hired barracks, the greatcoats were dark grey with small cuffs. The money that was paid from the British military chest was mostly silver and formed a considerable portion of circulating species (…) as the banks issued only paper money at that time their payment being suspended during the period of rebellion –

(…)

1841-1855 – The volunteers of Canada continued to drill without pay during peace, unless embodied for duty during that period, supplied their own uniform clothing, and were allowed much latitude in the description worn by each arm, but having chosen a uniform with approval from Head Quarters were compelled to continue to wear that pattern until changed by authority, the cavalrymen continued to supply their own horses and much of their saddlery & appointments. Corps of all arms were requested as far as possible to conform to similarity in colour to that worn by their own arm in the regular forces of Great Britain (either full dress or undress as could be most suitably procured), but the similarity in colours worn not universally maintained as, grey (scarlet facings) was a favourite colour and could be provided in Canada." (…) (LAC, Powell's Scrapbooks Vol. 1.)

Appendix 3

Dress Regulations 1848-1852

"Canada West Militia General Order

Adjutant General's Office, Montreal, 25 August 1848

G.O.

His Excellency the Governor General has been pleased to order that the Uniform to be worn on Service by the Militia of Canada West, shall be that recommended at Toronto in 1838 and approved by His Excellency the Lieutenant Governor of Upper Canada by the General Order of the 23rd April 1838, viz:—

For the General Staff.

The Uniform prescribed for the Staff of the Regular Service, excepting that the epaulette, embroidery, lace and mountings of sword belt are to be silver.

For the Artillery.

The Uniform prescribed for the Royal Regiment of Artillery, excepting that the epaulettes and lace are to be of silver, and that the chako is to be of blue cloth of the regulation pattern, with a mohair band two inches wide, having on the front a Crown, embroidered with Silver and the Royal Cypher beneath.

For undress the chaco is to be worn with an oilskin cover, buttoned at the sides.

For the Infantry.

Full dress. The same as prescribed for the Militia of Great Britain, with the exception of the chaco which is to be the same as the Artillery already described. — Undress. A blue frock coat single breasted with nine buttons down the front and a Prussian collar. The shoulder straps of silver cord one eight of an inch in diameter doubled.

The Glengarry Battalions are to wear the Highland bonnet, tartan scarf and trousers as heretofore.

By Command,

D. Macdonell, Lt. Colonel

Deputy Adjutant General Militia"

(LAC, RG9, IC4, Vol. 4)

The 24 August 1848 Canada West order was reissued word for word for all of Canada by the "Canada Militia General Order" from the Adjutant General's Office, Quebec, 16 April 1852. On 16 June 1852, a General Order was added as follows:

"The 3rd & 4th Battalions of Stormont Militia, shall wear the Highland Uniform, as prescribed for the Glengarry Battalions in the General Orders of 16th April 1852. Their Commanding Officers to select the pattern of the Tartan for their Battalions." (LAC, RG9, IC4, Vol. 2)

Appendix 4

1860 Dress Regulation

Militia General Order, Adjutant General's Office, Quebec, 19 March 1860

…the following pattern of all clothing to be furnished in future by the Volunteer force of the province: viz:

Field batteries and foot companies of artillery: - Blue tunics and trousers, the same as the Royal Artillery.

Cavalry:- Blue tunics, single breasted with scarlet facings and white cord – blue trousers with white stripe down the legs – officers to wear silver lace.

Rifle corps or companies – Rifle green tunic, single breasted with scarlet facings and black cord shoulder strap; collar and cuffs slightly braided.

Rifle green trousers with two stripes of black braid on a scarlet stripe down the leg; the Highland companies are recommended to wear tunics, or jackets and trews the same as usual in the regular service, the material and facing of the tunic or jacket to be in uniform with other rifle corps.

The buttons of the tunic of each arm of the force to be of the same description as those worn in the regular service encircled with the words "Volunteer Militia, Canada".

Head dress – His Excellency is pleased to leave the description of future head dress to be decided by the force themselves; but with a view to uniformity at each station…

(*The British North American Almanac and Record for the year 1864*, p 113.)

Appendix 5

Regulations and Orders for the Active Militia...of the Dominion of Canada [1870]

"Clothing

215 The Uniform of the militia is similar to that worm by the Regular Army, with the exception that in the Infantry, silver lace is substituted for gold- and the facings of Infantry corps clothed in scarlet are blue, those of Rifle corps, scarlet.

216. During the performance of the annual drill, and on all occasions of actual service, the undress uniform maybe worn by officers, and forage caps instead of chacos by non-commissioned officers and men.

217. All Clothing and Great Coats, issued at any time from the stores of the Dominion, must be considered the property of the Government, and as being issued only for use by the Corps, as provided by Law. See sec. 43 Militia Law."

(*Canada Gazette*, 4 March 1870)

Appendix 6

Militia Dress Regulations, 4 March 1876

Badges of Rank

The several ranks of officers are designated by embroidered badges, as follows:-

Colonels wear a crown and star; Lieutenant Colonels, a crown: and Majors, a star.

And similarly:-

Captains wear a crown and star: Lieutenants, a crown; and Sub Lieutenants, a star.

These badges, when on the saddle cloth, 2 1/2 inches high.

Regimental officers having Brevet Ran wear the badges of their regimental rank.

Boots and Spurs

Mounted officers, except where otherwise specified wear when on mounted duties, high boots with crane necked spurs fastened with straps and buckles . Officers who are not mounted wear Wellington ankle boots. The high boots must be, as regards shape, in accordance with the sealed pattern, but the height will depend upon the length of the leg and the relative height of the calf. The boot which is sloped down at the back, should reach at the back just to the top of the calf, generally about 4 1/4 or 5 inches from the top of the knee.

Braid, Buttons, Etc

Unless otherwise specified loops. frogs.and buttons on the front of tunics, etc, will be at equal distances. Buttons, generally, will be an inch. When loops of lace or cord are worn across the breast, the top loops will reach to the sleeve seams, and those at the waist will be 4 inches long.

Depth of Skirts

The skirts of tunics for officers 5 feet 9 inches high will be- ‘For Colonels on the Staff, Artillery, Engineers and Infantry regiments, 10 inches; For General Staff, Cavalry and Rifle regiments, 9 inches; The skirts of frocks will be 17 inches deep for all officers 5 feet.9 inches high. The proportionate variation for each inch of difference in height is about 1/4 inch in the skirts of frocks and 1/8 inch in the skirts of tunics.

Forage Caps

Forage caps will be 2 5/8 inches high.

Gloves

Gloves, except when otherwise stated, will be of white leather.

Patrol jackets

Patrol jackets, except those otherwise specified, will be made according to the following description:- Blue cloth. 28 inches long from the bottom of the collar behind, for an officer 5 feet 9 inches in height, with a proportionate variation for any difference in height, rounded in front, and edged with inch mohair braid all round and up the sides. On each side in front, four double drop loops of ¼ flat plait, with eyes in the centre of each loop, the top loops reaching to the sleeve seams, and the bottom ones 4 inches long; four netted olivets on the right side to fasten through the loops on the left. On each sleeve, an Austrian knot of lfat plait 7 inches high from the bottom of the cuff. Double flat plait on each back seam, with crow's foot at the top and bottom and two double eyes at equal distances, Pocket fitted with flaps in and out. Hooks and eyes in front.

Sashes

Sashes will be worn diagonally over the left shoulder and under the left shoulder cord; the ends will be crossed though a runner at the waist, and the sash will be of such length that the ends of the tassels shall just reach the bottom of the skirt of the tunic.

STAFF

Deputy Adjutant-General

Tunic

Scarlet cloth, edged all round, except the collar, with round-back silver cord, Blue cloth collar and cuffs, the collar ornamented with half inch lace round the top, silver Russia braid along the bottom, and a rich tracing in double silver braid below the lace, at each end the badges of rank embroidered in gold; the cuffs pointed with half-inch lace round the top and a rich tracing of in double silver braid above and blow the lace, extending to 7 inches from the bottom of the cuff, On each side of the breast, four loops of round-back silver cord, with caps and drops, fastening with gold-worked olivets, On three eyes at the top, passing under a netted cap at the waist, below which it is doubled, and ending in and Austrian knot, reaching to the bottom of the skirt. On each shoulder, a silver cord loop, with a small button. The skirt rounded off in front , closed behind, and lined with white.

Lace

Silver. Staff pattern.

Buttons

Silver, frosted, with burnished laurel round the edge.

Dress Trousers

Blue cloth, with 1 3/4 inch lace down the side seams.

Spurs

Brass.

Cocked Hat

Of black beaver, or silk; and of the following dimensions :—Time left side 7 inches high, the right side 6 1/4 inches, each corner 4 3/4 inches long. There will be a black silk cockade on the right side, with a loop and button over it; and at each corner a bullion tassel, 1 3/4 inch long, exclusive of the head, with loop of 3/4 inch lace ; tassels, netted, silver purl head; eight small silver bullions, with seven crimson bullions over them.

Plume.

White swan feathers, drooping outwards, 6 inches long, with red feathers under them long enough to reach the ends of the white ones; feathered stem 3 inches long.

Sword

Half-basket hilt, and back piece of steel or gilt metal, black fish-skin grip, with silver wire, when the hilt is of steel, or with gift wire when the hilt is of gilt metal; slightly-curved blade, grooved and spear-pointed. Full Size—blade 35 inches long and 1 1/8-inch wide at the shoulder; extreme length, including the hilt, 41 inches weight. without scabbard 22 pounds. Second size - blade 33 inches long.and an inch wide at the shoulder ; extreme length. including hilt, 38 1/2 inches; weight, without scabbard , 1 pound,12 ounces.

Scabbard

Brass.

Sword Knot.

Silver and crimson cord, with silver and crimson acorn.

Sword Belt.

Russia leather, 1 1/4 inch wide, with slings an inch wide: two stripes of silver embroidery on belt and slings : a silver hook to hook up the sword.

Waist Belt.

Round, silver clasp, with royal cypher and crown In the centre, and a laurel wreath on the outer circle.

Shoulder Belt

Silver lace, 1 3/4 inch wide, with crimson morocco leather lining and edging; silver ornamented buckle, tip and slide.

Telescope Case

Black patent leather, to hold a binocular field glass: solid leather flap. reaching to the lower edge of the case, ornamented with royal cypher and crown in silver metal.

Frock

Blue cloth, double-breasted rolling collar, without badges of rank. The front and collar edged with 3/4 inch black mohair lace, an Austrian knot of black Russia braid on each sleeve, reaching to 6 inches from the bottom of the cuff; 5 loops of Black Russia braid on each side of the breast, fastening with black olivets; 2 olivets at the waist height. The skirts lined with black.

Waistcoats

Scarlet cloth, without collar, edged with silver Russia braid, and fastening with hooks and eyes.

Undress Trousers

Blue cloth, with scarlet stripes 1 1/4 inch wide down the side seams.

Undress Sword Knot

Russia leather cord and acorn,

Pantaloons

Blue cloth, with scarlet stripes… Mounted officers, except where otherwise specified wear when on mounted duties, high boots with crane-necked spur fastened with straps and buckles. The high boots must be as regards shape in accordance with the same patterns, but the height will depend upon the length of the leg and the relative height of the calf. The boot which is sloped down at the back…just to the top of the calf, generally about 4 or 5 inches from the top of the knee.

Forage Cap

Blue cloth, with silver embroidered peak, and band of 1 3/4 inch lace; silver purl button and braided figure on the crown.

Staff Jacket

Scarlet cloth edged all around, including the collar, with 1/2 inch lace, forming barrels at the bottom of the back seams. Blue cloth collar and cuffs; the badges of rank, embroidered in gold at each end of the collar (by field Officers only), and a silver braid along the bottom, with an eye in the centre; the ends pointed, with lace and braiding of the same pattern as on the cuffs of the tunic. A silver cord loop, with a silver stud on each shoulder. A row of silver studs in front, on the left side. Scarlet lining, hooks and eyes in front, and a loop of silver braid on the bottom of the collar, to fasten across the neck,

CAVALRY

Tunics

Blue cloth, edged all round with gold chain gimp, except the collar, which has three–quarter inch lace round the top with the badges of rank, embroidered in silver, at each end. On each side of the breast, six loops of gold chain gimp, with caps and drops, fastening with gold worked olivets. On each back seam a double line of the same gimp forming three eyes at the top, under a netted cap at the waist and ending in an Austrian knot reaching to the bottom of the skirt; with a tracing of gold braid all round the gimp. An Austrian knot: gold chain gimp on each sleeve, reaching to 8 inches from the bottom of the cuff. The skirt rounded off in front closed behind, and lined with black. The badges of rank, embroidered in silver, at each end at the collar.

Field officers have figured braiding below the lace on the collar. and figured braiding on the sleeve, round the Austrian knot, extending to 11 inches from the bottom of the cuff. Captains have a row of braided eyes on the collar bellow the lace; and a tracing of braided eyes round the knot on the sleeve nine inches deep. Lieutenants have a tracing of plain braid only below the lace on the collar and round the knot o the sleeve 8 inches deep.

Collars

Buff cloth.

Lace

Gold.

Trousers, Etc

Blue cloth with two stripes three quarter inch lace, one-quarter inch apart down each side seam; Wellington boots and brass spurs. Blue cloth, with stripes as on trousers ; over boots and steel spurs, as described in Section 1, of these regulations.

Busby

Black sable fur; outside measurement. 7 5/4 inches high in front, 8 inches at the sides and 9 inches at the back; top half inch less in diameter than the bottom; back shaped to fit the head. A gold, gimp oval cockade 2 Inches deep and half inch wide, in the centre in front, the top on a level with the top the cap. A spring socket, behind the cockade A buff cloth covering the top of the cap, and falling down the right side to within an inch of the bottom; a line of gold braid along the seam of the bag, and down the centre, with a gold gimp button at the bottom. A gilt hook at the top, on the right, to hook up the chain.

Sword

Half basket steel hilt, with two fluted bars on the outside; black fish-skin grip, bound with silver wire; slightly curved blade 35 1/2inches long, and 1 1/4 inch wide at the shoulder, grooved and spear-pointed.

Scabbard.

Steel, with a large shoe at the bottom, and a trumpet shaped mouth.

Sword-Knot

Gold and. crimson cord. with gold acorn.

Sword Belt

Gold lace 1 1/4 inch wide, with sword slings of the same width and a 1/4 inch buff stripe in the centre of the sword; pocket, and tache-slings 1/2 inch wide; buff morocco leather lining and edging, gilt snake fastening.

Sabretache

Buff cloth face with lace 2 1/2 inches wide all round, within 1/4 inch of the edge; a 1/2 inch silk stripe (buff) in the centre of the Lace. Embroidered regimental device in the centre. Buff morooco leather pocket. The sabretache not to hang below the calf of the leg.

Pouch Belt

Gold lace 1 1/4 inch wide, with sword slings of the same width and a 1/4 inch buff stripe in the centre of the sword; pocket, and tache-slings 1/2 inch wide; buff Morocco leather lining and edging gilt snake fastening.

Pouch

Black leather, silver flap, and gilt ornaments.

Undress

Frock

Blue cloth, single breasted. The collar edged with 3/4 inch black braid, and with figuring in narrow braid. A braided figure on each sleeve extending to 10 inches from the bottom of the cuff. Six loops of inch braid across the breast, w'th four rows of olivets. The back seams and back skirts trimmed with inch braid, traced round with narrow braid, and with olivets and tassels. The skirts lined with black.

Trousers

The same as in full dress, except that the spurs worn with the trousers are steel instead of brass; and that the double stripes on the trousers and pantaloons are of white cloth.

Forage Cap

Blue cloth, with band of 1 3/4 inch gold lace; gold purl button and braided figure on the crown, and a line of gold braid round the crown-seam.

Stable Jacket

Blue cloth, with olivet and lace, or cord. Field Officers wear the badges of their rank, embroidered in silver on the collar. Collar,—buff.

Mess Waistcoat

According to regimental pattern.

Sabretache

Black patent leather. The other articles as in full dress.

Cloak and Cape

Blue cloth, of the same pattern as for Officers of Dragoons, with scarlet lining.

Sub-Lieutenant

Sub-Lieutenants wear the same uniform as Lieutenants, except that the badge of rank is a star. They are not to provide themselves with horse furniture.

COBOURG CAVALRY

Tunic

Scarlet cloth, with collars and cuffs of yellow velvet. The collar ornamented with 3/4 inch lace all round for Field Officers, but round the top only for Captains and Lieutenants; with the badges of rank, embroidered in silver at each end. The cuffs pointed, and edged with round black gold cord, forming for Field Officers, a triple Austrian knot traced round with gold Russia braid, and extending to 11 inches from the bottom of the cuff for Captains a double Austrian knot similarly traced, 9 inches deep; and for Lieutenants, a single Austrian knot, 7 inches deep, eight buttons in front, and two at the waist behind; and a twisted gold cord loop, with a small button on each shoulder. A scarlet flap on each skirt behind, with three buttons. and edged with round-back gold cord. The front edged with the same material and colour as the facings, and the skirts lined with white.

Helmet

Gilt brass, according to sealed pattern.

Plume

Black and red horse-hair.

Gloves

White leather gauntlets. The remainder of the accoutrements are the same as the Cavalry.

GOVERNOR GENERAL'S BODY GUARD

Tunic

Blue cloth, edged all round, including the top and bottom of the collar, with round-back gold cord. Collar and cuffs of white cloth; the collar laced within the cord, with 3/4 inch lace, all round for Field Officers but round the top only for Captains and Lieutenants, with the badges of rank embroidered in silver at each end like those of the other regiments; the cuffs pointed with 1 1/2 inch lace round the and top and figured braiding extending to 11 inches from the bottom of the cuff, for Field Officers; with an Austrian knot of gold cord tracing of braid in the form of eyes, 8 inches deep, for Captains; and with a similar knot and a tracing of plain braid, i7 1/2 inches deep, for Lieutenants; eight buttons in front and two at the waist behind, and a twisted gold cord loop, with a small button, on each shoulder. A scarlet flap on each skirt behind with three buttons, and edged with round-back gold cord. The font edged with the same material and odour as the facings, and the skirts lined with white. [*Note: the gold cord was in fact silver and the silver badges of rank in fact gold. This seems to have been an undetected error*] .

Helmet

According to pattern.

Plume

White horse hair with rose at top, rising 6 inches above the top of the helmet, and then falling as far as the bottom.

The rest of the uniform as already prescribed for cavalry.

FIELD AND GARRISON ARTILLERY.

Full Dress

Tunic

Blue cloth, with scarlet cloth collar. the collar and sleeves laced and braided according to rank, as mentioned below. The skirt rounded in front, closed behind, with a plait at each side and lined with black. Buttons sown the front 2 1/4 inches apart, and two at the waist behind; and a gold cord loop, with a small button on each shoulder. Scarlet cloth edging all round, except the collar and up the skirt-plaits.

Field officers have 5/8 inch lace all round the collar, within the cord; and a chevron of 1 1/2 inch lace on each cuff, with figured braiding above and below the lace, extending to 11 inches from the bottom of the cuff.

Captains and Lieutenants have a lace round the top only of the collar; and an Austrian knot of gold cord on each sleeve, 7 inches deep, traced round with gold braid 8 inches deep and figured for Captains; 7 1/2 inches deep and plain for Lieutenants.

Plume

White goat's hair, 7 inches from top of busby, with gilt socket and ring; to be worn on the top of the busby.

Cap-Lines (For Field Brigades)

Gold cord with an acorn at each end, passing round the cap diagonally three times, then round the neck and looped on the left breast.

Dress Sword-Belt

Gold lace 1 1/2 inch wide, lined with blue Morocco leather; gilt S hook fastening, with two oval gilt plates bearing the Royal crest. Sword slings (and tache slings for mounted officers) fastened to a flat steel bar covered with blue Morocco leather, and attached to the inside of the belt by four flat steel hooks.

Sabretache (Full Dress.)

Blue Morocco leather, faced with blue cloth; 1 1/2 inch lace round the face, 1/4 inch from the edge. An embroidered device within the lace of the Royal arms above, and a gun below, with an oak and laurel wreath.

Sabretache(Undress)

Black patent leather, with regimental badge, in gilt metal.

Undress Sword-Belt

White patent leather,1 7-10 in wide, with sword slings (and tache slings for mounted officers) gilt frosted plated, with regimental device.

Undress Pouch-Belt

White patent leather, 2 inches wide.

Lace.

Gold.

Buttons

Gilt, burnished, with a gun and crown.

Trousers, &c

Blue cloth, with 1 3/4 inch lace down the side seem: Wellington boots and brass spurs.

Busby

Black sable skin, 7 1/2 inches high in front, 8 1/2 inches at the back, and 23 inches round the top, outside. A scarlet cloth bag, covering the top of the cap and falling down the right side to within an inch of the bottom, a spring socket at the top in front. Black leather chin-strap and brass buckle.

Sword

Half-basket steel hilt, with two fluted bars on the outside; black fish-skin grip, bound with silver wire; slightly curved blade, 35 1/2 inches long and 1 1/4 inch wide, grooved and spear-pointed.

Scabbard

Steel, with a large shoe at the bottom and a trumpet shaped mouth.

Sword-knot

Gold cord, with a gold acorn.

Pouch.

Blue Morocco leather collapsing pouch, with two pockets, the leaf 5 5/8 inches long and 2 3/4 inches deep, covered with blue cloth and edged with 3/4 inch last.

Pouch Belt.

Gold lace; 2 inches wide. lined with blue Morocco leather; gilt ornamented buckle and slide, and a grenade, encircled with a wreath, at the end.

Undress

Waistcoat

Scarlet cloth, without collar, edged all round with gold Russia braid; the pockets edged with the same braid, forming crow's feet at the end and centre.

Patrol jacket for Officers under the rank of Regimental Colonel.

Blue cloth, rounded in front, and edged with inch black mohair braid all round and up the openings at the sides; five loops of flat plait on each side in front, fastening with olivets, and with crow's feet and badges of rank. The sleeves ornamented with flat plait, forming crow's feet, 6 inches from the bottom of the cuffs. Double flat plait on each back seem, with crow's feet at top and bottom, and two eyes at equal distances. Pockets edged with flat plait, forming crow's feet and eyes.

The jacket to be long enough to reach the saddle when the officer mounted, and loose enough to be worn over the stable jacket.

Trousers

Blue cloth, with scarlet stripes two inches wide down the side seams; Wellington boots and steel spurs.

Pantaloon,&c, for Mounted Duties

Blue cloth, with stripes as on the trousers, over boots and steel spurs, as described in section 1 of these regulations.

Forage Cap

Blue cloth, with band of 1 5/8 inch gold lace, gold button and braided figure, of special pattern, on the crown. The cap to be 3 inches high.

Stable Jacket

Blue edged all around with gold Russia braid, forming a figure 8 at the bottom of each back seam, Scarlet cloth collar, edged all round with gold braid, and lined with black. Small buttons down the front 1 3/4 inch apart, and a gold cord loop, with a small button, on each shoulder. Scarlet alpaca lining.

Field officers wear the badges of their rank, embroidered in silver, on their collar; and a chevron of inch lace on each cuff, with braided eyes above and below the lace, extending to 6 inches from the bottom of the cuff.

Captains and Lieutenants have an Austrian knot of gold cord on each sleeve, 6 inches deep, traced round with gold braid, 7 inches deep, and figured for Captains; 6 1/2 inches dep, and plain, for Lieutenants.

Sword-Knot.

White buff leather, 1/2 inch wide, with runner and gold acorn.

Sabretache (For Mounted Officers only.)

Black patent leather, with regimental badge, in gilt metal.

Pouch

Black patent leather collapsing pouch, with two pockets 5 1/2 inches long, 2 1/2 inches deep. A gun in gilt metal…

Cloak and Cap

Blue cloth, with sleeves, standard fall collar, with three black gook and eyes in front, and three small gilt buttons at the bottom to fasten the cape. Round loose cuffs, 6 inches deep. A pocket in each side seam outside, and one on the left breast, inside; four buttons down the front. A cloth back strap to fasten with a large flat button at the top of each pocket; a similar button in front on the right to hold the end of the back strap when it is not buttoned across behind. White shaloon lining. The cloak to reach within 8 inches of the ground.

Blue cloth cape, 32 inches deep, lined with white shaloon. A cloth band round the top, to fasten with a black strap and buckle; and a fly inside the band, with three button holes, for attaching cape to cloak; three buttons down the front.

Horse Furniture.

Black sheepskin and wallet (leather).

Valise

Blue cloth, as described for Cavalry regiments.

ENGINEERS

Regimental Officers.

Tunics

Scarlet cloth with collar and cuffs of Garter blue velvet. The collar edged all around with round back gold cord; 3/4 inch lace all round within the cord for field Officers, round the top only for Captains and Lieutenants; at each end the badges of rank, embroidered in silver. The cuffs pointed and ornamented as described below, according to rank; nine buttons in front and two at the waist behind. A shoulder knot of treble twisted round-back gold cord on each shoulder, with a small button, near the neck; and a grenade, embroidered in silver, at the lower end. The skirt rounded of in front, closed behind, with a plait at each side, and lined with white. The front, skirt, and plaits edged with carter blue velvet.

Field Officers have 1 1/2 inch lace round the top of their cuff; and figured braiding above and below the lace, extending to 1 1/2 inches from the bottom of the cuff.

Captains have an Austrian knot of round-back gold cord on each sleeve, traced all round with braided eyes, the braid extending to 8 inches from the bottom of the cuff.

Lieutenants have a similar knot, but without the figured braiding.

Lace and Buttons

Gold, of regimental patterns.

Dress Trousers

Oxford mixture cloth, with 1 3/4 inch lace down the side seams; Wellington boots, and brass spurs.

Cocked Hat

As described in Section 1 of these Regulations, with loop of inch lace, button and black watered silk cockade; gold bullion tassels.

Plume

White cock's feathers, drooping outwards, 5 inches long.

Busby

Picked seal skin, 5 1/4 inches high in front and 8 inches at the back, on a pliable cork body, with ventilating cork inside. An ornament of round-back gold cord, hooked up at each side, and curved downwards across the front and back, the front plaited, and the back plain doubled, with flounders and eggs at the left side. Gilt, burnished chain, lined with black velvet, with a gilt lion's head hook at the back to fasten it up. A gold gimp boss in front 1 1/2 inch long.

Plume

White goat's hair, 6 1/2 inches long, with bright blue feather, 2 1/2 inches long, round the bottom. Plume socket, a gilt fuse and bill.

Sword

As described in section 1 of these Regulations, with hilt of gilt metal, pierced and engraved according to special pattern.

Scabbard

For Field Officers, brass; for other officers, steel.

Sword Knot

Gold cord, with gold acorn; for regimental duty, Russia leather and gold acorn.

Sword Belt

Russia leather, 1 1 /2 inch wide, with slings and inch wide, the front sling lined with Russia leather; two stripes of gold embroidery on belt and slings. Gilt burnished plated, with regimental device in sliver.

Shoulder Belt

Russia leather, 2 inches wide, with three stripes of gold embroider, the centre one wavy, the others straight; gilt engraved buckle, tip and slide.

Telescope Case

Black patent leather, to hold a binocular field glass; solid leather flap, with gilt regimental badge.

Frock for Regimental Field Officers

Blue cloth, single breasted, with rolling collar; ornaments on sleeve as on the tunic, but in black mohair braid, traced; eight loops of 1/4 inch black braid down the front, with barrel buttons placed according to regimental pattern. The front edges, collar. back and sleeve seams, and back slits trimmed with 7/8 inch black braid traced. Hooks and eyes in front. The skirt lined with black.

Patrol Jacket, for Officers under the rank of Regimental Colonel

Blue cloth 28 inches long from the bottom of the collar behind, for an officer 5 feet 9 inches in height, with a proportionate variation for any difference in height, rounded in front, and edged with inch black mohair braid all round and up the openings at the sides. Stand and fall collar and cuffs of blue velvet. Five loops of flat plait on each side in front, with crow's feet at the ends and three rows of olivets. Crow's foot on each sleeve, ?1/2 inches deep from the bottom of the cuff. Pockets fitted with flaps in and out. Hooks and eyes in front.

Field Officers wear the badge of their rank embroidered in gold on the collar.

Undress Trousers, &c.

Oxford mixture cloth, with scarlet stripes, 2 inches wide down the side seams; brass spurs for Field Officers, steel spurs for other mounted Officers.

Pantaloons, &c for Mounted Duties.

Oxford mixture cloth, with stripes, as on the trousers; over-boots and brass or brass steel spurs, as described in Section 1 of these Regulations.

Forage Cap

Blue cloth with band of 1 3/4 inch gold lace, a scarlet welt round the crown, and a gold netted button in the centre Black patent leather peak and chin-strap.

Shell Jacket

Scarlet cloth, with garter blue velvet and cuffs. Gold braid all round the jacket and along the bottom of the front, and a crow's foot at the centre of collar seam and of waist., Hook and eyes and gilt studs down the front; and a twisted gold cord loop, with a small button on end shoulder. Pointed cuffs, 5 inches deep: with inch lace and a tracing of braid round he top, for field officers with a double line of small braided eyes for Captains; and with plain braid edging for Lieutenants, scarlet silk serge lining.

Field Officers wear the badges of their rank, embroidered in gold, on the collar.

Waistcoat

Scarlet cloth with hooks and eyes and gilt studs down the front, and edging of gold braid all round and on collar seam. Pockets edged with gold braid forming crow's feet at ends and centre.

INFANTRY OF THE LINE.

Tunic

Scarlet cloth, with blue collar and cuffs. The collar ornamented with half-inch lace along the top and silver Russia braid at the bottom ; with the badges of rank, embroidered in gold at each end. The cuffs pointed, with 1/2 inch lace round the top, and a tracing in silver Russia braid, quarter inch above and below the lace, the lower braid having a crow's foot and eye, and the upper and Austrian knot, at the top, eight buttons in front, and two at the waist behind; and a silver square cord loop, with a small button on each shoulder. The skirt closed behind, with a plait at each side, and lined with white. The front collar and skirt plaits edged with white cloth quarter inch wide.

Field Officers have a row of braided eyes, below the lace, on the collar; two bars of lace along the top of the cuff showing 1/4 inch of the facings between the bars; and the braiding of the sleeve is in the form of eyes, above and below the lace for Colonels and Lieutenant-Colonels, and above the lace only for Majors. the lace on the sleeve extends to 8, and the Austrian knot to 10 inches, form the bottom of the cuff.

Captains have no braided eyes on the collar. The lace and braiding on the sleeves are the same as those of Field Officers, except that the tracing is plain, without eyes.

Lieutenants have one bar of lace extending to 7 1/2 and the Austrian knot to 9 1/2 inches, from the button of the cuff. In other particulars, the lace and braiding are the same as those of Captains.

Lace

Silver.

Buttons

Silver.

Trousers

Blue cloth with a scarlet welt 1/4 inch wide down each side seam; in summer, blue tartan, with similar stripes. On occasions and at balls, blue cloth with silver lace 1 1/8 inch wide; and with 1/8 inch crimson silk stripe in the centre, down the side seam.

Pantaloons, &c, for Mounted Duties

Blue cloth with stripes as on the trousers; over boots and spurs as described above under "Staff".

Spurs, with Trousers or Pantaloons

For Field Officers, brass; for Adjutants and Musketry Instructors, steel.

Chaco

Blue cloth, 4 inches high in front and 6 1/2 inches at the back, the crown 6 inches long, and 5 1/2 inches across. Silver braid 1/4 inch wide round the bottom, up the sides and back, and in two lines round the top 1/8 inch apart. Silver cap plate with the number in the centre within a garter bearing the royal motto, a laurel wreath round, and a crown above. Silver burnished chain 1/2 inch wide, lined with black velvet silver rose fastenings at the sides and a silver lion's head hook at the back.

Colonels and Lieutenant Colonels have two lines of 1/2 inch lace round the top of the cap, instead of braid. Majors have a line of 1/2 inch lace instead of the upper line of braid.

Tuft

Worsted ball: two third white and one third red, the red at the bottom; silver socket.

Sword

As described above under "Staff": The hilt of gilt metal, with device of Royal cypher and crown, and lined with black patent leather.

Scabbard

For Field Officers, brass; for other officers, steel.

Sword Knot

Silver and crimson strap, with silver acorn.

Undress Sword-Knot.

White buff leather.

Sword Belt

White enamelled leather 1 1/2 inch wide, with slings an inch wide, flap, and silver hook. On state occasions and at balls, silver lace, of the same pattern as on the full dress trousers, lined with crimson morocco leather; slings of similar lace, 3/4 inch wide.

Waist Plate

Round gilt clasp, with the number of the regiment surrounded by a crown, in silver, on the centre piece, and the regimental title in silver on the outer circle.

Sash

Crimson silk net on ordinary occasions. On state occasions and at balls, silver and crimson net 2 1/2 inches wide in 1/2 inch stripes of silver and crimson silk, alternately; silver and crimson runner and tassels.

Blue Patrol Jacket

Blue cloth 28 inches long from the bottom of the collar behind, for an officer 5 feet 9 inches in height, with a proportionate variation for any differences in height, rounded in front and edged with inch black mohair braid all round and up the opening at the sides. On each side to front, four double drop loops of 1/4 inch flat plait, with eyes in the centre of each loop, the top loops reaching to the sleeve seams… Four netted olivets on the right side to fasten through the loops on the left. On each sleeve an Austrian knot flat plait, 7 inches high from the bottom of the cuff. Double flat plait on each back seam with crow's foot at the top and bottom, and two double eyes at equal distances. Pockets fitted with flaps in and out. Hook and eyes in front.

Field Officers wear the badges of their rank embroidered in gold on the collar.

Forage Cap

Blue cloth, with black leather peak and chin strap; black netted button and braided figure on the crown. Band 1 1/4 inch wide, of red cloth. The number of the regiment in silver embroidered figures, 1 1/4 inch high, on the band in front.

Shell Jacket

Scarlet cloth, with collar and pointed, cuffs of blue facings; silver braid edging all round, including the top and bottom of the collar; a loop of silver braid at bottom of collar to fasten across the neck; shoulder cords as on the tunic, a row of silver studs and hooks and eyes down the front, scarlet lining.

Field Officers have a row of braided eyes on the collar below the upper line of braid, and the badges of rank embroidered in gold at each end. Colonels and Lieutenant-Colonels have two chevrons of braid on each sleeve, 3/4 inch apart, the upper forming an Austrian knot extending to 10 inches from the bottom of the cuff, and the lower braid, a crow's foot and eye; a row of braided eyes above and below the chevrons, as on the tunic. Majors have same braiding on the sleeve, omitting the lower row of braided eyes.

Captains have similar braiding, but without the braided eyes; the Austrian knot extends to 9 inches only.

Lieutenants have a single chevron of braid forming an Austrian knot, 8 inches high, and a crow's foot and eye below it.

Mess Waistcoat.

Blue cloth, silver braid edging round the top, down the front, and along the bottom to the side seams; the pocket-edged with braid forming crow's feet and eyes, a row of silver studs and hooks and eyes down the front.

RIFLES

Tunic

Rifle green cloth, edged all round, except the collar, with black square cord, collar and cuffs, scarlet; the collar edged with 1/2 inch black lace, and with the badges of rank, embroidered in black silk at each end; the cuffs pointed and ornamented as described below, according to rank; the skirt rounded off in front, closed behind, and lined with black; on each side of the breast, five loops of black square cord, with netted caps and drops, fastening with black olivets. On each back-seam, a line of the same cord forming three eyes at the tip, passing under a netted cap at the waist, below which it is doubled, and ending in an Austrian knot reaching to the bottom of the skirt; on each shoulder, a square cord loop with a black netted button.

Field officers have a figured braiding below the lace on the collar; and 1 1/2 inch black lace round the top of the cuff, with figured braiding above and below the lace, extending to 11 inches from the bottom of the cuff.

Captains have a row of braided eyes below the lace on the collar, and Austrian knot of black square cord on the sleeve, with a tracing of braided eyes all round it, extending to 8 inches from the bottom of the cuff.

Lieutenants have a tracing of plain braid only below the lace on the collar, and an Austrian knot on the sleeve, with a tracing of plain braid round it, extending to 7 inches only from the bottom of the cuff.

Lace

Black mohair, of special pattern.

Buttons

Bronze, with bugle and crown.

Trousers &c.

Rifle green cloth with 2 inch black lace down the side seam; in summer, rifle green tartan, without stripes. Wellington boots and steel spurs. for mounted officers.

Pantaloons. &c, for Mounted Duties

Rifle green cloth with stripes as on the cloth trousers; overboots, as described above, under "Staff" with steel spurs.

Sword.

As described above, under "Staff" ; steel hilt with device of bugle and crown.

Scabbard

Steel, for all ranks.

Sword Knot

Black leather strap and acorn.

Sword Belt

Black enamelled leather, 1 1/4 inch wide, with slings an inch wide; silver shake clasp and mountings.

Pouch Belt

Black enamelled leather, 2 1/2 inches wide, with silver breast ornament whistle and chain, of regimental patterns.

Pouch

Black enamelled leather, with a silver bugle on the flap.

Gloves

Black leather.

Patrol Jacket

Rifle green cloth, of the size and shape prescribed for the infantry, with collar and pointed cuffs of scarlet as on the tunic. Black mohair inch braid, traced with black Russia, at the top and bottom of collar, showing scarlet facings. The tracing forms a crow's foot on the top of the cuff, a plume at the top of the back of the jacket, and a crow's foot at the bottom. On each side to front, five loops of black round cord, with caps and drops, fastening with black olivets. Field Officers wear the badges of their rank on the collar as on the tunic.

Forage Cap.

Rifle green cloth, with band of 1 1/2 inch black lace, black netted button and braided figure on the crown and black leather chin strap. No peak.

Shell Jacket

Rifle green cloth edged all round with inch black lace collar and pointed cuffs of the regiment. Facings (scarlet), laced and braided according to regimental patterns. Black silk lacings.

Mess Waistcoat.

Rifle green cloth, edged braided round according to regimental patterns.

COLONEL ON THE STAFF

Tunic

Scarlet cloth, with blue cloth collar and cuffs. The collar laced round the top and bottom with 1/2 inch lace. The cuffs round, with 2 bars 1/2 inch lace round the top, showing 1/2 inch of blue cloth between the bars. Scarlet flap on each sleeve…edged with 1/2 inch skirt and 2 1/2 inch wide, on each skirt behind. the back skirts edged with 1/2 inch lace. Eight buttons down the front, three on each flap; two at the waist behind and loop of silver cord, with a small button on each shoulder. The collar, cuffs and flaps, edged with white cloth…and the skirts lined with white. On each side of the collar a crown and star, embroidered in gold.

Lace

Silver, staff pattern.

Buttons

Silver, frosted, with burnished laurel round the edge.

Dress Trousers

Blue cloth, with 1 1/4 inch lace down the side seams.

Spurs

Brass.

Cocked Hat

As described above, under "Staff", with loop of 3/4 inch lace; tassels, netted silver, pearl heal, eight small silver buttons with seven crimson bullions under them.

Plume

White swan feathers, drooping outwards, 8 inches long, with red feathers under them long enough to reach the ends of the white ones, feathered stem 3 inches long.

Sword

As described above, under "Staff", the hilt of gilt metal, with device of word and baton crossed encircled with laurel leaves, surmounted by a crown.

Scabbard

Brass.

Sword Knot

Silver and crimson cord with silver and crimson acorn.

Sword-Belt

Russia Leather, 1 1/4 inches wide with slings an inch wide; 2 stripes of silver embroidery on belt and slings, a silver hook to hook up the sword. Waist Plate, Sliver, rectangular, burnished plate, with a device in silver of the Royal cypher and motto, with the crown above, and an oak branch at each side.

Sash

Silver and crimson net, 2 1/2 inches wide, in 1/2 inch striped of silver and crimson silk alternately; runner of plaited silver and crimson; tassels of silver and crimson fringe, 9 inches long.

Frock

Blue cloth, double breasted. Rolling collar; the front and collar edged with 1/4 inch black mohair lace. An Austrian knot of black Russia braid on each sleeve, reaching to 6 inches form the bottom of the cuff; five loops of black Russia braid on each side of the breast, fastening with black olivets; two olivets at the waist behind. The skirts lined with black.

Waistcoat

Scarlet cloth, without collar edged with silver Russian braid, and fastening with hook and eyes.

Undress Trousers

Blue cloth, with scarlet stripes 1 1/4 inch wide. down the side seams.

Pantaloons, &c., for Mounted Duties

Blue cloth, with scarlet stripes as on trousers. Over-boots as described above, under "Staff", with brass spurs.

Forage Cap

Blue cloth, with silver embroidered peak, and band of 1 3/4 inch lace; silver pearl button and braided figure on the crown.

Undress Sword Knot

Russia leather cord and acorn.

Shell Jacket

Scarlet cloth, edged all round, including the collar, with 1/2 inch lace, forming laurels at the bottom of the back-seams. Blue cloth collar and cuffs; the badges or rank, embroidered in gold at each end of the collar, and a line of silver braid along the bottom, with an eye in the centre; the cuffs round, 3 inches deep, with two bars of 1/2 inch lace round the top, 1/4 inch apart. A silver cord loop, with a silver stud, on each shoulder. A row of silver studs in front, on the left side. Scarlet silk lining; hooks and eyes in front, and a loop of silver braid at the bottom of the collar, to fasten across the neck.

UNATTACHED OFFICERS

Tunic

Scarlet cloth, with blue cloth collar and cuffs; of the same pattern as for officers of Infantry of the Line, except that there is no braiding above the lace on the cuffs, and that the bars of lace, of which there are three for Field Officers. extend as far up the sleeves as the braiding does on the Infantry Officers tunics.

Lace

Silver; two-vellum pattern.

Buttons

Silver, burnished with crown and scalloped edge. All the other articles of uniform as for Officers of Infantry.

STOREKEEPERS

Tunic.

Blue cloth; the skirt rounded off in front, and closed behind. Blue velvet collar and cuffs; the collar laced round the top and bottom with inch Lace, eight buttons down the front and two at the waist behind; a silver square cord loop with a small button on each shoulder, The front collar, and skirt plaits edged with white cloth, 1/4 inch wide; and the skirts lined with black silk.

Lace

Silver.

Buttons

Silver, with "Canada Militia" on it.

Trousers

Blue cloth, with two stripes of white, 1/4 inch wide and 1/4 inch apart down each side seam. On state occasions and at balls, blue cloth with 2 inch lace down the side seam.

Forage Cap

Blue cloth, with silver embroidered horizontal peak, and band of 1/2 inch lace.

Cocked Hat

Of black beaver, or silk, the left side 7 inches high, the right side 6 1/4 inches, each corner 4 3/4 inches long. There will be a black silk cockade on the right side, with a loop and button over it; and at each corner a silver tassel 1 3/4 inch long, exclusive of the head. (At Officer's discretion)

Plume

White swan feathers, drooping outwards 4 inches long. (At officer's discretion)

Sword: Same as Infantry. Scabbard: Same as Infantry. Sword-Knot: Same as Infantry. Sword-Belt: Same as Infantry

Waist-Plate

Round silver clasp, with the royal crest on the centre piece.

Pouch Belt.

Brown untanned leather, 2 1/2 inches wide; silver buckle, tip and slide.

Pouch

Brown untanned leather, to hold writing material.

Great Coat and Cape

The same as officers of the Infantry

Frock

Blue cloth, single breasted; blue velvet rolling collar and pointed cuffs; at each end of the collar the badges of rank embroidered in gold; the cuffs ornamented with black braid of the same pattern as the cuffs of the tunic. The collar, front and back skirts edged with3/4 inch black mohair braid; five loops of the same braid on each side in front, with two olivets on each loop; two olivets at the waist behind encircled with braided crow's foot the skirts lined with black.

MEDICAL OFFICERS

The medical Officers wear the uniforms of their regiments, except in the following particulars:

Cocked Hat.

The same as the Staff.

Plume

Black cock's tail feather, drooping outwards, 4 inches long.

Shoulder Belt and Instrument Case.

Brown untanned leather.

…the foregoing Militia Dress Regulations were submitted to and approved by His Excellency the Governor General to Council on the 10th day of February instant.

(*Canada Gazette* 4 March 1876)

Appendix 7

Dress Regulations for Officers of the Militia of Canada, 1 May 1886

Memo. Officers at present possessing uniforms will be allowed three years from this time for the wearing of them out. Any new uniforms obtained are to be of the new pattern, as below described.

GENERAL INSTRUCTIONS.

Badges of Rank.

The several ranks of officers are designated by silver badges on the shoulder straps of the tunic, mess jackets and stable jackets, and gold for patrol jacket, frock coat, great coat and cloak. Rifles wear bronze badges.

Embroidery or metal optional, as follows :

Colonels—Crown and two stars;
Lieutenant-Colonels—Crown and one star;
Majors—Crown;
Captains—Two stars;
Lieutenants—One star;
And 2nd Lieutenants—No badge.

The badges to be an inch high, if embroidered; 3/4 inch if in metal.
Regimental officers having Brevet Rank wear the badges of their rank in the Militia, and Departmental Officers wear the badges of their honorary or relative rank.

Boots and Spurs.

Mounted officers, except where otherwise specified, wear, when on mounted duties, knee boots with crane-necked spurs fastened with straps and buckles. Officers who are not mounted wear Wellington, or ankle boots. The knee boots must be, as regards shape, in accordance with the sealed pattern, but the height will depend upon the length of the leg and the relative height of the calf. The boot, which is sloped down at the back, should reach in front to about 4 inches from the top of the knee, and at the back just to the top of the calf.

Braid, Buttons, &c.,

Unless otherwise specified, loops, frogs and buttons on the front of tunics, &c., will be at equal distances. Buttons, generally, will be an inch in diameter and 'small» buttons inch. When loops of lace or cord are worn across the breast, the top loops will reach to the sleeve seams, and those at the waist will be 4' inches long.

Cocked Hats.

Of black beaver or silk; and of the following dimensions: The left side 7 inches high, the right side 6 1/4 inches, each corner 4 3/4 inches long. There will be a black silk cockade on the right side, with a loop of 3/4 inch gold lace and button over it ; and at each corner a bullion tassel, 1 3/4 inch long, exclusive of the head ; tassels, netted, gold purl head, eight small gold bullions, with seven crimson bullions under them.

Collars and Cuffs.

The collars of all tunics, frock-coat and jackets, except those with rolling collars, will be rounded at the top in front. The height of the collar of the tunic for cavalry is not to exceed 2 inches.

Depth of Skirts.

The skirts of tunics for officers 5 feet 9 inches high will be-

For Artillery, Engineers and Infantry Regiments, 10 inches. For General Staff, Cavalry and Rifle Regiments, 9 inches. The skirts of frocks will be 17 inches deep for all officers 5 feet 9 inches high. The proportionate variation for each inch of difference in height is about 1/4 inch in the skirts of frocks and 1/8 inch in the skirts of tunics.

Forage Caps.

Round forage caps with peaks, will be 3 inches high, without peaks 2 5/8 inches high.

Gloves.

Gloves, except where otherwise stated, will be of white leather.

Great Coat.

Milled cloth, double-breasted, to reach within a foot of the ground. Stand and fall collar, four and a half inches deep, with a fly to cover the band of the cape when buttoned on. Loose round cuffs, six inches deep. Two pockets with flaps at the waist in front, two openings behind at the side seams, with pointed flaps eleven inches long; and a breast pocket on the left inside. A slit in the left side for hilt of sword to pass through. An opening behind, nineteen inches long, with a fly. Two rows of buttons down the front, six in each row, the top buttons six inches apart, the bottom ones four inches; three buttons on each skirt flap, the centre one to close the pocket; four small buttons at the opening behind; and five flat buttons under the fly at the collar. A cloth back-strap, attached to the top button of the skirt flap, to confine the coat at the waist; two hooks and eyes at the collar. Cape of the same cloth as the coat to button on, and long enough to cover the knuckles; with four small buttons in front, to fasten at the neck with leader strap, runner and buckle. Four cloth tabs with button holes in the lining at the bottom, one *on* either side in front and two *in* the rear, so, as to secure the cape to the bottom buttons of the coat in front and to the top buttons on the flap behind.

In the case of the Mounted Officers, for whom the above pattern of coat is authorised, the following modifications will be made:— The opening of the coat behind will be long enough to reach to the cantle of the saddle, and a gusset will be introduced commencing at the top of the slit and extending down-wards to about 24 inches, with about 19 inches width at the bottom. A tab, with button hole at the bottom of the gusset, to close it when the coat is worn on foot. A small pocket with a flap at the back of the left sleeve. On the inside of each skirt a cloth band with button to secure the skirts over the knees when the coat is worn on mounted duties. The four buttons at the opening behind are omitted. The coat to reach to the ankles when worn on foot.

Great Coats. (Winter Wear.) (Optional.)

Milled cloth (blue for General and Staff Officers' and Officers of Artillery, grey for other officers); double-breasted, with lapels, 4 1/2 inches wide at the top, and 2 1/2 inches at the bottom; the shape as for frock-coat; to reach within a foot of the ground. For mounted officers the coat will reach to the ankles. The body, skirt, and sleeves lined with real or artificial fur, optional; the collar and cuffs on the outside and the lapels on the inside lined with black Astrachan fur for General and staff Officers, Artillery and Rifles, grey Astrachan for other officers. The Astrachan forms an edging 1 inch wide on the inside of the collar and down the lapels, on the outside, to the waist. Four loops of 1/4 inch black flat cord down the front on each side. Each loop forms an eye in the centre at the top and bottom and a crow's foot at the end; a black netted olivet on each loop near the crow's feet, two olivets at the waist behind. Pocket inside the left breast. Pockets in the plaits of the skirts behind. Three hooks and eyes on the collar. Shoulder straps of the same material as the garment; a black netted button at the top. Badges of rank in gold for winter wear. Either description of great coat may be selected provided those worn by the officers of each corps are alike.

Patrol Jackets.

Patrol jackets, as far as measurement is concerned, will be made according to the following description: 28 inches long from the bottom of the collar behind, for an officer 5 feet, 9 inches in height, with a proportionate variation for any difference in height.

Sashes.

Sashes will be worn diagonally over the left shoulder and under the left shoulder cord, and over the sword belt. The ends will be crossed through a runner at the waist, and the sash will be of such length that the ends of the tassels shall just reach the bottom of the skirt of the tunic.

Sabretache.

Staff officers will or may wear, when mounted, Russia leather sabretaches and three slings of special pattern.

Officers of Dragoons and Mounted Officers of Artillery Engineers and Infantry Regiments, wear when on mounted duties, black leather sabretaches of similar pattern, and with slings to match their sword belts.

No device or badge to be worn on the sabretache of Mounted Infantry Officers. Mounted Officers of Rifle Regiments wear a silver bugle and strings.

Shoulder Straps.

Same as Imperial army.

Swords.

Unless otherwise described, will be of the following description: Half-basket hilt, and back piece of steel or gilt metal, black fish skin grip, bound with silver wire when the hilt is of steel, or with gilt wire when the hilt is of gilt metal; slightly curved blade, grooved and spear pointed. Full size blade 35 inches long and 13-inch wide at the shoulder; extreme length, including the hilt, 41 inches; weight, without scabbard, 2 pounds. Second size—blade 33 inches long, and an inch wide at the shoulder ; extreme length, including hilt, 38 1/2 inches; without scabbard, 1 pound 12 ounces.

Tunics and Jackets.

All tunics and jackets will be single-breasted.

Lace and Buttons.

Gold lace, maple leaf pattern, gilt buttons.

Waterproof Cloak.

Length to vary according to height and size of wearer. This cloak may be worn by the Staff, and in camp and quarters off parade by all other officers.

Regimental Staff

The Adjutants wear the uniform of their rank.

The Medical Officers wear the regimental uniform of their relative rank. except the head dress and belts instead of which the following are worn:

Cocked Hat.

As described in these regulations, bound with black silk lace, 2 inches wide ; gold lace loop, tassels of gold crape fringe, with crimson crape fringe underneath; plume of black cock's tail feathers, drooping from a feathered stem 3 inches long.

Sword Belt, Pouch Belt, Pouch.

As for Army Medical Department, Imperial service but with waist plates and devices on pouches of regimental patterns.

Quartermasters wear regimental uniform of their honorary or relative rank, except

Sword Belt, Pouch Belt, Pouch.

Black leather, waist plates and devices on pouches of regimental pattern.

Paymasters.

Wear the regimental uniform of their honorary or relative rank with following exceptions, cocked hat as for Imperial Army Pay Department.

Undress Sword Belt and Pouch Belt.

Black leather with slings of same material. Pouch belt of black leather, 2 inches wide, with pouch of black leather, waist plate and devices on pouch belt of regimental pattern.

Full dress Sword and Pouch Belt.

As for Imperial Army Pay Department, but with waist plates and devices on pouch of regimental pattern.

STAFF

Adjutant-General

Tunic.

Scarlet cloth, edged all around, except the collar, with round-back gold cord. Blue cloth collar and cuffs; the collar ornamented with half-inch lace round the top, gold Russia braid along the bottom, and a rich tracing in double gold braid below the lace; silver, the cuffs pointed, with half-inch lace round the top, and a rich tracing in double gold braid above and below the lace, extending to 9 inches from the bottom of the cuffs. On each side of the breast, four loops of round-back gold cord, with caps and drops, fastening with worked olivets. On each back seam, a line of the same cord forming three eyes at the top, passing under a netted cap at the waist, below which it is doubled, and ending in an Austrian knot, reaching to the bottom of the skirt. The skirt rounded off in front, closed behind, and lined with white. Shoulder straps of worsted gold cord, universal pattern, lined with scarlet; a small button at top; badges of rank in silver.

Lace.

Gold, Staff pattern, maple leaf.

Buttons.

Gilt, frosted, with burnished laurel round the edge

Dress Trousers.

Blue cloth, with 1 3/4 inch lace down the side seams.

Dress Pantaloons.

Blue cloth with gold lace as on trousers.

Spurs

Brass

Cocked Hat.

As described in General Instructions of these Regulations.

Plume.

White swan feathers, drooping outwards, 6 inches long, with red feathers under them long enough to reach the ends of the white ones; feathered stem 3 inches long

Helmet.

Imperial. White.

Sword.

As described in General Instructions with gilt hilt.

Scabbard.

Brass.

Sword Knot.

Gold and crimson cord, with gold and crimson acorn.

Sword Belt.

Russia leather, 1 1/4 inch wide, slings an inch wide; two stripes of gold embroidery on belt and slings; a gold hook to hook up the sword.

Sabretache.

Imperial.

Waist Plate.

Round, gilt clasp, with Royal Cypher and Crown in the centre, and a laurel wreath on the outer circle.

Shoulder Belt.

Gold lace, 1 3/4 inch wide, with crimson Morocco leather lining and edging; gold ornamented buckle, tip and slide.

Telescope Case.

Black patent leather, to hold a binocular field-glass; solid leather flap, reaching to the lower edge of the case, ornamented with Royal Cypher and Crown in gilt metal.

Frock.

Blue cloth, double-breasted. Rolling collar. The front and collar edged with 3/4 inch black mohair braid. An Austrian knot of black Russia braid on each sleeve, reaching to 6 inches from the bottom of the cuff; 5 loops of black Russia braid on each side of the breast, fastening with black olivets; 2 olivets at the waist behind. The skirts lined with black. Badges of rank in gold on shoulder straps, which are of same material as garment.

Patrol Jacket for service in the Field.

Blue cloth, blue cloth stand up collar, height not to exceed two inches, rounded in front 1/2 inch mohair braid at top and bottom; inch mohair braid traced with Russia braid all round, up the slits and along the back seams. The tracing forms an eye at each angle of the braid, except at the top of the slits and back seams, where it forms a crow's foot 1 inch in length, and at the bottom in the centre, where it forms a long crow's foot 1 1/2 inches in length. Five loops of inch mohair braid at equal distances down the front on each side, with two olivets on each loop, the top loops extend to the shoulder seams and the bottom to 4 inches. Blue cloth cuffs, pointed with inch mohair braid traced with black Russia braid forming an Austrian knot above and below the mohair braid. The mohair braid reaches to 7 inches from bottom of cuffs and the Austrian knot at the top to 9 inches. Pockets in front edged at the bottom with inch mohair braid; black silk lining; pocket inside left breast. Shoulder-straps of the same material as the garment, edged with 1/2 inch black mohair braid, except at the base; black netted button at top. Badges of rank in gold.

Waistcoat.

Scarlet cloth, without collar, edged with gold Russia braid, and fastening with hooks and eyes, a pocket on each side.

Undress Trousers.

Blue cloth, with scarlet stripes 1 3/4; inch wide down the side seams.

Pantaloons, etc., for Mounted Duties.

Blue cloth, with scarlet stripes as on the trousers. Mounted officers, except where otherwise specified, wear, when on mounted duties, high boots with crane-necked brass spurs, fastened with straps and buckles.

Forage Cap.

Blue cloth, with gold embroidered drooping peak, and band of 1 3/4 inch gold lace; gold button and braided figure on the crown.

Shell Jacket.

Scarlet cloth, edged all round, including the collar, with inch gold lace, forming barrels at the bottom of the back seams. Blue cloth collar and cuffs: and a line of gold braid along the bottom, with an eye in the centre; the cuffs pointed, with lace and braiding of the same pattern as on the cuffs of the tunic. A row of gold studs in front, on the left side. Scarlet lining; hooks and eyes in front, and a loop of gold braid at the bottom of collar, to fasten across the neck. Shoulder straps and badges of rank as for tunic.

Mess Waistcoat.

Blue cloth; pattern as for infantry. Gold Russia braid edging round the top, down the front, and along the bottom to the side seam; at an interval of 1/2 inch, Russia braid forming small eyes 3/8 of an inch apart. The edges of the pockets to be braided in a similar manner. A row of gold studs and hooks and eyes down the front.

Horse Furniture for Adjutant General

Saddle.

Hunting, with plain stirrups and blue girths.

Saddle Cloth.

Blue, 3 feet long at the bottom and 2 feet deep with gold lace an inch wide, and scarlet cloth beading all round ; the badges of rank embroidered in silver on the hind corners.

Bridle and Breast Plate.

Brown leather, according to sealed pattern ; bent branch bit, with gold bosses, bearing the Royal Cypher within a garter and a Crown above ; blue front and rosettes ; steel chain reins.

Wallets.

Brown leather ; with brown leather covers.

Deputy Adjutant General

The uniform and horse furniture are the same as those of the Adjutant-General, except that the braided figures on the collar and sleeves are smaller, the latter extending to seven inches only from the bottom of the cuff.

Brigade Major

The uniform, &c., are the same as those of a Deputy Adjutant-General, except that on the tunic there is only a tracing of small eyes in single braid on the collar, and only a line of single braid above and below the lace on the cuff of the tunic and shell jacket, the braid on the sleeve extending to 5 inches only from the bottom of the cuff.

Inspector of Artillery.

Uniform, dress and undress, and horse furniture, in every respect as for Deputy Adjutant-General, except that the tunic and jacket are blue with scarlet cloth collar and cuffs, the sword and knot, cloak and cape of regimental pattern, the shoulder straps, lace on tunic and dress trousers, forage cap and belts are of Artillery pattern and width. The full dress Artillery pouch belt and pouch will be worn in both dress and undress uniform. A telescope case with regimental badge may be worn instead of pouch in undress uniform. The peak to forage cap as for Deputy Adjutant-General. Sabretache as for Staff Officers with Artillery pattern slings.

Assistant Inspector of Artillery.

Uniform, &c., are the same as those of a Brigade Major with the above exceptions.

CAVALRY.

Hussars—Full Dress

Tunic.

Blue cloth, edged all round with gold chain gimp, except the collar, which has 3/4 inch lace round the top. On each side of the breast, 6 loops of gold chain gimp, with cap and drops, fastening with gold worked olivets. On each back seam, a double line of the same gimp, forming 3 eyes at the top, passing under a netted cap at the waist, and ending in an Austrian knot reaching to the bottom of the skirt; with a tracing of gold braid all round the gimp. An Austrian knot of gold chain gimp on each sleeve, reaching to 8 inches from the bottom of the cuff. The skirt rounded off in front, closed behind, and lined with black. Shoulder straps of plated gold chain gimp, lined with blue, a small button at the top; badges of rank in silver.

Field Officers have figured braiding below the lace on the collar, and figured braiding on the sleeve, round the Austrian knot, extending to 11 inches from the bottom of the cuff.

Captains have a row of braided eyes on the collar, below the lace ; and a tracing of braided eyes round the knot on the sleeve, 9 inches deep.

Lieutenants have a tracing of plain braid only below the lace on the collar; and round the knot on the sleeve, 8 Inches deep.

Collars.

Buff cloth.

Lace.

Gold, universal.

Trousers, &c.

Blue cloth, with 2 stripes of 3/4 inch lace, 1/4 inch apart, down each side seam; Wellington boots and brass spurs.

Pantaloons, &c., for Mounted Duties.

Blue cloth, with stripes as on trousers; knee boots and steel spurs as described under "Staff."

Plume

White egret feathers, nine inches high from the top of the cap; with a gilt ring, and a gilt, corded, ball socket with four upright leaves.

Cap Chain.

Dead-and bright gilt corded chain ; lined with black Morocco leather.

Cap-Line.

Gold purl cord, with sliders and olive ends ; encircling the cap diagonally three times, passing through a gilt ring under the bag, then round the body and looped on the breast.

Sword.

Half-basket steel hilt, with two fluted bars on the outside; black fish-skin grip, bound with silver wire ; slightly curved blade, 35; inches long and 1; inch wide at the shoulder, grooved and spear-pointed.

Scabbard.

Steel, with a large shoe at the bottom, and trumpet-shaped mouth.

Sword-Snot.

Gold and crimson cord, with gold acorn. Sword-Belt.

Sword -Belt

Gold lace, 1 1/4 inch wide, with sword slings of the same width, Buff Morocco leather lining and edging ; gilt snake fastening.

Sabretache.

Buff cloth face, with lace, 1 3/4 inches wide all round, within 1/4 inch of the edge. Embroidered regimental device in the centre. Buff Morocco leather pocket. The sabretache not to hang below the calf of the leg.

Pouch-Belt.

Gold lace, 1 1/2 inch wide, with 1/2 inch buff silk stripe in the centre. Morocco leather lining and edging of the same colour as the sabretache (butf). Silver ornamented buckle tip and slide.

Pouch.

Black leather, silver flap, and silver ornaments.

Undress

Frock.

Blue cloth, single-breasted. The collar edged with 3/4 inch black braid, and with figuring in narrow braid. A braided figure on each sleeve, extending to 10 inches from the bottom of the cuff: Six loops of inch braid across the breast, with four rows of olivets. The back seams and back skirts trimmed with inch braid, traced round with narrow braid, and with olivets and tassels. The skirts lined with black. Shoulder straps of same material as garment, edged with 1/2 black mohair braid, except at base, badges of rank in gold. Either metal or embroidered.

Trousers.

The same as in full dress, except that the spurs worn with the trousers are steel instead of brass and that the double stripes on the trousers and pantaloons are of white cloth.

Forage Cap.

Blue cloth, with band of 1 3/4 inch gold lace; gold purl button and braided figure on the crown, and a line of gold braid round the crown-seam.

Stable-Jacket

Blue cloth, with olivets and lace, or cord. Shoulder straps and badges of rank as for tunic, Collar,—buff.

Sword-Belt

White leather, with slings an inch wide, gilt mountings ; waist plate as in full dress.

Pouch-Belt

White leather, with brass buckle, tip, and slide.

Mess Waistcoat

According to Regimental pattern.

Sabretache

Black-patent leather. The other articles as in full dress.

Cloak and Cape

Blue cloth, of the same pattern as for Officers of Dragoons, with scarlet lining.

Gloves

White leather.

Horse Furniture

Throat Ornament.

White horse hair, 18 inches long, brass ball and socket.

Valise.

Blue cloth, twenty-seven inches long, hollowed at centre; ends six and a half inches in diameter, with regimental number and initial letters embroidered in gold.

Dragoons wearing Scarlet Tunics.

Tunic.

Scarlet cloth, with collar and cuffs of yellow velvet. The collar ornamented with 1/4 inch lace all round for Field Officers, but round the top only for Captains and Lieutenants. The cuffs pointed and edged with round back gold cord, forming, for Field Officers, a triple Austrian knot traced round with gold Russia braid, and extending to 11 inches from the bottom of the cuffs; for Captains, a double Austrian knot similarly traced, 9 inches deep ; and for Lieutenants, a single Austrian knot, 7 inches deep, eight buttons in front, and two at the waist behind and a twisted gold cord loop, with a small button on each shoulder. A scarlet flap on each skirt behind, with three buttons and edged with round-back gold cord, traced inside and out with gold Russia braid. The front edged with the same material and colour as the facings, and the skirts lined with white. Shoulder straps of plaited flat gold cord, lined with scarlet ; small button at the top. Badges of rank in silver.

Helmet.

Gilt brass, according to sealed pattern.

Plume.

Black and red horse hair.

Gloves.

White leather gauntlets, to be worn only on mounted parades, with tunics. On all dismounted parades, short gloves will be worn.

Dragoons wearing Blue Tunics.

Tunic.

Blue cloth, edged all round, including the top and bottom of the collar, with round-back gold cord. Collar and cuffs, of white cloth; the collar laced within the cord, with 3/4 inch lace all round for Field Officers, but round the top only for Captains and Lieutenants; the cuffs pointed with 1 1/2 inch lace round the top, and figured braiding extending to 11 inches from the bottom of the cuff, for Field Officers; with an Austrian knot of round-back gold cord, and tracing of braid in the form of eyes, 8 inches deep for Captains; and with a similar knot and a tracing of plain braid, 7 1/2 inches deep, for Lieutenants; eight buttons in front and two buttons at the waist behind. A scarlet, white, or blue flap on each skirt behind, with three buttons, and edged with round-back gold cord. Shoulder straps of plaited flat gold cord, lined with blue; a small button at top. Badges of rank in silver.

Trousers blue cloth, with 1 3/4 inch lace down the side seams, for corps wearing scarlet tunics, and two stripes ofd inch lace, 1 inch apart, down each side seam, for corps wearing blue tunics. Wellington boots, and brass spurs.

Pantaloons for mounted duties, blue cloth, with stripes as on trousers ; knee boots and steel spurs, as described in general regulations, except that there is a V cut on top, in front of the knee boots.

Helmet.

According to Regimental pattern.

Plume.

White horse hair with rose at top, rising 6 inches above the top of the helmet, and then falling as far as the bottom.

Sword.

Three-quarter-basket steel hilt, with pierced scroll -work guard; black fish-skin grip, bound with silver wire, slightly curved blade, 36 inches long-grooved and spear-pointed.

Scabbard.

Steel, with a large shoe at the bottom, and a trumpet shaped mouth.

Sword-Knot.

White leather strap, with gold acorn.

Sword-Belt.

Gold lace, 1 1/2 inches wide, with slings an inch wide; Morocco leather lining, and velvet edging of the same colour as the facings.

Waist-Plate.

Gilt rectangular plate, frosted, with burnished rim; in the centre the Royal Cypher and Crown, encircled with oak leaves in silver.

Sabretache.

See general instructions.

Pouch-Belt

Gold lace, 2 1/2 inches wide, with the same lining and edging as the sword-belt; silver buckle, tip, and slide, and breast ornament.

Pouch.

Black leather, with gold embroidered edging round the top; solid silver flap, 7 1/2 inches long and 2 3/4 inches deep, engraved round the edges; in the centre, the Royal Cypher and Crown, in gilt metal.

Gloves.

White leather gauntlets.

Patrol Jacket.

Blue cloth, stand up collar, rounded in front, half inch mohair lace at top and bottom of collar, figured braiding in centre. Inch mohair braid, traced with Russia braid, all round, up the slits, and along the back seams. Five loops of inch mohair braid, at equal distances, down the front on each side with two olivets on each loop, the top loops extend to the shoulder seams and the bottom to 4 inches. The cuffs pointed with inch mohair braid, traced with Russia braid, and figured braiding at the top and bottom. The mohair braid reaches to 5 inches from bottom of cuff, and the figured braiding at the top to 8 inches. Pockets in front edged all round with inch mohair braid. Figured braiding at the top of slits, and at the top of the shoulder seams. Figured braiding in the centre of the back, at the bottom of the collar; and at the bottom of the jacket, figured braiding on the right, left, and between the back seams. Hooks and eyes in front. Black lining. Pocket inside left breast. Shoulder straps of the same material as the garment, edged with half-inch black mohair braid, except at toe base; a black netted button at the top. Badges of rank in gold, on shoulder straps.

Stable Jacket.

Cloth of the colour of tunics, edged all round, including the collar, with inch gold lace of the same pattern as that on the tunic. Collar and cuffs of the colour of facings of tunics ; the cuffs pointed with inch lace round the top. Shoulder straps and badges of rank as for tunic.

Forage Cap.

Blue cloth; with band of 1 3/4 inch gold lace; gold purl button and braided figure on the crown; blank patent leather chin-strap.

The other articles, except the gauntlets, as in full dress.

Cloak and Cape.

Blue cloth, of the same pattern as for rank and file; to reach to the ankles when worn on foot. Collar of cape of white cloth. Gilt buttons. Badges of rank in gold on shoulder straps, which are of same material as garment.

Regimental Staff Officers

The Adjutant and Riding Master wear the same uniform as the other officers of their rank.

The Quartermaster wears the same uniform as the other officers of his honorary or relative rank, with the following exceptions:

Cocked-Hat. (Instead of the Helmet).

As described in general instructions, with loop of four-fold gold chain gimp ; gold bullion tassels.

Plume.

White cock's feathers, drooping outwards, 5 inches long.

Horse Furniture

Saddle.

Hussar pattern, with brass head and cantle; blue girths; brown leather wallets.

Bridle.

Brown leather, with brass whole buckles; bent branch bit, with pads and plain bent bar; link-and tee bridoon; plain leather head-collar; bit-head and bridoon rein sewn on; bosses on bit and ear bosses.

Chain.

Steel, with swivel, rings, and spring hook.

Breastplate and Crupper.

Brown leather, with brass whole buckles, and bosses as on bit.

Dress Lambskin.

Black Ukraine lambskin, 3 feet 6 inches long and 13 inches deep, edged with white cloth, and lined with moleskin.

Undress Lambskin.

Black Ukraine lambskin, with black leather seat and large flap to open for wallets, edged and lined as the dress lambskin.

FIELD AND GARRISON ARTILLERY.

Full Dress.

Tunic.

Blue cloth, with scarlet cloth collar. The collar and sleeves laced and braided according to rank, as mentioned below. The skirt rounded in front, closed behind, with a plait at each side, and lined with black. Buttons down the front, 2 1/4 inches apart, and two at the waist behind. Scarlet cloth edging all round, except the collar, and up the skirt-plaits.

Field Officers have 5/8 inch lace all round the collar, within the cord; and a chevron of 1 1/2 inch lace on each cuff, with figured braiding above and below the lace, extending to 11 inches from the bottom of the cuff. Shoulder straps of plaited gold wire, and lined with blue. A small button of regimental pat-tern at top. Badges of rank embroidered in silver.

Captains and Lieutenants have lace round the top only of the collar; and an Austrian knot of gold cord on each sleeve, 7 inches deep, traced round with gold braid 8 inches deep and figured for Captains; 7 1/2 inches deep and plain for Lieutenants.

Sword-Belt.

Gold lace 1 1/2 inch wide, lined with blue Morocco; gilt S hook fastening, with two oval gilt plates bearing the Royal crest. Sword slings (and tache slings for mounted officers) fastened to a flat steel bar covered with blue Morocco leather, and attached to the inside of the belt by four flat steel hooks.

Sabretache. (Optional).

Blue Morocco leather, faced with blue cloth 1 1/2 inch lace round the face, 1/4 inch from the edge. An embroidered device within the lace of the Royal arms above, and a gun below.

Lace.

Gold, of regimental pattern.

Buttons.

Gilt, burnished, with a gun and Crown.

Trousers, &c.

Blue cloth, with 1 3/4 inch lace down the side seam; Wellington boots and brass spurs.

Fur Cap

Wedge shape. Persian lamb with scarlet cloth bag on the right side, falling to within an inch of the button. Gold embroidered grenade 2 1/2 inches long in front.

Helmet.

The same pattern as for Royal Artillery, abroad.

Sword.

Half-basket steel hilt, with two fluted bars on the outside; black fish-skin grip, bound with silver wire; slightly curved blade, 35 1/2 inches long, and 1 1/4 inch wide, grooved and spear-pointed.

Scabbard.

Steel, with a large shoe at the bottom and a trumpet-shaped mouth.

Sword-Knot

Gold cord, with a gold acorn.

Pouch.

Blue Morocco Leather collapsing pouch, with two pockets, the leaf 5 5/8 inches ling and 2 3/4 inches deep, covered with blue cloth and edged with 3/4 inch lace. An embroidered device within the lace similar to that on the sabretache.

Pouch Belt

Gold lace, 2 inches wide, lined with blue Morocco leather; gilt ornamented buckle and slide, and a grenade, encircled with a wreath, at the end.

Undress

Mess Waistcoat.

Scarlet cloth with collar edged all round with half-inch gold lace, regimental pattern, including collar—pocket edged with gold Russia tracing braid, crow's feet with figure eight at each end, with crow's feet in centre, to fasten with hooks and eyes, small gilt studs up front.

Patrol Jacket

Blue cloth, rounded in front, and edged with inch black mohair braid all round and up the openings at the sides; 5 loops of flat plait on each side in front, fastening with netted olivets, and with crow's feet and olivets at the end. Stand-and-flat collar. The sleeves ornamented with flat plait, forming crow's feet, 6 inches form the bottom of the cuffs. Double flat plait on each back seam with crow's foot at top and bottom, and 2 eyes et equal distances. Pockets edged with flat plait, forming crow's feet and eyes. Shoulder straps, of the same material as the garment, edged with half inch black mohair braid, except at the base; black netted button at the top. Badges of rank in gilt metal.

Patrol jacket for Active Service, Camps and duties in Barracks

Fine blue serge; welted seams; stand up collar, square in front; fastened with one hook and eye, a grenade 2 1/4 inches long, in gold embroidery at each end. Shoulder straps of the same material as the garment, fastened at the top with a small black netted button, 1/2 inch in diameter, badges of rank embroidered in gold; five gilt ball-buttons down the front; a slit on each side, sleeves ornamented with flat plait, forming crow's feet, 6 inches form bottom of the cuffs; two inside breast pockets, and watch pocket.

Trousers

Blue cloth, with scarlet stripes 1 3/4 inches wide down the sides seams; Wellington boots and steel spurs.

Pantaloons, &c., for Mounted Duties.

Blue cloth, with stripes as on the trousers; overboots and steel spurs as described under "Staff».

Forage Cap.

Blue cloth, with band of 1 5/8 inch gold lace, gold button and braided figure, of special pattern, on the crown. The cap to be 2 5/8 inches high.

Forage Cap for Active Service, Camps, and duties in Barracks.

Blue cloth folding cap, 4 1/2 inches high, with blue side flaps, 4 inches deep, to turn down when required. Gold French braid welts on cap and flaps, and at front and back seams, a grenade embroidered in gold in front.

Stable Jacket.

Blue cloth, with scarlet collar, and pointed blue cuffs, laced all round with 3/4 inch gold lace, regimental pattern, to form bull's eye at the bottom of each back seam- small gold tracing on collar seam to fasten with hooks snd eyes, small gilt studs up the front, sdarlet lining shoulder straps and badges as for tunic.

Field Officers to wear a flat chevron of inch lace, extending to six inches form the bottom of the cuff, with braided eyes above and below the lace, the bottom of the braiding to reach just over the top of the scarlet cuff.

Captains and Lieutenants have an Austrian knot of gold cord on each sleeve, 6 inches deep, traced round with gold braid, 7 inches deep, and figured for Captains; 6 1/2 inches deep, and plain, for Lieutenants.

Sword-Knot

White buff leather, 1/2 inch wide, white runner and gold acorn.

Sabretache (For Mounted Officers only.)

Black patent leather, with regimental badge, in gilt metal.

Pouch-Belt

White patent leather, 2 inches wide.

Pouch

Black patent leather collapsing pouch, with two pockets 5 3/4 inches long, 2 3/5 inches deep, A gun in gilt metal, on the leaf or telescope case with gun ornament.

Cloak and Cape.

Blue cloth, with sleeves, stand-and-fall collar, with three black hooks and eyes in fornt, and three small flat buttons at the bottom to fasten the cape. Round loose cuffs, 6 inches deep. A pocket in each side seam outside, and one in the left breast, inside; four buttons down the front. A cloth back strap, to fasten with a large flat button at the top of each pocket; a similar button in front on the right to hold the end of the back strap when it is not buttoned across behind. White shalloon lining. Shoulder straps of same material as garment; a small button of regimental pattern on top. Badges of rank in gilt metal.

Blue cloth cape, 32 inches deep, lined with white shalloon. A cloth band round the top, to fasten with a strap and buckle; and a fly inside the band, with three button holes, for attaching cape to cloak; three buttons down the front.

Horse Furniture

Black sheepskin and wallet (leather).

Valise

Blue cloth, as described for Cavalry regiments.

Winter coat for Permantnt Corps.

Same pattern as Patrol jacket, edged all round with black Persian lamb, and on cuffs below the tracing plait, ornaments of rank in shoulder straps, in gold metal.

Winter Cap.

Wedge shape, Persian lamb.

Winter Gloves.

Gauntlet shape, Persian lamb.

ENGINEERS

Regimental Officers

Tunic

Scarlet cloth with collar and cuffs of Garter blue velvet. The collar edged all round with round-back gold cord; 2/4 inch lace all round within the cord for Field Officers, round the top only for Captains and Lieutenants; at each end a grenade embroidered in silver.The cuffs pointed and ornamented as described below, according to rank; nine buttons in front and two at the waist behind. Shoulder knots of treble twisted round back gold cord, lined with scarlet. A small button of regimental pattern at the top. Badge of rank in silver. The skirt rounded off in front, closed behind, with a plait at each side, and lined with white. The front, skirt, and plaits edged with Garter blue velvet.

Field Officers have 1 1/2 inch lace round the top of the cuff; and figured braiding above and below the lace, extending to 11 inches from the bottom of the cuff.

Captains have an Austrian knot of round-back gold cord on each sleeve, traced all round with braided eyes, the braid extending to 8 inches from the bottom of the cuff.

Lieutenants have a similar knot, but without the figured braiding.

Lace and Buttons.

Gold, of regimental pattern.

Dress Trousers.

Blue cloth, with 1 3/4 inch gold lace down the side seams; Wellington boots and brass spurs.

Helmet

White as for Artillery with spike except badge, which is to be wreath of maple leaves enclosing a grenade.

Sword,

As described in General Instructions of these Regulations, with hilt of gilt metal, pierced and engraved according to special pattern.

Scabbard.

For Field Officers, brass; for other Officers, steel.

Sword Knot.

Gold cord, with gold acorn; for regimental duty, bridle leather with bridled leather runners and acorn,

Sword Belt.

Russia leather, 1 1/2 inch whide, with slings an inch wide, the front sling lined with Russia leather; two stripes of gold embroidered on belt and slings. Gilt burnished plate, with regimental device in silver.

Shoulder Belt.

Russia leather, 2 inches wide, with three stripes of gold embroidery, the centre one wavy, the others straight; gilt engraved buckle, tip and slide.

Telescope Case.

Black patent leather, to hold a binocular field glass; solid leather flap, with gilt regimental badge.

Frock, for Regimental Field Officers.

Blue cloth, single breasted, with rolling collar; ornaments on sleeve as on the tunic, but in black mohair braid, traced; eight loops of 3/4 inch black braid down the front, with barrel buttons placed according to regimental pattern. The front edges, collar, back and sleeve seams, and back skirts trimmed with 7/8 inch black braid, traced. Hooks and eyes in front. The skirt lined with black. Shoulder straps of same material as garment, edged with 1/2 inch Black mohair braid, except at base. Badges of rank in gold.

Patrol Jacket.

Scarlet cloth with Garter blue facings on collar and cuffs. To be similar in shape as for Infantry, with dark blue piping all round, except the collar, the same as tunic. Shoulder straps of same material as garment, with small regimental button at the top. Badges of rank in gold.

Undress Trousers, &c.

Blue cloth, with scarlet stripes, 2 inches wide, down the side seams; brass spurs for Field Officers, steel spurs for other mounted officers.

Pantaloons, &c., for Mounted Duties.

Blue cloth, with scarlet stripes, as on the trousers; knee boots and brass or steel spurs, as described under “Staff.»

Forage Cap. -

Blue cloth, with band of 1 3/4 inch gold lace, a blue cloth welt round the crown, and a gold netted button in the centre. Black patent leather drooping peak and chin-strap. The peak ornamented with 1/2 inch full and gold embroidery.

Shell Jacket.

Scarlet cloth, with Garter blue velvet collar and cuffs. Gold Russia braid,1/8 inch wide, all round the jacket and along the bottom of the collar, with small eyes at the ends of the collar and at the bottom of the front, and a crow's foot at the centre of collar seam and of waist. Hooks and eyes and gilt studs down in front; and a twisted gold cord loop, with a small button on each shoulder. Pointed cuffs, 5 inches deep; with inch lace and a tracing of braid round the top, for field officers; with a double line of small braided eyes for Captains; and with plain edging for Lieutenants, scarlet silk serge lining, Shoulder knots, with badges of rank as for tunic

Waistcoat.

Scarlet cloth, with hooks and eyes and gilt studs down the front, and edging of gold braid all round and on collar seam. Pockets edged with gold braid forming crow's feet at end and centre.

Great Coat and Cape.

Blue cloth, of the pattern described in General Instructions of these Regulations, lined with scarlet shalloon, the collar lined with Garter blue velvet.

Horse Furniture.

Saddle.—Hunting, with plain stirrups and white girths. Bridle and wallets as for General Staff, but with Regimental bosses.

INFANTRY

Governor General's Foot Guards.

Dress Regulations for Officers.

The Badge and Device of the Regiment which is to be worn on the appointments of Officers, non-commissioned Officers and Men shall be a six-pointed star, with the initial letters of the following Provinces upon the points of the star, namely: O for Ontario, Q for Quebec, M for Manitoba, N.S. for Nova Scotia, B.C., for British Columbia, and N.B. for New Brunswick, surrounded by a garter bearing thereon the following motto "Civitas et Princeps cura nostra" surmounted by a Royal Crown, and having underneath a scroll bearing the words "Governor General's Foot Guards."

Tunic.

Scarlet cloth; blue cloth collar and cuffs; the collar embroidered in front and round the top ; at each end, a star embroidered in silver 1 1/16 inch long, by 15/16 wide, six pointed. Oval medallion in centre of star, with a blue cross (St. George) in the oval, cross 9/16 by 6/16; the cuffs round, 2 3/4 inches deep, embroidered round the top. Blue flaps on each sleeve, 5 1/2 inches long and 2 1/2inches wide; scarlet flaps at the skirt plaits, reaching within a half inch of the bottom of the skirt; ten buttons in front, two and two, and four bars of embroidery, two and two, on each skirt and sleeve flap, 2 buttons at the waist behind about three inches apart. Blue cloth shoulder straps of regimental pattern, embroidered with two rows of purl embroidery, except at the base, small button at top. The front, collar, cult's, and flaps edged with white cloth 1/4 inch wide ; and the skirts lined with white. Badges of rank in silver.

The Field Officers and Captains are distinguished by embroidery round the bottom of the collar and round the skirt and sleeve flaps, and by a second bar of embroidery round the cuff.

Embroidery.

Gold, of special pattern ; that round the collar, cults and flaps to be a half inch wide.

Lace.

Gold, of regimental pattern.

Buttons.

Gilt, of regimental pattern.

Trousers.

At levees, drawing rooms and in the evening, blue cloth, with 1 1/2 inch gold lace down the side seams ; at mess and on other occasions blue cloth, with scarlet stripes 1 1/2 inch wide.

Pantaloons (for mounted duties.)

Blue cloth, with scarlet stripes 1 1/2, inch wide.

Cap.

Black bearskin, 8 inches high, fastened under the chin by a plain gilt taper chain.

Plume.

Scarlet goat's hair, six inches long, on the left side.

Winter Cap.

Grey Astrachan fur of regimental pattern. Bag of blue cloth on left side, with tracing of gold braid 1/4 inch from edge, and fastened at the lower end by regimental button.

Winter Gloves.

Gauntlets, of grey Astrachan fur.

Sword

As described in General Instructions of Militia Regulations; steel hilt with regimental device pierced and chased in the guard; black fish-skin grip. bound with gilt.

Scabbard

Steel.

Sword Knot.

Gold cord, with gold acorn, in full dress; on other occasions white cord with gold acorn.

Sword Belt.

On state occasions, gold lace, lined with crimson Morocco leather, 1 1/2 inch wide, with slings an inch wide; for ordinary use, white patent leather of the same dimensions.

Waist Plate.

Round, gilt clasps, with regimental badge on centre-piece.

Sash.

On state occasions, blue and gold net; at other times crimson silk. To be worn only with the tunic (or summer patrol jacket.)

Spurs

Brass.

Sabretache.

Field and Staff Officers will wear, when on mounted duties, black leather sabretaches, of regimental pattern; with slings of an inch wide to match their sword belt. Metal ornaments of regimental pattern to be worn on the flaps.

Patrol Jacket.

Blue cloth, braided according to regimental pat-tern. Shoulder straps of same material as Jacket, edged with 1/2 inch black mohair braid, except at base; black netted button at top. Badges of rank embroidered in gold.

Summer Patrol Jacket.

White Duck, of regimental pattern. Shoulder straps of same material as Jacket. Gilt metal rank badges.

Forage Cap.

Blue cloth, of special pattern, gold cord round top, with drooping peak ornamented with 1/4 inch gold embroidery. Black lace band 1 1/2 inch wide, with embroidered regimental badge of six pointed star in front. Note —The use of the Field Service Cap (Guards' Pattern) with Regimental Badge, upon occasions, other than those of drill, is optional.

Shell Jacket.

Scarlet cloth edged all round, including the collar with gold braid, with crow's foot in centre of back. Blue cloth collar and cuffs. The collar rolled, with six pointed stars similar to that on collar of tunic on each end, 5 inches from the shoulder seam— the cults pointed. Field officers have on each sleeve three rows of gold braid, Captains two rows, and Subalterns one row, terminating in Austrian knot over point of cuff.

Mess Waistcoat.

Blue cloth, gold braid edging round the top, down the front, and along bottom to side seams--edges of pockets similarly braided—crow's foot at each end, and in centre above and below, row of gilt studs and hooks and eyes down the front.

Boots.

For mounted duties, Riding boots according to pattern, for others, Wellington or ankle boots. During the winter season, knee boots of regulation pattern.

Great Coat.

Grey cloth, of Regimental pattern, with gilt Regimental buttons. Shoulder straps of same material as the garment ; a small button of regimental pattern at the top. Badges of rank in gold. During winter season, a square cut detachable collar of grey Astrachan fur is worn with the overcoat. The sword belt is worn outside the coat.

Regimental Staff

The Adjutant wears the uniform of his rank. The Medical Officers wear the regimental uniform of their relative rank except the bearskin cap, sash and sword belt, instead of which- the following are worn :

Cocked Hat, As described in the General Instructions of Militia; Regulations, button of regimental pattern, tassels of gold crape fringe, with crimson crape fringe underneath ; plumes of black cock's tail feathers, drooping from a feathered stem, 3 inches long.

Sword Belt, Pouch Belt, Dress Pouch, Field Pouch.

Same as for the Imperial Army Medical Department, but with waist plates and devices on pouches of regimental pattern.

The Quarter Master wears the regimental uniform of his honorary or relative rank with the following exceptions:

Cocked Hat, as for the Medical Officers, button of regimental pattern, but with upright red feather 5 inches long.

Sword Belt, black leather, with slings of same material.

Sash, crimson silk net. At balls, levees, and on state occasions, he will wear the gold lace belt, and blue and gold net sash.

The Paymaster wears, the regimental uniform of his honorary or relative rank with the following exceptions:

Cocked Hat as for the Medical Officers, but with loop of1/2 inch lace and regimental button over it: tassels, flat gold,—worked head, 6 gold bullions with 5 gold bullions under them. Plume, mushroom shaped, of white cock's tails feathers 4 inches long, with yellow feathers under them long enough to reach 2 inches below the white.

Undress Sword Belt and Pouch Belt.

Black leather with slings of same material. Pouch Belt of black leather, 2 inches wide, with pouch of black patent leather. Waist plate and devices on pouch belt of regimental pattern.

Full Dress Sword and Pouch Belt.

As for Imperial Army Pay Department, but with waist plates and devices on pouch belt and pouch, of regimental pattern.

Horse Furniture

Saddle.

Hunting, with plain stirrups and blue girths.

Holsters.

Brown leather, with brown leather covers.

Saddle Cloth.

Blue cloth, edged with gold lace an inch wide; three feet long and two feet deep. The field officers are distinguished by a second stripe of lace, and the badges of rank, embroidered in silver, at each hind corner.

Bridle.

Brown leather, cavalry pattern, with gilt bosses bearing regimental device, brown leather breast-plate, and steel chains reins.

INFANTRY [OF THE LINE]

Tunic.

Scarlet cloth, with blue cloth collar and cuffs. The collar ornamented with 5/8 inch lace along the top, and Russia braid at the bottom. The cuffs pointed with 1/2 inch, lace round the top, and a tracing in gold Russia braid, quarter inch above and below the lace, the lower braid having a crow's foot and eye, and the upper an Austrian knot at the top, eight buttons in front, and two at the waist behind; and a gold square cord loop, with a small button on each shoulder. The skirt closed behind, with a plait at each side, and lined with white. The front, collar, and skirt plaits edged with white cloth quarter inch wide. A hook and eye inside the bottom of the collar. Shoulder straps of twisted round gold cord, universal pattern, lined with scarlet. A small button of regimental pattern at the top. Badges of rank in silver.

Field Officers have a row of braided eyes below the lace on the collar; two bars of lace along the top of the cuff, showing 1/4 inch of the facings between the bars ; and the braiding on the sleeve is in the form of eyes, above and below the lace for lieutenant-Colonels, and above the lace only for Majors. The lace on the sleeve extends to 8, and the Austrian knot to 10 inches, from the bottom of the cuff.

Captains have no braided eyes on the collar. The lace and braiding on the sleeves are the same as those of Field Officers, except that the tracing is plain, without eyes.

Lieutenants have one bar of lace extending to 7 1/2 and the Austrian knot to 9 1/2 inches, from the bottom of the cuff. In other particulars, the lace and braiding are the same as those of Captains.

Lace.

Gold, maple leaf pattern.

Buttons.

Gilt.

Trousers.

Blue cloth with a scarlet welt 1/4 inch wide down each side seam ; in summer, blue tartan, with similar stripes. On state occasions and at balls, blue cloth, with gold lace 1/8 inch wide; and with 7/8 inch crimson silk stripe in the centre, down the side seam.

Pantaloons, &c., for Mounted Duties.

Blue cloth with stripes as on the trousers ; knee-boots and spurs as described under " Staff."

Spurs, with Trousers, or Pantaloons.

For Field Officers, brass for Adjutants, steel.

Helmet.

White, gilt chain, and small gilt badge: wreath of maple leaves, surmounted by Crown, enclosing numeral. A light ventilator on top in place of spike. Pattern similar in other respects as to material as that now used in Her Majesty's regular army. The peak of back and the front somewhat shorter.

Plate.

In gilt metal, wreath of maple leaves, surmounted by a Crown, enclosing a numeral.

Sword.

As described in General Instructions of these Regulations; the hilt of gilt metal, with device of Royal Cypher and Crown, and lined with black patent leather.

Scabbard.

For Field Officers, brass; for other officers, steel.

Sword Knot.

Gold and crimson strap, with gold acorn.

Undress Sword Knot.

White buff leather.

Sword-Belt.

White enamelled leather, 1 1/2 inch wide, with slings an inch wide, flap, and gilt book. On state occasions and at balls, gold lace, of the same pattern as on the full dress trousers, lined with crimson Morocco leather; slings of similar lace 3/4 inch wide.

Waist Plate.

Round gilt clasp, on which is a wreath of maple leaves surmounted by a Crown, in silver. If a Battalion the number also in silver, in the centre.

Cross or Shoulder Belt.

White enamelled leather, 2 1/2 inches wide, with a plain gilt buckle tip and slide.

Field Glass Pouch.

Black patent leather, ornamented with Royal Cypher and Crown, in gilt metal.

Sash.

On state occasions and at balls, gold and crimson net, 2 1/2 inches wide, in 1/2 inch stripes of gold and crimson silk, alternately; gold and crimson runner and tassels.

Blue Patrol Jacket.

Blue cloth of blue serge, 28 inches long from the bottom of the collar behind, for an officer 5 feet 9 inches in height with a proportional variation for any difference in height, rounded in front and edged with inch black mohair braid all round and up the opening sat the side. On each side in front, 4 double drop loops of 1/4 inch flat plait, with eyes in the centre of each loop; the top loops extend to the shoulder scams and the bottom to 4 inches; 4 netted olivets on the right side to fasten through the loops on the left. On each sleeve an Austrian knot of flat plait

7 inches high from the bottom of the cuff. Double flat plait at each back seam with crow's foot at top and bottom and 2 double eyes at equal distances. Pockets fitted with flaps in and out. Hooks and eyes in front. Shoulder straps of the same material as the garment, edged with inch black mohair braid, except at the base, black netted button at the top. Badges of rank in gold.

Scarlet Patrol Jacket.

Scarlet cloth, 28 inches long from the bottom of the collar behind, for an officer 5 feet 9 inches in height, with a proportionate variation for any difference in height, rounded in front and edged with white cloth piping all round and up the openings at the sides. Five gilt buttons; blue cloth collar and cuffs, white piping same as tunics. Collar and cuffs trimmed, with one row of gold tracing braid at top of each.

Field Officers to wear an additional row of 1/4 inch braid: For Major, one; For Lieutenant-Colonel, two;

Two outside breast pockets.

The badges of rank embroidered in gold on the shoulder straps which are of same material as garment.

This jacket may be used in place of the tunic on all occasions except those of state or ceremony, and in place of the blue patrol jacket, provided uniformity in Corps is observed.

Forage Cap.

Blue cloth, straight up 3 inches high with black leather drooping peak. The peak ornamented with 1/2 inch full gold embroidery; black netted button and braided figure on the crown. Band, 1 3/4 inch wide, of black lace, maple leaf pattern. (In corps designated "Royal" the band to be of scarlet cloth). Black netted button and braided figure on the crown. Regimental number in gold embroidered figures, 1 1/4 inch high, on the band in front, chin strap, black leather. Field Officers will have a gold braid welt round the top of the cap.

In Corps in which the Glengarry cap is worn by the men, officers will wear the same.

Shell Jacket.

Scarlet cloth, with collar and pointed cuffs of blue cloth; gold braid edging all round, including the top and bottom of the collar, a loop of gold braid at bottom of collar to fasten across the neck; shoulder cord, as on the tunic, a row of gold studs and hooks and eyes down the front, scarlet lining. Shoulder straps and badges of rank as for tunic.

Field Officers have a row of braided eyes on the collar below the upper line of braid. Lieutenant-Colonels have two chevrons of braid on each sleeve, 3/4 inch apart, the upper forming an Austrian knot extending to 10 inches from the bottom of the cuff, and the lower braid, a crow's feet and eye; a row of braided eyes above and below the chevrons, as on the tunic. Majors have the same braiding on the sleeve, omitting the lower row of braided eyes

Captains have similar braiding, but without the braided eyes; the Austrian knot extends to 9 inches only.

Lieutenants have a single chevron of braid forming an Austrian knot, 8inches high, and a crow's foot and eye below it.

Mess Waistcoat.

Blue cloth, gold braid edging round the top, down the front, and along the bottom to the side seams; the pockets edged with braid forming crow's feet and eves, a row of gilt studs and hooks and eyes down the front.

Whistle.

See General Instructions. White metal; pattern as for sergeants; to be carried in a pocket in the officers dress.

Great Coat and Cape.

Grey cloth, as described in General Instructions of these Regulations.

Winter Cap.

Wedge shape, of black fur.

Regimental Staff Officers

Adjutants wear the uniform of their rank. The other regimental staff officers wear the uniform of their relative rank with the following exception: Black waist-belts and cross-belts, instead of white. Sashes are not worn.

At balls or levees, Paymasters of Infantry are authorised to wear gold lace sword-belts similar to those worn by other officers of the regiment.

Horse Furniture

Saddle.

Hunting, with plain stirrups and blue girths.

Bridle and Breast Plate.

Brown leather with gilt bosses bearing the maple leaf, and the Crown above, and " Infantry Mounted Officers" round. Blue front and rosettes, and steel chain reins.

Wallets.

Brown leather, with brown leather covers.

RIFLES.

Tunic.

Rifle green cloth, edged all round, except the collar, with black square cord, collar and cuffs, scarlet cloth ; the collar edged with 1/2 inch black braid, the cuffs pointed and ornamented as described below, according to rank; the skirts rounded off in front, closed behind, and lined with black; on each side of the breast, five loops of black square cord, with netted cap and drops, fastening with black olivets. On each back-seam, a line of the same cord forming three eyes at the top, passing under a netted cap at the waist, below which it is doubled, and ending in an Austrian knot reaching to the bottom of the skirt. Shoulder straps of black chain gimp, with small button of Regimental pattern at top. Badges of rank in bronze.

Field Officers have figured braiding below the lace on the collar; and 1 1/2 inch black lace round the top of the cuff; with figured braiding above and below the lace, extending to 11 inches from the bottom of the cuff.

Captains have a row of braided eyes below the lace on the collar, and an Austrian knot, of black square cord on the sleeve, with a tracing of braided eyes all round it, extending to 8 inches from the bottom of the cuff.

Lieutenants have a tracing of plain braid only below the lace on the collar, and an Austrian knot on the sleeve, with a tracing of plain braid round it, extending to 7 inches only from the bottom of the cuff.

Braid.

Black mohair, of special pattern.

Buttons.

Bronze, with bugle and crown.

Trousers, &c.

Rifle green cloth, with 2 inch black braid down the side seams. Wellington boots and stool spurs for mounted officers.

Pantaloons, ,&c, for Mounted Dutics.

Rifle green cloth with stripes as on the cloth trousers: knee-boots, as described under "Staff," with steel spurs.

Helmet.

White, same form and material as for Infantry ; bronze ornaments; chain chin strap Plate bronze, same pattern as for infantry.

Winter Cap.

Wedge shape, of black fur.

Sword.

As described in General Instructions, steel hilt, with device of bugle and Crown.

Scabbard.

Steel, for all ranks.

Sword Knot.

Black leather strap and acorn.

Sword Belt.

Goat skin leather, 1 1/4 inch wide, with slings an inch wide ; silver snake clasp and mountings.

Pouch Belt.

Goat skin leather, 2 1/2 inches wide, with silver breast ornament whistle and chain, of regimental patterns.

Pouch.

Goat skin leather, with a silver bugle on the flap. Shape for field glass.

Sabretache.

As laid down in General Instructions, device on flap: a bugle with strings in silver.

Gloves.

Black leather.

Patrol Jacket.

Rifle green cloth, of size and shape prescribed for Infantry, with collar and pointed cuffs of the same facings as on tunic. Black mohair inch braid, traced with black Russia braid, down the front, on the top of the cuffs, and along the back seams. Half inch mohair braid traced with black Russia braid at the top and bottom of collar, showing scarlet facings. The tracing forms a crow's foot and eye on the top of the cuff. Two outside pockets, trimmed with tracing braid. Shoulder straps as for tunic, badges of rank in bronze.

Serge Fatigue Jacket. (Optional.)

Of Regimental pattern.

Forage Cap.

Rifle green cloth, with band of 1 1/2 inch black mohair braid, black netted button and braided figure on the crown and black leather chin strap. No peak.

Forage cap for active service and peace manoeuvres. Green cloth, special pattern, ornamented in front with a silver bugle placed on a scarlet cord boss.

Shell Jacket.

Rifle green cloth, upright scarlet collar and scarlet cuffs. Black mohair braid, traced 1 1/4 inch wide, all round the body, forming barrels (or dummies) at bottom of back seams. Back seams trimmed with a double row of 1/4 inch mohair braid, forming crows-foot at top and finishing over the barrels (or dummies) at the bottom, pockets trimmed with 1/4 inch mohair . braid, forming a crow's foot at each end and in centre. Five waved loops of square cord in front, with four rows of knitted olivets, two olivets on each loop. Pointed culls of 1 inch mohair braid, with tracing of black Russia braid forming a row of small eyes on the outside and inside of cuffs, and extending 6 1/2 inches from the bottom of each cuff. Collar—Mohair braid, 1/2 inch wide all around; trimmed through centre with plumes, and row of small eyes along top edge; a loop at bottom of collar to fasten across the neck. Shoulder straps and badge of rank as for tunic.

Mess Waistcoat.

Rifle green cloth, single breasted, no collar, open halfway down. Hooks and eyes. Half inch mohair braid on edges, with a 1/4 inch braid down the front, one inch from the edge. Scarlet cloth between the two braids, with row of eyes of black Russia braid down front edge on the scarlet cloth. Pockets trimmed with 1/4 inch mohair braid, forming a crow's foot at each end, edged all round with scarlet cloth.

Great Coat and Cape.

Grey cloth, as described in General Instructions of these Regulations. Bronze buttons.

Regimental Staff Officers

Uniform as for the other Officers of their rank.

Horse Furniture

As prescribed for Infantry of the Line, with the following exceptions:

Girths.

Green.

Wallet covers (with undress).

Black goat skin leather, with silver studs.

Bridle, Breast Plate, &c.

Black leather. the head stall lined with scarlet cloth with scalloped edges, with silver whole buckles, and silver bit bosses green front and rosettes. Black and scarlet horse-hair throat ornament 18 inches long, with silver ball socket.

By Command,
Walker Powell, Colonel, Adjutant General of Militia, Canada
Headquarters, Ottawa, 1st May 1886

Appendix 8

GENERAL ORDERS, HQ, Ottawa, 24 March 1892

No. 1

FIELD SERVICE DRESS FOR OFFICERS.

The Serge Patrol Jacket of the pattern approved for the Imperial Army has been adopted for the officers of the Canadian Militia, and will be worn in marching, field day and drill order.

A detailed description is appended and sealed patterns will be issued to the Royal Schools of Instruction to secure uniformity.

Cavalry

Serge Patrol Jacket. Blue; (in Canadian Mounted Rifles and 3rd Prince of Wales' Dragoons, scarlet) of the same cut as the serge frock now issued at the Royal School of Cavalry, Quebec, for non-commissioned officers and men. Full in the chest, collars and cuffs of the same colour and material as the rest of the jacket. Shoulder-straps of cloth of the colour of the regimental facings, with a small regimental button at the top. Badges of rank in gold.

Artillery

Blue Serge; Welted seams; stand-up collar, square in front, fastened with one hook and eye, a grenade, two and one quarter inches long, in gold embroidery at each end; shoulder straps of the same material as the garment, fastened at the top with a small black netted button, half an inch in diameter, badges of rank embroidered in gold. Five gilt ball-buttons down the front; a slit on each side, sleeves ornamented with flat plait, forming crow's feet six inches from bottom of the cuffs; two inside breast pockets and watch pocket.

Infantry and Engineers

Scarlet Serge; Full in the chest. Collar, cuffs and shoulder-straps of cloth of the colour of the regimental facings. A small regimental button at the top of the shoulder-strap. Badges of rank in gold. Collar rounded in front with black enamelled leather tab and hook and eye. Two pleats on each side; on the left side an opening for the support of the sword belt. Five small regimental buttons down the front. A patch pocket with pointed flap and small button on each breast. Cuffs pointed five inches deep in front, and two inches deep behind. Scarlet lining, no collar badge.

Rifles.

Rifle Green Serge; Square in front, stand up collar with hook and eye and black silk tab. A body seam on each side, seven regimental horn buttons down the front. Two pockets on each side with pointed flaps. A small button with tab under each flap. A drawing string inside at the waist. Shoulder-straps of the same material as the garment, a small button at the top. Badges of rank in bronze. Collar and cuffs of same colour as the regimental facings.

No.2

UNIFORM OF OFFICERS ATTACHED TO SCHOOLS OF INSTRUCTION.

Officers of the Active Militia attached to Schools of Instruction will be required to provide themselves with the field service uniform, which will be worn at all instructional drills and parades.

The patrol jacket may be worn at mess by attached officers who are not provided with tunic or mess dress. It will likewise always be worn when performing the duties of orderly officer.

(*Canada Gazette*, 26 March 1892)

Index

www.ingramcontent.com/pod-product-compliance
Lightning Source LLC
LaVergne TN
LVHW070832220926
848486LV00036B/269